Geography and
soil properties

A. F. PITTY

Geography and soil properties

METHUEN & CO LTD

First published in 1979 by Methuen & Co Ltd
11 New Fetter Lane, London EC4P 4EE
© *1978 A. F. Pitty*
Printed in Great Britain at the
University Press, Cambridge

ISBN 0 416 75380 9 hardbound
ISBN 0 416 71540 0 paperback

This title is available in both hardbound and paperback editions. The paperback edition is sold subject to the condition that it shall not, by way of trade or otherwise, be lent, re-sold, hired out or otherwise circulated without the publisher's prior consent in any form of binding or cover other than that in which it is published and without a similar condition including this condition being imposed on the subsequent purchaser.

To EDWARD *and* ALICE

Contents

Acknowledgements	ix
Preface	xi

1	**Geography and soils**	**1**
1.1	Physical geography and soils	1
1.2	Soils and human geography	15
1.3	Soils and geographical method	33
1.4	Conclusions about soils and geography	41

2	**The mineral fraction of the soil**	**42**
2.1	The classification of particle sizes	42
2.2	The coarse fraction	44
2.3	Sand	45
2.4	Silt	51
2.5	Clays	55

3	**Soil organic matter**	**69**
3.1	Supply of organic matter to the soil	69
3.2	Breakdown of organic matter in the soil by macrofauna	76
3.3	Biochemical compounds in soil organic matter	78
3.4	Humus	89
3.5	Peat	97

4	**Soil structure and porosity**	**101**
4.1	Soil structure	101
4.2	Porosity	111

Contents

5	**Physical properties of the soil**	**115**
5.1	Introduction	115
5.2	Soil water	117
5.3	Soil temperatures	136
5.4	Soil air	143
6	**Chemical properties of the soil**	**151**
6.1	Oxidation-reduction processes	151
6.2	Ion exchange	154
6.3	Soil acidity	162
6.4	Interaction of organic matter with metals, oxides, and clays	166
6.5	Iron, aluminium, and silica	167
6.6	Potassium, calcium, and magnesium	174
6.7	Nitrogen, phosphorus and sulphur	178
6.8	Micronutrients	193
6.9	Sodium chloride and associated salts	197
7	**Soil mechanical properties**	**203**
7.1	Bulk density	203
7.2	Soil strength	207
7.3	Swelling and dispersion	210
7.4	Shrinkage and fissuring	213
7.5	Consistency or Atterberg limits	214
7.6	Compaction	216
7.7	Brittleness	218
7.8	Cementation of iron concretions	219
7.9	Induration	220
8	**Soil colour**	**224**
8.1	The influence of colour on other soil characteristics	224
8.2	Description of soil colour	224
8.3	Sources of pigments in soil colours	225
8.4	Influence of other soil properties on soil colour	226
8.5	Significance of soil colour	227
8.6	Soil colour and man in contact with the earth	228
	Bibliography	229
	Appendix: soil names	271
	Subject index	277

Acknowledgements

In writing this book, I've much appreciated the helpfulness and high professional skill of the Library staff in the University of Hull, and I am particularly grateful to Mr G. D. Weston. I am grateful to my former colleagues in the Department of Geography of the University of Hull, Mr K. Scurr, Mr A. Key, and Miss W. A. Wilkinson, for drawing the diagrams. Mr B. Fisher and Mr S. Moran prepared photographic reductions. I am also indebted to my former research students who incorporated a range of soils work in their investigations, including Dr R. R. Arnett, Dr I. Reid, Dr J. L. Ternan, Dr J. I. Pitman, Dr R. A. Halliwell, Dr P. A. Whittel, Mr R. G. Cooper, and Mr J. Crowther. Other former students to whom I'm grateful include Dr A. C. Imeson, Dr M. Bonell, and Mr J. A. Milne.

My awareness of the significance of soils in geographical studies was first sharpened when I was an undergraduate, whilst abstracting soils information from geological literature for Dr R. Webster and Dr P. H. T. Beckett's project on terrain analysis. More recently, I became increasingly indebted to Dr R. R. Arnett, with whom I shared several teaching commitments. I am also grateful to former colleagues and students of the Department of Geography in the University of Western Australia for their hospitality and advice during a year when my teaching experience was widened and my familiarity with the characteristics of field soils invaluably enlarged. I and the publishers would also like to thank the following for permission to reproduce copyright figures:

Academic Press for 6.1c and d; Athlone Press for 1.11; The Canada Department of Agriculture for 4.2a, b and c; Council for Scientific and Industrial Research for 2.1c; *The East Midland Geographer* for 1.10b;

x Acknowledgements

Economic Geology for 1.15a and b; Elsevier for 6.5a; *Geochimica et Cosmochimica Acta* for 3.7a; *Geoderma* for 1.3a, 3.1, 3.9, 4.3b, 5.3c, 5.7a, 6.10a and 7.5a, b and c; Dr D. M. Gray and the National Research Council of Canada for 1.1c; John Wiley and Sons for 1.2 and 1.4; *Journal of Applied Ecology* for 1.7a and b and 1.13d and e; *Journal of Ecology* for 1.8a, 1.13a and 7.7b; *Journal of Soil Science* for 1.6c, 1.9b, 1.17b and c, 2.1b, 2.2, 2.3b and c, 4.1a, b and d, 5.1d and e, 5.5a, 5.6, 5.7d, 6.3b, f and g, 6.6f, 6.7c, d, e and f, 6.9d, 7.4a, 7.6 and 7.7d; *Journal of the Science of Food Agriculture* for 1.10c and d; *Nature* for 1.10a; North Carolina Agricultural Experimental Station for 6.1a; Oxford University Press for 1.9; *Pedobiologia* for 3.2; *Pedologie* for 2.4a, b, c and e and 4.5; *Quarterly Journal of Engineering Geology* for 1.2b and 7.3a; *Revue d'Ecologie et de Biologie du Sol* for 3.2c and d; Rothamsted Experimental Station for 7.1a; School of Forestry, Yale University for 1.14a; *Scottish Geographical Magazine* for 1.9c; *Soil Biology and Biochemistry* for 5.5c; *Soil Science* for 3.7c, 3.8, 5.2a, 5.3a and b, 6.1b, 6.3a and c, 6.4a, b, c and d, 6.5c, 6.6c, d and e, 6.9b and c and 6.11a and c; Soil Science Society of America for 1.1b, 1.5a and b, 1.6a and b, 1.11b and c, 1.13b and c, 1.14b and c, 1.19a and b, 1.20, 2.1a, 2.3a, 4.2d, 4.3a, 5.1c, 5.2b, 5.3d, 5.7b and c, 6.6a and b, 6.7b, 6.8, 6.9a, 6.10f, 6.11b and d, 7.1b and c, 7.2a, c and d, 7.3c and 7.4b; *Soviet Soil Science* for 1.17a and d, 1.18, 1.19d, 2.1d, 3.7b, 4.1c, 4.4, 5.1a and b, 5.2c, 5.5b, 6.3e, 6.5b, 7.2b, 7.3b and 7.7a; United States Government Printing Office for 1.16; University of California Press for 3.3; Van Nostrand Reinhold for 1.15c.

ALISTAIR F. PITTY

Preface

Soils attract emotional responses. Titles like 'Mother Earth' and 'The rape of the soil' have gone forth and multiplied. One of John Steinbeck's characters loved the soil so desperately that he physically embraced it. The cooler desire to control inspires research in laboratories where plans to order and control the soil are conceived. Conversely, soil may invoke revulsion, for we are advised that there are certain matters with which we should not soil our hands. However, the soil is neither a person nor an electronic circuit and is increasingly vital as the natural filter which removes noxious contaminations.

We cannot jump straight to conclusions and to lay plans about crops, vegetation, and water supply either by deep feelings and intuition or by clinically applying the principles and methods of chemistry or physics directly to plant physiology. Furthermore, the characteristics of soils merit study in their own right and need not always be viewed as means to anthropocentrically defined ends.

The fundamental expression of soils is essentially geographical. The continuous distribution of soils at the earth's surface and their conspicuous regional differences at all scales from the continent to the hillslope make soils perhaps the most intrinsically geographical phenomenon. Certainly, in soil mapping, one sees, possibly, the clearest contemporary expression of the essential unit of study in classical geography, the 'natural region'. Equally, in the development of the other complementary strand of geography, the man-land inter-relationships, the ground trodden underfoot is of immediate interest. However, in the following pages, only passing reference is made to either the regional or the environment themes. The approach is essentially systematic, simply because the level of appreciation of the two fundamental themes of geography depends entirely on the degree of pre-existing, systematic knowledge which the observer can apply. The outward expression of regional differences in soils and of the substantial artificial modifications of the soil are so obviously germane and central to geography that they can too readily invite interest onwards before a basic knowledge of 'baseline' properties of soils has been acquired as a fundamental part of the preliminaries in a geographer's training.

It is not possible to carry the argument for the geographer's systematic study of the soil simply by bold assertion. This is evident from the themes of preceding texts for the soil geographer, in which a regional approach has persisted probably longer than in any other branch of the discipline. Therefore, a lengthy introductory statement follows, exemplifying the innumerable instances within systematic geography where intricacies can only be understood if an understanding of soil characteristics and processes is more than superficial. Otherwise, within his urban-built environment and talking of grass-roots philosophies over his ploughman's lunch, the geographer is in danger of losing touch with the interdependencies of reality.

1 Geography and soils

1.1 Physical geography and soils

1.1.1 HYDROLOGY AND SOILS

(a) *Soils and the hydrological cycle*
At the critical interface in the hydrological cycle, where precipitation meets the soil, water is either absorbed into the soil or it runs off on the surface. When absorbed in sufficient quantities, soil water favours the growth of economically valuable crops or contributes to groundwater reserves. If absorbed in smaller amounts, much moisture may be lost in evapotranspiration and an economically valueless crop may result. If precipitation runs off on the surface, it may be channelled for irrigation or impounded for urban and industrial use. Alternatively, runoff may mount into destructive floods, constitute an erosion risk, and as such create sedimentation problems downchannel and in estuaries.

Whether precipitation takes beneficial or destructive routes is determined largely by the properties of the surface horizon of the soil. Infiltration rates depend on the water physical properties of the soil, especially its permeability. For instance, infiltration decreases with increasing clay content and increases with greater porosity in the soil (fig. 1.1a). A major factor is also simply the amount of water already occupying the soil pores prior to a given precipitation event. Permeability may change with depth. If subsoils are less permeable than surface horizons, the lower horizons control infiltration rates and the upper horizons may become saturated rapidly (fig. 1.1b). Alternatively, a surface horizon or crust may develop and be less permeable than the subsoil if rainbeat

2 Geography and soil properties

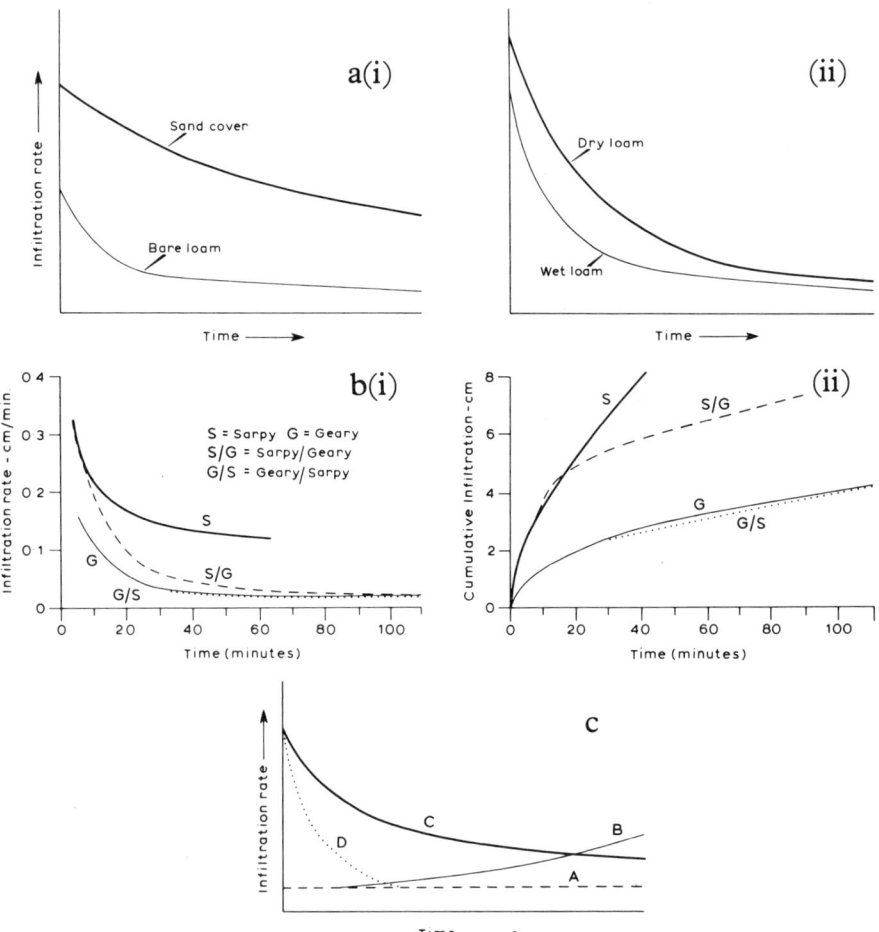

Figure 1.1 Effect of soil texture, layering and moisture content, and temperature on infiltration.
a Generalized variation in infiltration curves, related to soil texture (i) and to preceding moisture content of the soil (ii).
b Predicted influence of layered combinations of soils and uniform soils on infiltration (Hanks and Bowers, 1962). The coarse-over-fine layered soil (Sarpy/Geary) was identical with the coarser soil until the wetting front reached the finer soil.
c Generalized infiltration curves for frozen soils (Gray, 1970).
  A Soil frozen when saturated, or persistence of impervious ice layer during melting.
  B Soil frozen at high moisture content, with meltwater progressively melting ice-filled pores.
  C Soil frozen at low moisture contents and at temperatures near freezing, then thawed by meltwater.
  D Soil frozen at low moisture contents but at temperatures well below freezing; water entering pores is frozen and infiltration is inhibited.

compacts a poorly structured soil and washes fine particles into the pores of the surface layers. This occurrence may reduce permeability within the course of one storm.

Soil temperatures of field soils of different textures concern the hydrologist. If clays crack on drying, the first ensuing rains infiltrate rapidly. Where freezing occurs, apart from depth and persistence of frost in the soil, the changes which may take place within the frozen layer during freeze–thaw cycles are particularly significant. The type of response depends on such factors as the degree to which a soil's pores are saturated with moisture prior to freezing or on thawing, or on the degree of freezing when water first enters a dry soil (fig. 1.1c). For example, the common spring floods in the drift-free area of the upper Mississippi valley result primarily from restricted infiltration into frozen soils.

(b) *Soils and water quality*
It is increasingly doubted whether the soil can retain sufficient of added chemicals to keep their levels low in nearby fresh water bodies. In particular, nitrification occurs rapidly in most soils and excessive N and P are a major cause of increased growth of undesirable aquatic vegetation in lakes and streams. Deleterious side-effects also include algal blooms, fish kills, filter clogging, and undesirable taste and smell in drinking waters. Aesthetic qualities in the environment may also deteriorate. Methema-globinemia in animals and young infants has been attributed to excessive nitrates in drinking water. In semi-arid and desert soils, irrigation is usually responsible for secondary salinization processes. The suitability of waters for irrigation is, therefore, being increasingly evaluated in conjunction with studies of soil characteristics. Otherwise, solutes in soil water may create saline, sodic or toxic levels which are hazardous for crop growth or to animals and man consuming these crops (Bettenay *et al.* 1964).

The soil has been for ages a major medium for the disposal of a variety of wastes, owing to its colloidal organic matter and other related properties. Organic compounds and natural wastes added to the soil are usually destroyed in the upper zones or are retained in the surface horizons long enough for them to be entirely mineralized. Nonetheless, even the slow release of a toxic compound from the soil affects the quality of water for drinking and for agricultural and industrial purposes. Deterioration of this type has become more marked with the extension of urban residential areas and with the increased use of biocides made of synthetic compounds which are not readily degraded by soil processes.

Clearly, to comprehend quantities and qualities of water as a prime natural resource, the geographer must understand the role and significance of several characteristics of the soil.

4   Geography and soil properties

1.1.2.   GEOMORPHOLOGY AND SOILS

The study of the hydrophysical characteristics of soils and their response to precipitation in particular, enlightens the geomorphologist about the nature of geomorphological processes (Reid 1973). Furthermore, soil profile characteristics increasingly afford precise information on the spans of time which must elapse for such processes to have some visible

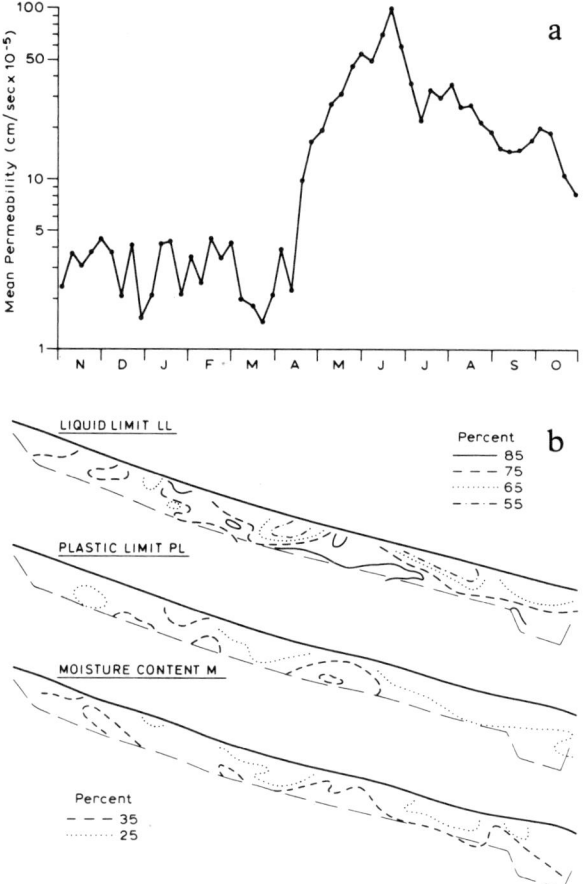

Figure 1.2   Geomorphological significance of soil moisture.
a Differences between topsoil and subsoil permeability, due to seasonal swelling of the clay fraction (Arnett, 1976). This contrast appears to account for much of the variations in interflow regimes in these soils, developed on Jurassic strata in the North York Moors.
b Moisture content and the Atterberg limits (plastic limit and liquid limit) which describe the engineering stability of the soil in a coastal landslip area near Charmouth, Dorset (Denness *et al.*, 1975). The profile is 100m long.

Geography and soils 5

if not measurable effect on landsurface evolution. In fact, it can be claimed that the geomorphological history of the earth's landsurface and the history of soil formation have many similar or parallel or even identical characteristics (Daniels *et al.* 1971).

(a) *Contemporary weathering processes*
In studying water movement through the soil, like the hydrologist, the geomorphologist is concerned with porosity and permeability and the degree to which their indices decline with finer particle sizes or are affected by other soil characteristics such as aggregation, compaction, or the dispersion or swelling characteristics of the clay fraction. The degree to which porosity and permeability decrease with depth in the soil is a particularly significant geomorphological factor (fig. 1.2a). In addition, soil water is the fundamental control on certain mechanical characteristics of soils. In areas where mass-movement of the soil or landslipping occurs, the amount of water in the soil may reach such levels that the soil behaves like a plastic material or even as a liquid (fig. 1.2b). Like such soil mechanics investigations by the civil engineer, studies of soil erosion by the agricultural engineer have a critical relevance for the geomorphologist. Soil temperatures are significant where their changes induce freezing and thawing or wetting and drying in the surface, with knowledge about the case of permanently frozen ground being essential. Data describing soil air are equally relevant as the carbon dioxide released by biochemical decomposition of organic matter and by plant root respiration in the soil greatly increases the potential solutional activity of soil water when it encounters rock fragments or bedrock.

(b) *Relict landsurfaces and ancient soils*
One of the most precise statements which W. M. Davis made was that weathered soils to a depth of 50 ft might characterize the surface of a peneplain. Indeed, pedological evidence supports the interpretation of the landsurface of South Limburg as part of a Tertiary and Late-Tertiary peneplain (fig. 1.3b). Its characteristics include thin local deposits of much weathered and pitted sand and gravel, strongly weathered outcrops of bedrock, deep relict soils and extensive surface silification. Australia in Late-Tertiary times was characterized by a general planation surface with minor relief and with limited areas of higher land (fig. 1.3a).

The characteristics of ancient soils do not necessarily corroborate interpretations of landsurfaces as former peneplains. For instance, soil mapping and associated laboratory studies on samples from various parts of the Chalk outcrop in southern England illustrate the importance of continued subaerial weathering during the Quaternary in the evolution of the landsurface (Catt and Hodgson 1976). However, where they

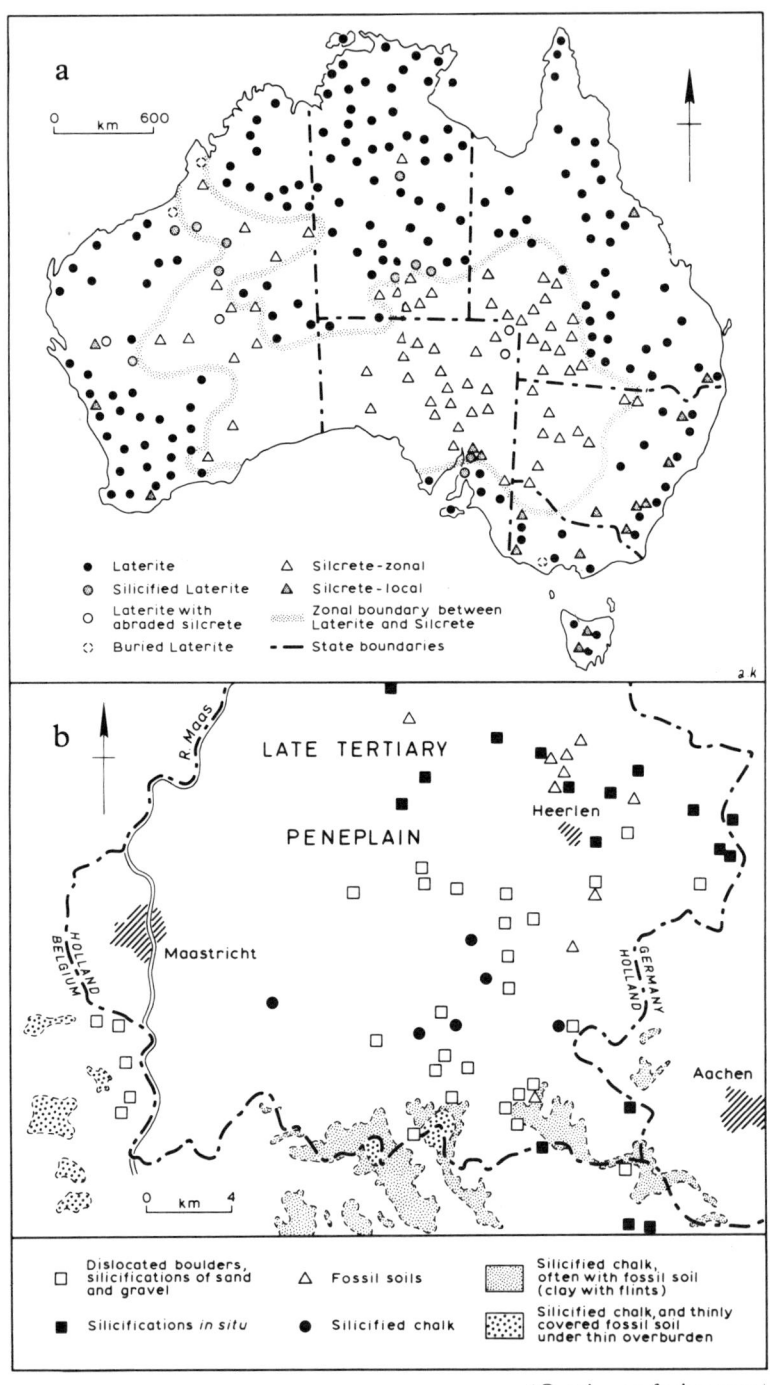

(Caption on facing page)

occur, landsurfaces capped by laterite, calcrete, or silcrete demand the geomorphologist's particular attention and invite his understanding of these distinctive horizons as extreme cases in the expression of soil weathering processes (Lepsch et al. 1977).

(c) *Depositional landsurfaces and soils*
Floodplain soils reflect the various fluvial processes of former river bed, floodplain, delta or estuary. These are discernible in a specific zonality and pattern of change in the soil cover. Studies of contemporary processes show that coarse-textured soils develop in the portion of the floodplain adjacent to the river (fig. 1.4). Soils developed in slack water sediments are widely distributed and may be particularly significant for agriculture, as noted in the Lower Ohio River floodplain. Chemical processes in floodplain soils are also an expression of distinctive local conditions. These are related to changes in oxidation–reduction conditions, a change in soil reaction, or to the formation of organomineral compounds. In particular, the accumulation of iron and its geochemical associates characterizes certain zones, past and present in floodplains. In sub-humid climates, the formation of a solonchak water regime is associated with virtually enclosed drainage areas in floodplains. Surface and ground-waters in such hollows are almost entirely lost by evapotranspiration.

Soil characteristics are useful in differentiating patterns of glacial deposition and outwash, particularly where periglacial conditions favoured wind-transport too. Organic soils frequently have developed in water-logged depressions, but distinctive, very soft and highly compressible soils indicate where post-glacial lakes were impounded, as observable in soils of lowland Ayrshire, or in the glacial landsurface of Wisconsin (fig. 1.5).

(d) *Buried soils*
Buried soils are particularly common in areas where the summer thaw of frozen ground leads to lower slopes being covered with solifluction lobes. Where buried soils occur in fluvial depositional zones, their presence indicates that the deposition rate and the associated erosion and soil development rates have been periodic. For example, in New Mexico in the Jornada del Muerto basin, gullies in an alluvial fan expose a succession of four major sediments, each of which has a distinct soil profile. In such

---

Figure 1.3   Relict weathering features on ancient landsurfaces.
a   Distribution of laterite and silcrete in Australia (Stephens, 1971).
b   Distribution of silicifications and fossil soils in South Limberg (Broek and Waals, 1967).

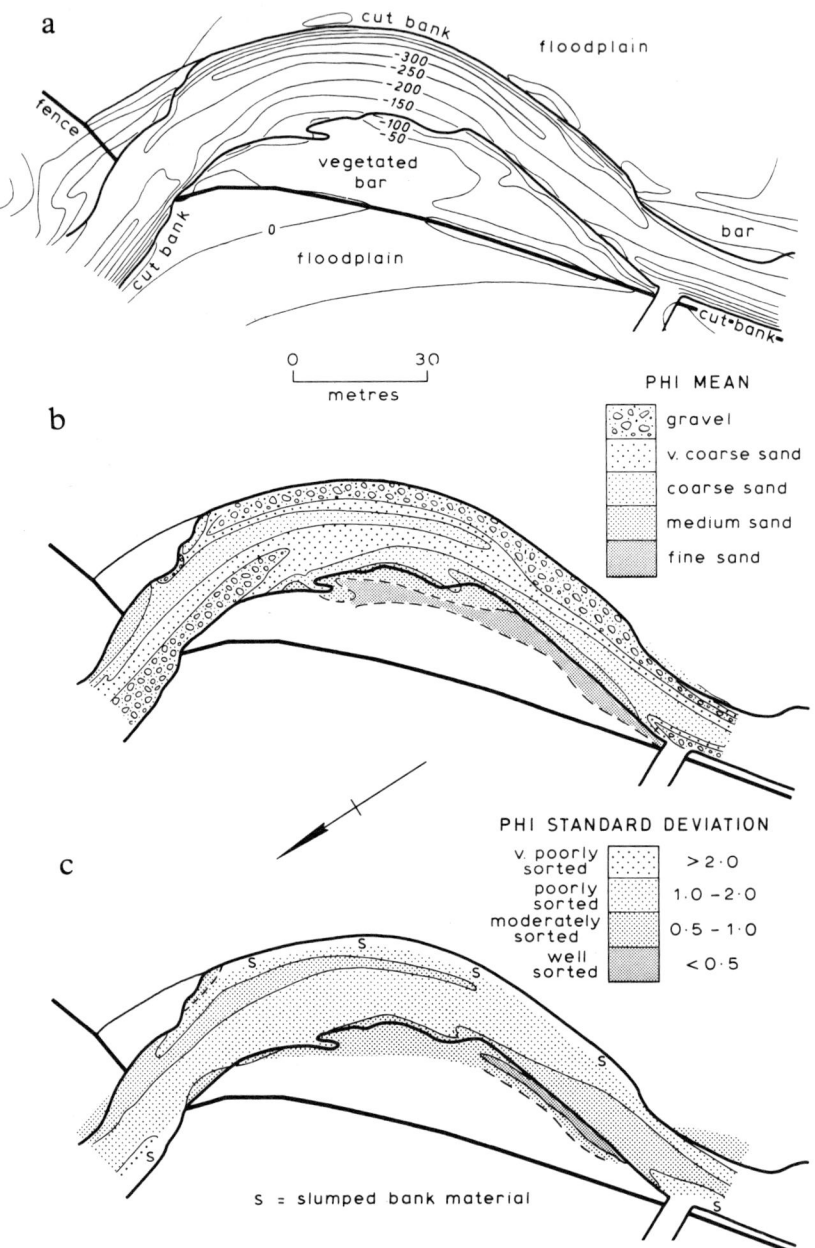

Figure 1.4 Particle characteristics of a dynamic depositional landform (Bridge and Jarvis, 1976), a meander in the floodplain of the South Esk, Glen Clova, Scotland.

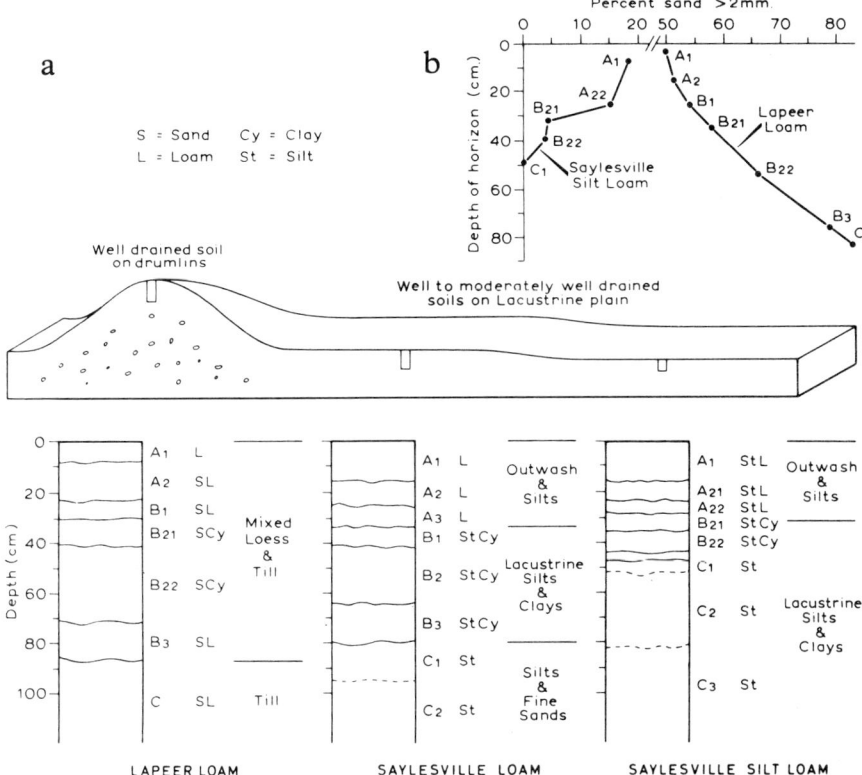

Figure 1.5 Soil particle and profile characteristics of relict depositional landforms Borchardt et al., 1968), in a glaciated landscape of south-eastern Wisconsin.
a The moderately dolomitic sandy Lapeer loam on the glacial till of the drumlins merges downslope into the highly dolomitic Saylesville silty clay loam of the glacio-lacustrine plains.
b Sand content decreases with proximity to the soil surface in the Lapeer soil, but increases in the Saylesville soil. There is an apparent dilution of sand of glacial till by aeolian silt and clay. One quarter of the 86 cm deep soil incorporates leached loess which would have been laid down as a 22 cm thick cover.

semi-arid areas, the buried soil usually includes a carbonate-enriched horizon (fig. 1.6a). Nearby, buried charcoal horizons were found, indicating that several deposits of alluvium are present, ranging in age from less than 1100 years to over 5000 years B.P. Climatic change, causing a decrease in vegetative cover, could have started the pronounced erosion of Pleistocene soils, resulting in the deposition of these highly calcareous Recent sediments (fig. 1.6b).

Radiocarbon datings of buried soils are of particular interest to the geographer, as are the findings of palynology. In the case of a buried

10 Geography and soil properties

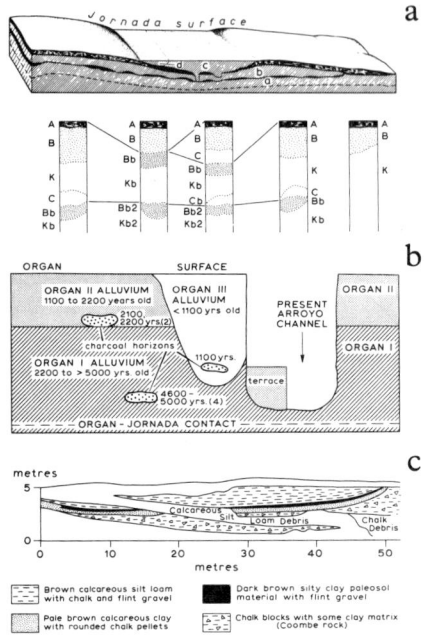

Figure 1.6 Soils of buried landsurfaces.
a Periodicity in the landsurface and desert soil development in southern New Mexico (Gile and Hawley, 1966). The block diagram shows buried sediments (a and b), the former channel (c) and the present arroyo channel (d). The profile sections show A and B horizons (stippled) and high-carbonate K horizons (blank).
b Detail of recent periodicity in southern New Mexico (Gile and Hawley, 1968).
c Typical buried soil at the base of chalk escarpments and dry valleys in southern England (Valentine and Dalrymple, 1976). Moderately decalcified soil materials are buried by a series of highly calcareous layers, up to 3 m in depth, due to late Neolithic/early Bronze Age slopewash and ploughwash.

soil in a scarp foot valley in Buckinghamshire, a radiocarbon date of 3910 years B.P. indicated that the soil was buried by colluvium which was slopewash and ploughwash initiated in late Neolithic or early Bronze Age, by man's clearance of the woodland and cultivation of the soil (fig. 1.6c).

1.1.3 ECOLOGY AND SOILS

Ecology has long been recognized as the study of the responses of organisms to their habitats and their adjustments to environmental changes. Today, however, vegetation covers are so altered that some geographies of organisms can only be delineated with the aid of a microscope trained on samples of fossil pollen. Man has replaced a variety of grass-eating animals by one or two types of stock animal and has virtually eliminated many fur-bearing animals from woodlands and forest. Despite man's regimentation of organisms, ecological patterns persist in several notable contexts which reflect a distinctive influence of soil properties, the edaphic factor.

On larger scales, as in the distribution of the elements in the British Isles' flora, climate plays the primary ecological role. At smaller scales, the secondary role of the edaphic factor emerges, with soil properties measurably influencing the characteristic chemical composition of a plant species. Thus, within regions of relatively uniform climate, the broader effects of soil properties on tree growth often appear fairly obvious, and many investigators correlate wood production of a tree species with selected, measurable soil properties (Gersmehl 1976). For example, the distribution and vigour of Australian species of *Eucalyptus* are controlled by available soil nutrients, particularly phosphorus (Moore 1959).

(a) *The edaphic factor and desert flora*

Desert flora illustrate well the important fact that even on local scales, a vegetation mosaic is not invariably a function of soil properties. In an Egyptian wild flora in a desert area west of Alexandria, P-rich species of shallow rooting annuals or of herbaceous varieties were mainly confined to the soils on alluvial deposits, where both the total and the available P were relatively high, being 200–300 and 10–25 ppm, respectively (El-Ghonemy *et al.* 1977). In contrast, the Ca- and Mg-rich species were perennial shrubs with no specific relationship with the edaphic factor. Similarly, in the Intermontane region of western U.S.A., where the shadscale vegetation zone provides some 17 m ha. of important winter forage for sheep and cattle, the sharply distinct vegetation patterns are not linked rigidly with soil properties (Gates *et al.* 1956). Of 5 species studied, sagebrush *(Artemisia tridentata)* and winterfat *(Eurotia lenata)* appeared least tolerant of high salt content, whereas the Nuttall saltbush *(Atriplex nuttallii)* grew on soils with the most salt. Nonetheless, all five species were found growing in pure stands in soils of similar salt content. All appeared to have a wide range of tolerance, and winterfat, in particular, is apparently a poor indicator of soil properties. Abrupt changes in soil chemistry do not occur concomitantly with changes in vegetation type. Later studies in the shadscale zone revealed surprisingly uniform soil physical properties at these boundaries also, both within and between communities (Mitchell *et al.* 1966). The fact that edaphic differences between soils occupied by various species is not expressed obviously and as a rigid control, only adds to the ecological interest in specifying and comprehending the exact nature of soil properties.

12 Geography and soil properties

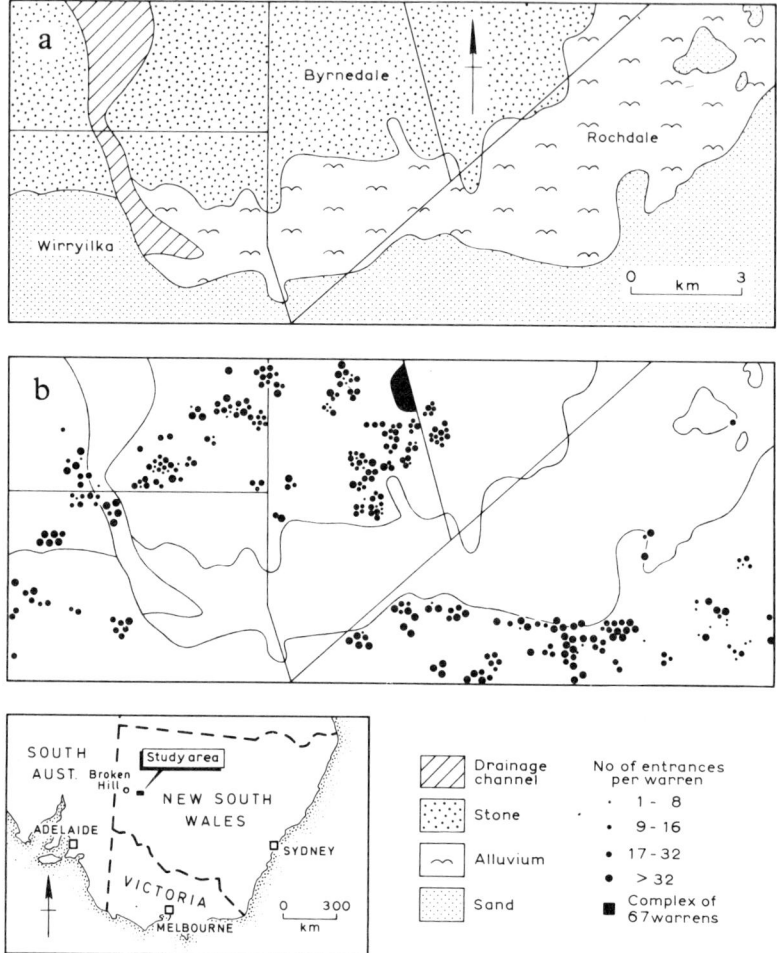

Figure 1.7 Soil types and the distribution of rabbit warrens (Parker et al., 1976), on grazing properties 90 km east of Broken Hill.
a Soil parent materials of Quaternary deposits of angular, poorly sorted sand and gravel, and small outcrops of Tertiary silicified conglomerates.
b Distribution and size of rabbit warrens in 1965.

(b) *Soil as a shelter for animals*

The link between the soil and the animals it shelters is made obvious by the macrofauna and 95 per cent of all micro-organisms spend part of their existence in the soil. However, even the most obvious case of the rabbit warren is not necessarily as simple a response to soil conditions as might be anticipated. The specially favoured habitat of sand dunes may become too hot, collapse, or become drifted over in very dry spells.

The long-term maintenance of a rabbit population may then depend on stony soils with greater cohesion than sand and sited near to damp localities. For example, on the Byrnedale station (fig. 1.7a), rabbit warrens as plotted in 1965 are seen to be as prolific on the stony ground as in the more easily excavated sand ridges whilst the alluvial plain of clay loam is largely avoided (fig. 1.7b).

(c) *Edaphic factors in semi-natural environments*

In areas remote from farming and afforestation, the physical, chemical and mechanical properties of the soil are prominent influences on vegetation patterns which are vital for scientific enquiries and of immeasurable aesthetic value. Animals, by migrating, may appear to avoid direct environmental influences. The food chain, however, inseparably links animals with green plants and hence indirectly to edaphic and other environmental controls. The natural link from soil to plant and on to animal is clearly visible in the exceptional caldera habitat of the Ngorongoro Crater in the Crater Highlands of northern Tanzania. This enclosed basin contains a permanent concentration of what are elsewhere strongly migratory animals. Most of the crater is clothed with grasslands which is conveniently grouped into four categories according to height. Several species of herbivorous game animals graze on these fodders and a variety of predators, notably lion and hyena, feed on these ungulates. There are marked changes in soil properties (fig. 1.8) from the lava and tuff outcrops on the crater walls, through ash and scoria slopes, to the alluvial, waterlogged crater floor. Of these soil properties, texture is one of the main influences in differentiating vegetation. Its affect on water-holding capacity regulates rooting pattern, percentage foliage cover and height of grassland. In addition, salinity is associated with poor drainage, and drainage also controls the weathering–leaching balance and hence the chemical character of the soil water on the slopes. For grazing, the Thompson's gazelle prefer the very short grassland during the rains but move on to the medium-height grassland after it has been well-grazed by other game. As with zebra and wildebeest, preference for short grass, firmness of footing and clear vision of predators are considered the main reasons for this preference. The kongoni and buffalo stay mainly on the taller *Themeda* grasslands where the fodder may also be too fibrous and unpalatable for other species. Soils also have a direct effect on faunal patterns, as the more sticky vertisols of depressions are avoided until late in the dry season since these soils adhere to animals' feet when wet (Anderson and Herlocker 1973). This example shows the striking influence of soils on vegetation types and the caldera also illustrates the indirect dependence of an animal population on a diversity of soils and the patterns, balance, and associations in a semi-

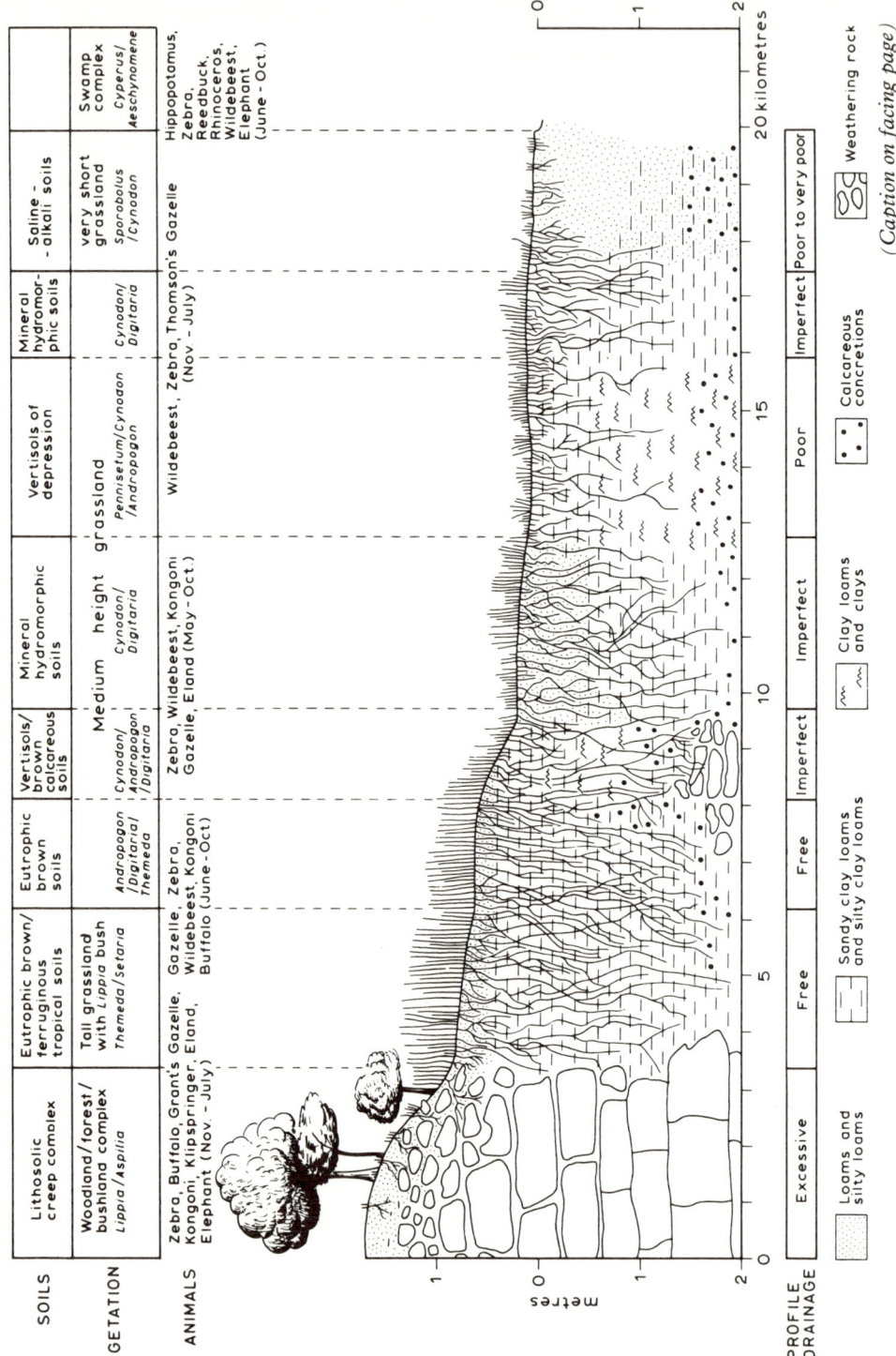

(Caption on facing page)

natural environment which encompasses also the domestic livestock of the Masai pastoralists.

1.2 Soils and human geography

1.2.1 SOILS AND HUMAN HISTORY

From the drier, short-grass savannahs of the Rift Valley which they occupy in the wet season, many of the Masai migrate during the drought to taller grasses of the higher plateaus. Traditionally, they set up temporary kraals during such seasonal migrations and are devoted to the interests of the cattle. Where and when man changed from hunting and nomadic herding to a fixed mode of agriculture and began cultivating the soil, the delicate balance between plant cover and the erosive forces of wind and water was disrupted. Thus for man the soil affords problems as well as providing indirectly shelter, food, and water.

(a) *Shelter and communications*
Soils have sheltered man, as in the ancient loess villages excavated into the side of ravines at Yang Shao Tsun in China or in contemporary dwellings cut into sun-baked clays near Matmata in southern Tunisia. More widely, soil materials were daubed on the sides of wooden frameworks and early settlers in many lands have lived beneath a layer of sod. With more sophisticated buildings, the soil in which their foundations settle is a critical factor. At Tikal in Guatemala, Maya ruins cluster on silty upland soils, with limestone bedrock at shallow depth affording good foundations for the most impressive architecture of pre-European civilization in North and Central America. The clays in the lowland seasonal swamps are sticky and shrink and swell with drying and wetting. These were avoided as construction sites (Olson and Puleston 1972). Similarly, implicit in the durability of the very heavy structures of the Roman Empire is their understanding of the safe bearing capacity and performance of soil under the action of load and water. Various types of sandy, mud, or marshy soil have impeded travellers and have been decisive factors in some military strategies.

Early settlers sought well-drained soils for their permanent settlements. Thus, in the Upper Thames Valley, Anglo-Saxon populations concentrated on dry-point sites slightly above the general level of the land, and these sites persist as the present-day villages (fig. 1.9a). Similarly, the Insch Valley was one of the most densely populated areas in Aberdeenshire in the 17th and 18th centuries, and at that time there was a remarkable correlation between distribution of this population and the deeper soils of the area (fig. 1.9c).

Figure 1.8 Association between slope, soil properties, vegetation and the distribution of wild animals (Anderson and Herlocker, 1973), in the Ngorongoro Crater of northern Tanzania.

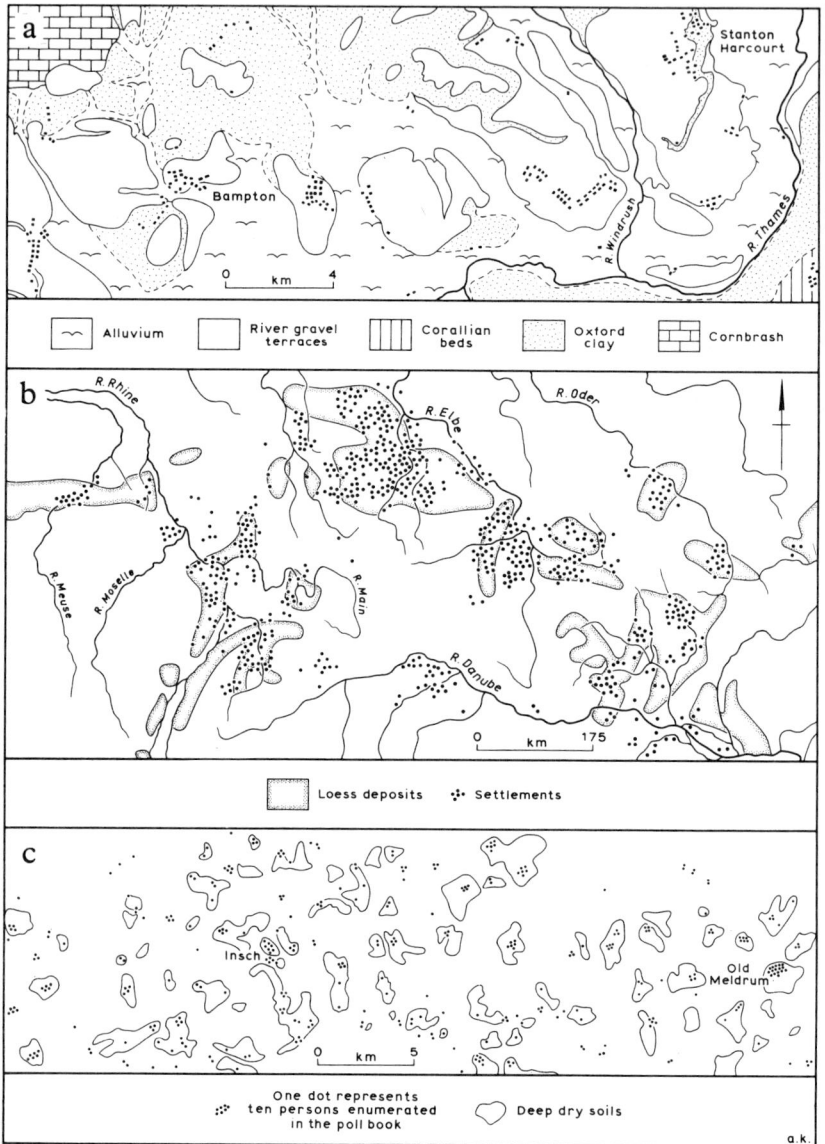

Figure 1.9 Association between early human colonization and soil properties of transported parent materials.
  a Concentration of Anglo-Saxon settlements at dry-point sites on river terrace gravels (Martin and Steel, 1954), in the Upper Thames Valley.
  b Relation between Central European settlement of the Danube culture and wind-blown loess deposits, about 4000 B.C. (Vink, 1963).
  c Seventeenth-century population sites in Aberdeenshire, in relation to deep dry morainic soils (Walton, 1950), mantling south-east facing slopes, the leeside in relation to the last ice movement in the area.

(b) *Attractiveness of fertile soils*
The first Stone Age farmer who started to look for a suitable piece of land for scattering some seeds was also the first soil surveyor (fig. 1.9b). The ancient farmers of Classical Greece and Rome recognized the variety of soils, and writings of the scholars record how soil fertility was judged by the colour, texture, and water content of the soil. Seed was sown thickly or scantily according to the adjudged quality of the soil (Semple 1928, pp. 66–9). In the Yucutan peninsula there was a specific vocabulary for distinctive soil types, such as very stony land or black earth.

More modern examples of initial colonizations are particularly instructive because documentation is more detailed. For instance, in the comparatively recent colonization of Western Australia from the 1830s onwards, the pattern of agricultural land use was established within the first few years in very close accordance with soil types. The sands of the coastal strip were passed and most colonists chose sites at the foot of the Darling Ranges where alluvial soils had been left where the rivers emerged from the Ranges. The Bassendean sands, between these alluvial soils of the east and the coastal sands and limestones were left almost untouched and remained so until well into the present century. Inland, early agriculture was confined to the finer-textured soils, characteristic of most of the broad valleys of what became the Wheat Belt. Large areas of sand-covered laterites and other, coarse-textured soils were avoided.

(c) *Inherent and induced limitations on the agricultural use of soils*
(fig. 1.10)
In the Old World and in the East, many soils are, in part, the creation of patient effort by successive generations. For example, in some parts of the west of Ireland, the soils have been built up by carting seaweed and sand on to ill-drained clays. Patches of peaty humus have been placed between larger rocks, whilst smaller rocks have been removed and stacked in countless walls. In contrast, areas which were famous for oaks, wheat fields and well-nourished herds now conceal once-powerful cities of Greek and Roman culture beneath metres of debris eroded from adjacent, now-bare hills. In speculating about the collapse of the old Maya Empire about A.D. 600, many writers have considered the role of soil exhaustion and erosion in accounting for this inexplicable decline. A sharp drop in the population's stature in A.D. 600–800 has been attributed to malnutrition, and the exceptionally large area enclosed by defensive earthworks at Tikel attributed to declining yields. Certainly, some soils have been removed down to bedrock and some metres' depth of sediment has accumulated in reservoirs and in earthwork ditches.

Erosion of red soils featured in the Phoenicians Festival of Adonis, but the annual reddening of the River Adonis in the Byblus area was

18 Geography and soil properties

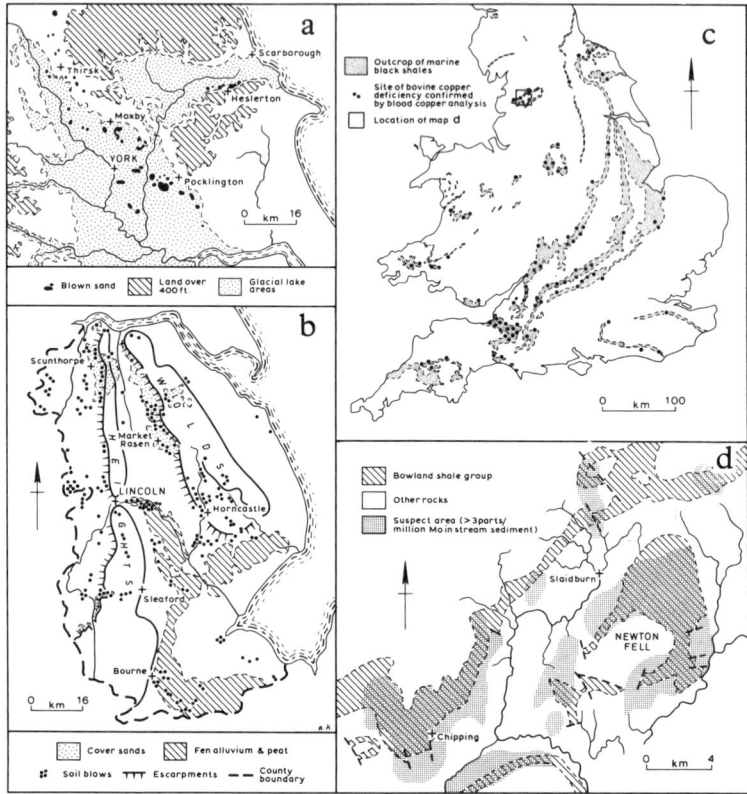

Figure 1.10 Imbalances between soil properties and contemporary agricultural practices.
a Areas of sand blows in Yorkshire in 1967 (Radley and Simms, 1967), as related to areas of former glacial lakes.
b Observed soil blows in Lincolnshire in March 1968 (Robinson, 1968), due to intensification of arable cropping.
c and d Bovine hypocuprosis induced by molydeniferous soils (Thomson *et al.*, 1972). Distribution of known copper deficiencies as related to outcrops of molybdeniferous shales (c), and extension of areas affected by high Mo concentrations in stream sediments (d).

attributed to the wounding to death of the god, annually, by a wild boar on Mt. Lebanon. Clearly, the Babylonians, Phoenicians, and other ancient peoples did not understand the importance of protecting their cultivated and grazing lands from the kinetic energy of raindrop impact. Today, stupendous Roman buildings remain, towering over desert sands and eroded soils in North Africa. Here knowledge of soil properties was sufficient to ensure that urban edifices endured, but was insufficient to prevent an irreplaceable natural resource slipping through fingers.

1.2.2 SOILS AND AGRICULTURE

(a) *The association between soils and crops*
Certainly, there are many examples of a broad association between crops and soils and statistical correlations between crop yields and productivity indices, based on an evaluation of soil properties, can be obtained (Clarke 1951). However, progressive mechanization and technological advances make soil just one of many variables in the complicated interaction between plants, animals, their environment, and economics. Improved food production is sought in improvements in seed supplies, crop varieties, insect and disease control, and in the wider use of irrigation and fertilizer practices (Coppock 1976). The relative significance perceived for soil science in contemporary food production is reflected in the training of specialists. Of the PhD graduates in agriculture in the USA in 1971, 29 were employed in aspects of breeding and of plant physiology, 6 in soil physics and chemistry, and 6 in soils and soil fertility.

(b) *Rotations and crops*
One of the earliest advances which transferred yields from a close control by soil type towards stronger links with management practices was the concept of rotations. In Britain, average yields of wheat in the eighteenth century were increased $2\frac{1}{2}$-fold by 1840, following the introduction of the Norfolk Rotation. Subsequently, further rotations have contributed to increased yields in a wide range of agricultural systems and further potential improvements remain. For instance, quite recently the best yields of cotton in the Gezira area are obtained when this crop follows a legume such as lubin, in a newly devised rotation (Robinson 1971). Contemporary research on crop planting includes the study of the density of plant spacing, which influences both the most economical use of available moisture and affects the role of light intensity on dry matter production, according to the degree of shading. Shading is also related to plant density which affects nitrogen nutrition, thick planting being deleterious with plants accumulating relatively more of the carbon-containing substances and less of those containing nitrogen.

(c) *Plant physiology and crops*
Genetics, as it controls plant nutrition, provides continuing scope for progress in agriculture (fig. 1.11b,c). Recent dramatic improvements of rice strains have invited the description 'Green Revolution' (Johnson 1972). In pasture production, Australia used to provide a case of singular deficiency in indigenous legumes and grasses which could have provided a basis for improved pastures. Now, reliable tropical legumes and grasses are available, with a higher nutritive value (Hutton 1970). Genetics and

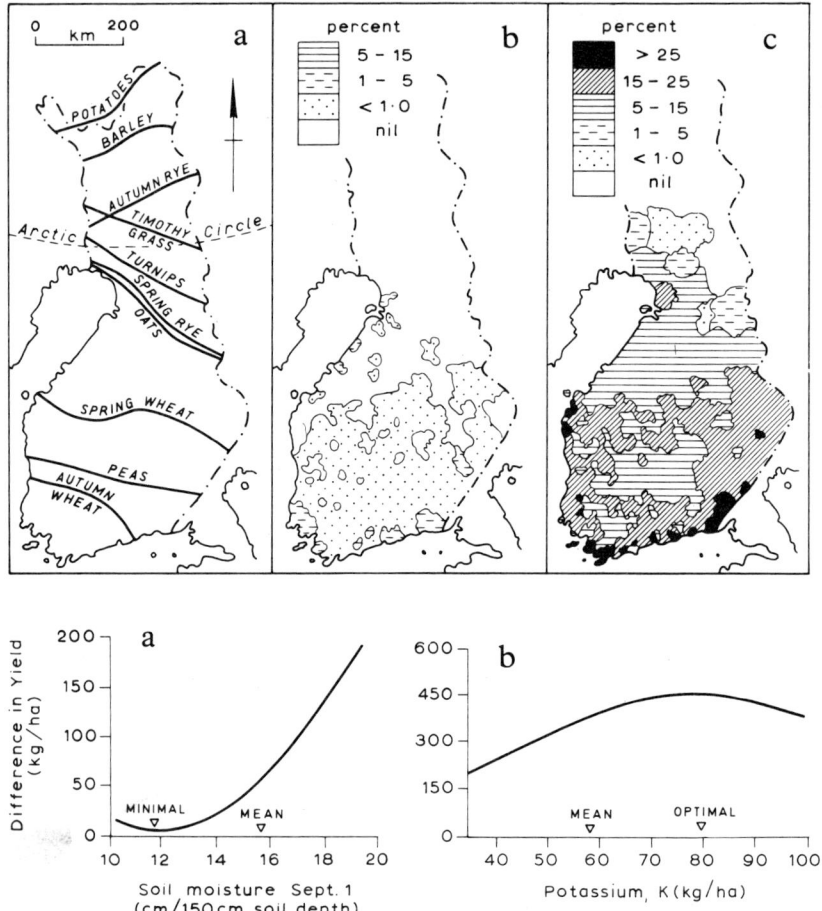

Figure 1.11   Some factors other than natural soil fertility which influence crop yield. (ABOVE) Patterns of crop boundaries in Finland, limited by climate, with extension and intensification due to plant breeding (Mead, 1953).
a Boundaries of crop cultivation.
b and c Spring wheat as a percentage of the total area in 1920 (b) and in 1945 (c).
(BELOW) Influence of water present and fertilizer additions on corn yield (Gross and Rust, 1972), in the fine loamy Clarion soils of Minnesota.
a Amounts of water in the soil above the mean value have a greater positive effect than the negative effect below the mean.
b Illustration of optimum amount of fertilizer application, in this case being 82 kg/ha.

plant physiology are also keys to the breeding of disease-resistant strains. In cereal production, for instance, a problem equal to any adverse soil property is the fact that no variety resistant to take-all disease has been developed. In addition, a study of the plants themselves, rather than soil tests, may indicate plant requirements for optimal growing conditions. For instance, iron deficiency may be due to the inactivation of iron within the plant and not necessarily be related to soil processes.

(d) *Fertilizers, biocides, and crops*

In China, labour intensive agriculture allows four vegetable and grain crops to be grown in a single year, with double and triple cropping common (Plambeck 1976). In most countries, however, labour costs are too high for such labour-intensive agriculture, even though agricultural labourers are often the lowest paid workers in the community. Instead, herbicides such as atrazine and paraquat control weeds normally destroyed by cultivation. This no-tillage approach has increased corn yields by 12–15 per cent in the U.S.A., where the area of cropland treated with herbicides had passed the 100 m. acre mark by 1966. Intensification is thus largely the work of the chemicals industry, with more than 100 m. tonnes of essential plant nutrients being added annually to world soils. With increased fertilizer application as the main feature of intensification

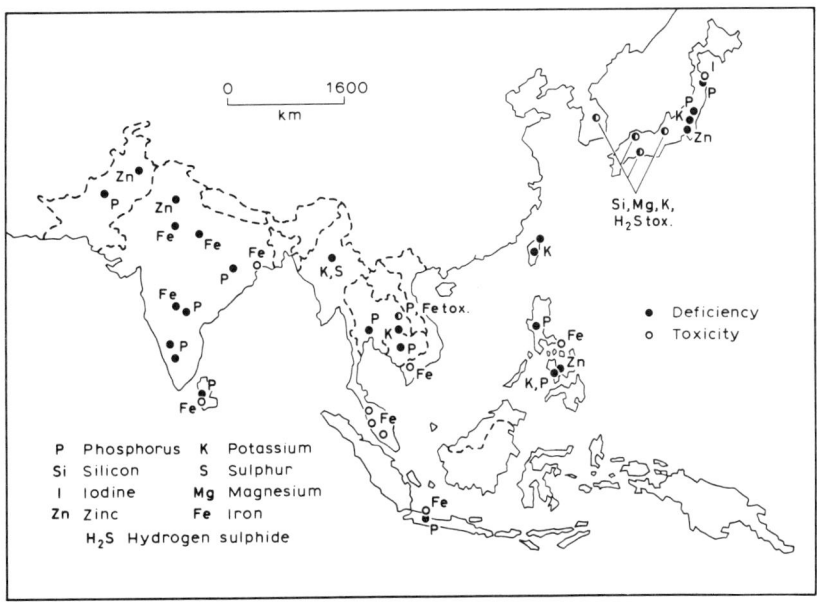

Figure 1.12 Fertility deficiencies and toxicities in natural soils, illustrated by nutritional disorders of rice in Asia (Tanaka and Yoshida, 1970).

of agriculture in the U.S.S.R., average annual grain production increased by 45–50 per cent between 1962 and 1972. Current plant and animal production levels are greatly dependent on additions on such scales (fig. 1.11). In addition, many soils are naturally deficient in certain vital micronutrients, and these have to be corrected artificially (fig. 1.12). Silicon nutrition increases the resistance of rice to disease and insect pests. Also, trace elements such as cobalt, selenium and copper have been applied to soils in order to correct deficiencies of these elements in the animals which consume the plants grown (Allaway 1968). Fertilizers can also be used in the build-up of soil fertility in general. Beneath subterranean clover pastures in South Australia, for example, residual fertilizer phosphorus and soil organic matter accumulate, which usually increases the amounts of available phosphorus, sulphur, nitrogen and trace elements, and augments the soil's nutrient-holding capacity (Williams and David 1976).

(e) *Irrigation and soils*

In drier areas, man's first agricultural task has always been the provision, where possible, of irrigation water. Although supply and quality of water are the primary concerns in irrigation agriculture, soil type influences the use of saline water. Salts leach out more readily from coarse-textured soils, but wilting occurs more readily due to the low water-retentiveness of such soils. The study of water-physical properties of soils is therefore necessary in irrigation schemes. The desired effects may not be realized if such studies are too few, as during the widespread growth of Soviet irrigation schemes during 1948–53 (Kirillina 1968). Regardless of climatic zone, such studies are particularly significant in predicting the effectiveness of dams and the influence of reservoirs on adjacent lands.

(f) *Weather and crops*

Compared with the general advance of yields linked with increased output from the chemical industry, crop productivity has always been markedly but unpredictably influenced by natural variability of weather, particularly rainfall (Williams *et al.* 1975). In the case of the subhumid northern Great Plains as little as 12.5 mm of soil water may provide the margin between a poor and a good crop yield. About three-quarters of the U.S.S.R. lies in regions with inadequate or fluctuating moisture, and droughts can inflict much damage over vast areas in the steppe and forest-steppe zones. Indeed, since the introduction of artificial fertilizers to counterbalance nutrient deficiencies, withering has become the main limitation in many, particularly sandy, soils. Specific measures of the detrimental or beneficial effects of high or low rainfall include the example of rye crops grown on unfertilized plots in the Netherlands, where 61 per cent of the variance in crop yield was explained by rainfall in the preceding November-February period (van der Paauw 1972). In the case of pasture nutritive values, particularly in the tropics, a major limitation is the natural stage of growth at which most of them are grazed. The seasonal nature and extreme variability of the rainfall and temperature prevents pastures being stocked to the optimum during periods of active growth. Year-to-year differences in temperatures and sunshine hours may be critical. For example, in Hungary toxic hydrogen sulphide forms in rice soils in cool and sunless summers. The tolls of drought are proverbial (Skaggs 1975).

Variations in weather complicate the examination of correlations between soil properties and yields and limits due to weather variability are largely inevitable and uncontrollable (Munro and Davies 1973). The amelioration of the deleterious effects of weather fluctuations therefore involves a knowledge of soil properties and how they might be regulated or manipulated to minimize the impact of capricious weather changes on crops and fodder. In the Canadian Prairies, for example, the influence of soil texture on water storage is a major control on cereal yields (Williams *et al.* 1975).

(g) *Significance of soils in technological agriculture*
From reviewing the influence of weather and climate, the application of chemicals and irrigation water, and the significance of physiology and crop rotations in farming practices, it is clear that the soil is only one of many highly intricate and inter-related factors. It is therefore logical to anticipate only approximate relationships between soils and the way in which they are used by the farmer and to accept that in many instances soil properties can be successfully over-ridden. Today, regardless of region or political affiliation, it is widely held that 'data on the expenditures for fertilizers and on the cost of tools and machines are the most accurate statistical expression of the degree of intensification of agriculture' (Unanyants 1970). Today, crops can be grown without soils. Sawdust and other soilless media, supplied with nutrient solutions, are not merely economically feasible for supporting certain greenhouse crops, but also avoid problems of low productivity resulting from compacted soils or soil-borne diseases. In British Columbia yield increases and cost reductions obtained by commercial growers of tomatoes are sufficient to promote a rapid conversion from soil to soil-less growing media. On farmland, the prospect of no-tillage agriculture moves forward, with chemicals removing residues from a previous crop as well as supplying many of the needs for the coming season. Yields are maintained by applications of chemicals on scales which appear to dwarf the significance of soil properties. However,

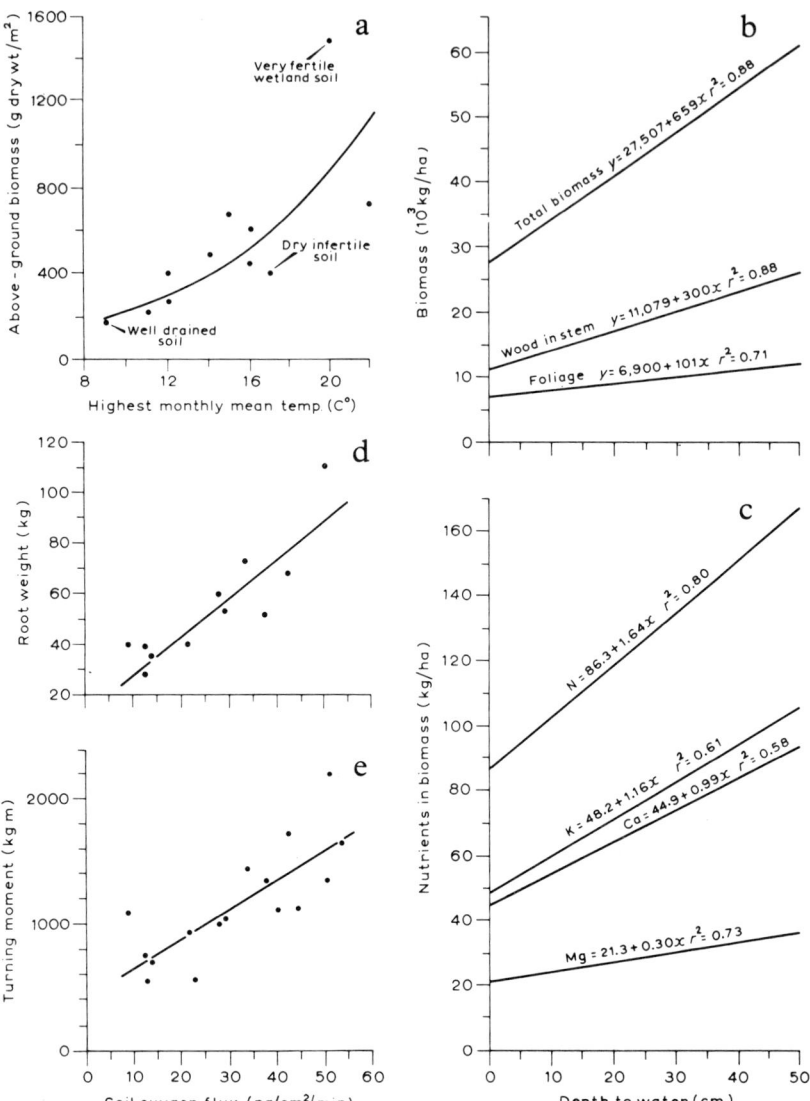

Figure 1.13 Environmental influences on plant and tree growth.
a Effect of temperature on growth for various sedges in northern and mid-latitudes (Gorham, 1974).
b and c Influence of winter root environment on the growth (b) and on nutrient uptake (c) for slash pine (*Pinus elliottii* Engelm. var. elliottii) in plantations on imperfectly drained soils in Louisiana (McKee and Shoulders, 1974). Depth to free water and associated redox potential of the soil in winter profoundly influence early growth, as the more extensive root systems cope better with summer droughts.

(Continued on facing page)

effective action of organic chemicals is governed by factors which include their capacity to be adsorbed by the soil. The nature of adsorption controls their uptake, persistence, mobility and toxicity, thus determining the optimum amount needed to be applied without deficiency or pollution effects (Bailey and White 1970). Thus, the progressive mechanization and chemicalization of agriculture requires a continuing revision of the standards used for the evaluation of soil suitability rather than an overriding of a soil's complexity. For instance, the mechanical harvesting of root crops and bulbs, traditionally grown on heavy clay soils, is often difficult and the use of lighter soils is now more preferable than formerly. Continuing re-evaluation of soil suitability is particularly relevant where practices from one major geographical region are transferred to another. In the case of pasture, particularly in the tropics where half the world's permanent pastures occur, many limiting factors are related to soil properties (Wilson and Haydock 1971). Therefore, an integral part of the progress of technological agriculture remains a systematic knowledge of soil properties and, in the final analysis, soils will always influence the cost of implementing any agricultural system which they are required to carry.

1.2.3 SOILS AND FORESTRY

Temperature is the major control on natural biomass production (fig. 1.13a) and by relegation in land-use priorities, usually forest soils are inherently infertile. However, intensive forest management is becoming increasingly vital to meet the growing demand for wood and wood products. Therefore, a comprehensive awareness of soils must include some knowledge about soil characteristics of special relevance to forestry (fig. 1.13b,c) as well as the long-standing association of soil study with agriculture. Increased felling rates characterize intensive forest management and accelerate the continual removal of nutrients and the role and problems of fertilization of forest soils (fig. 1.14). Forest management may also include responsibilities for the supply of game food for forest fauna. This supply usually relates to the organic fraction of forest soils and involves maintaining desirable forms of humus layers, ground-cover plants, insects, worms, and other organisms. Also facilities for visitors to forest parks are often influenced by soil properties (Lutz 1945).

---

d and e. The relationship between soil aeration and growth of Sitka spruce (*Picea sitchensis* (Bong.) Carr.) on upland peaty gleys (Armstrong *et al.*, 1976). The observations were made at 5–11 cm depth in April in the Kielder Forest in Northumberland.

26 Geography and soil properties

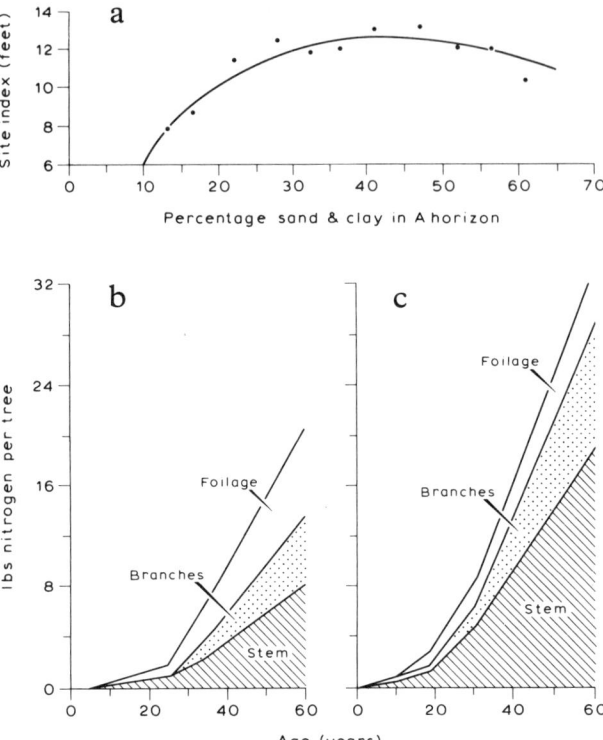

Figure 1.14 Influence of soil properties on growth rate and quality in forest stands.
a Lower growth rates at both low and high clay contents (Haig, 1929).
b and c. Increased nitrogen uptake by Loblolly Pine (*Pinus taeda* L.) at good quality soil sites (c) compared with poorer, podzolic soils (b) (Swincer *et al.*, 1966). The sites are in the Upper Coastal Plain of Mississippi and are Red Yellow Podzols (b) and Low-Humic Gleys (c).

Forest soils are particularly important in terms of scale. For instance, 32.5 per cent of the total area of the USSR has been classified as forest, and the proportion of Czechoslovakia occupied by forest is 36 per cent. In addition, skilful afforestation can control or regulate unfavourable natural phenomena and processes, such as exposure, drought, and erosion.

The fertilization of forest soils is likely to assume an important role in the near future. The intensified use of nitrogen in forestry has indicated that nitrogen loss is a problem requiring intensive study (fig. 1.14b,c). Also, it has been shown for *P.radiata* that on some soils it is possible for a single application to be sufficient to supply phosphorus at a suitable rate for the whole rotation period. *P.elliotti* (Engelm) has often shown a positive response to phosphate fertilizers, especially when planted outside its natural habitat. Information on water supply and availability is also frequently vital. Insufficient moisture is one of the main reasons for the difficulties involved in afforestation in drier environments, and early mortality on eroded sites may be attributed more to drought than to nutritional deficiencies of the soils. Both N availability and uptake may be restricted during drought periods and, as with agriculture, fertilizer application has to be synchronized with weather conditions. Up to 90 per cent of nitrogen added as ammonium sulphate has been observed to pass through a forest floor during a period of heavy rainfall. Clearly, successful forestry, like agriculture, often depends on factors other than soils. Nonetheless, critical limits placed by such other factors are inseparable from soils.

There are some distinctive features involved in forest management, such as fire risk. The morphology of the humus layers determines the susceptibility of forest stands to fire and the fertility status of burned-over soils. Raw humus ignites readily and burns persistently. Soils with raw humus lose nitrogen through volatilization and a major part of their nutrients through leaching of ash constituents. Conversely, soils with an efficient microbiota expose only a thin layer of litter to ground fires and nutrient losses are very small. Secondly, there is the time factor, with a crop taking 40–80 years to mature. Over this span of time the crops' requirements change. At seedling time, it is the selection of the soil according to its uppermost layers which is significant. In the longer term, fertility and water availability problems involve the root range down to depths of as much as 4 m or more. Trees cannot respond to fertilization like field crops, and in most instances, it is still not known how fertilization might affect tree growth over a 20–30 year period. Demands on water, however, are known to increase. During the first few years when the trees are still small, they can grow and tolerate insufficient moisture, but as they grow this insufficiency will have an increasingly adverse effect, with the ultimate risk of drying out at the crowns or withering. Also, owing to the time involved in monoculture, the properties of a forest soil may change progressively under the influence of a plantation, particularly where introduced species are grown and where the composition of forest litter may modify the direction of soil genesis.

Thirdly, in addition to being a resource, trees may be planted primarily to improve the environment or to arrest erosion induced by ill-advised agricultural practices. Afforestation is a major, versatile method of controlling water erosion by increasing both evapotranspiration and infiltration, which reduces surface runoff. The injurious or uncomfortable effects of winds can be reduced by shelterbelts and the moisture regime in leeward fields is improved. Again, it may be skilful management rather than soil type alone against which success should be measured.

For instance, it is now known that only shelterbelts with a few widely spaced and carefully managed rows will improve agriculture in sub-humid areas rather than barriers a kilometre wide like the 'State' shelterbelts of the Russian steppes.

Despite the many factors involved in the distinctive land use of successful silviculture, the interaction of trees and soils is one of the basic problems of forest science (Leaf 1956). In large measure, soils influence species composition and growth rate. An understanding of the inter-relationships of forest and soil is necessary before many practical problems can be appreciated, especially since forest soils are by designation marginal both geographically and in terms of natural fertility. Problems of regenerating and expanding forests have to be met in mountainous areas or in remote plains like the taiga regions of Siberia. The steep slopes of many forested areas means that the physical properties of soils, in relation to their resistance to pressure or slippage, have also to be taken into account. Otherwise, landslides may follow clear-cutting and logging. This adds to the expense of logging road construction and creates sedimentation problems in lowlands and estuaries downstream.

1.2.4 SOILS AND URBANIZATION

With the growing concentration of population in towns, agricultural soils gradually disappear beneath urban structures. A knowledge of soils, however, remains relevant to an understanding of the built environment. The patterns of excavation, reconstruction and re-vegetation on soils being converted for urban land-use reflects increasingly decisions made in the light of detailed knowledge about soil properties.

(a) *Engineering properties of soils*
Engineering works in the environment involve the effects and cost of loading the landsurface with artificial structures. If stresses in the uppermost soil layers cause local shearing and particle re-orientation, the reduction in hydraulic conductivity is substantial, particularly in clay and loam soils. Thus soil strength under buildings, roads, bridges, embankments, and dams is widely studied (Northey 1966), both before and after loading, with the aid of concepts and techniques developed by the specialized technology of soil mechanics (fig. 1.15). The highly compressible materials beneath Mexico City provide one of the most striking examples of the problems which arise in cities with inadequate foundation in their subsoil. The general process of subsidence of this area began in pre-Spanish and Spanish times when the area was first drained, and has accelerated with pumping of water from artesian wells. Since 1900 the central part of the city has sunk by about 5 m and continues

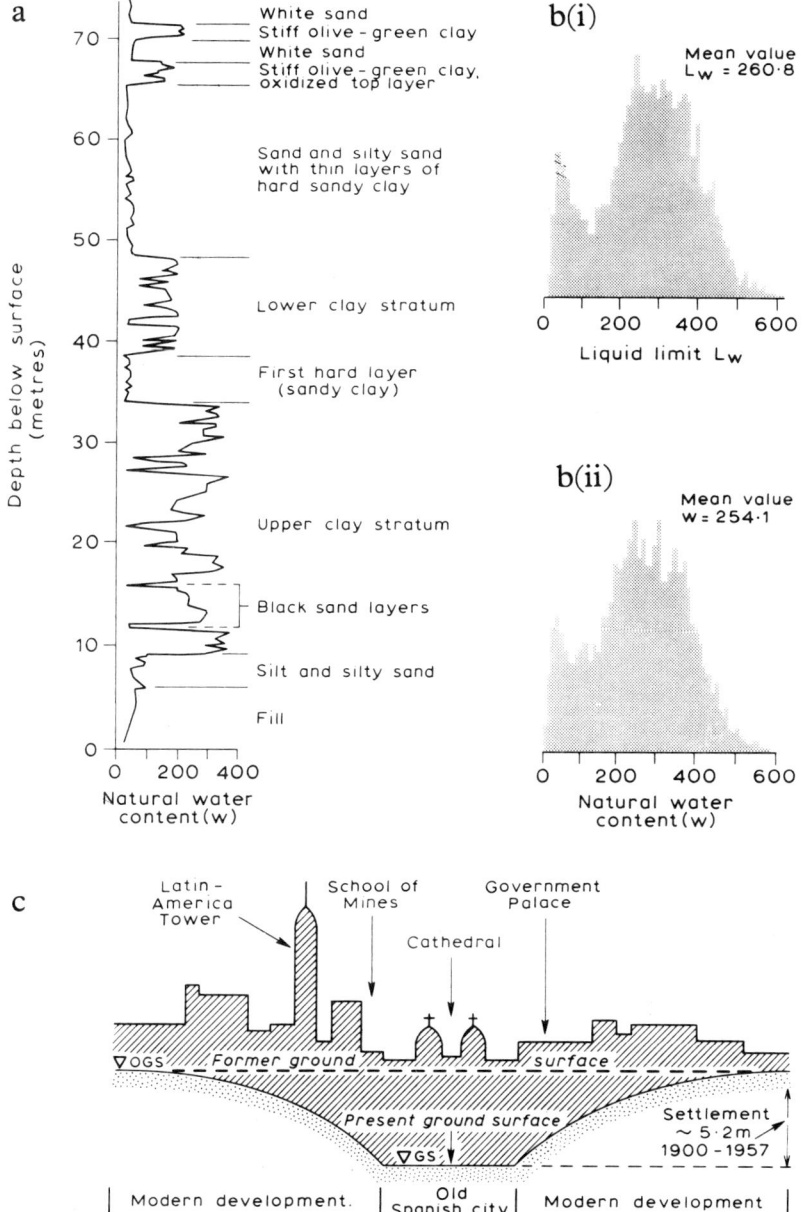

Figure 1.15 Significance of soil properties in a city built on sand.
a and b. Particle sizes of typical subsoil profile beneath Mexico City (a), and the approach of natural moisture content to the liquid limit of the soils (b) (Terzaghi, 1955).
c Settlement under the weight of Mexico City during the twentieth century (Jumikis, 1962).

to do so at a rate of 1 mm per day. The result is the tilting of monumental buildings, disastrous breakages in the city's sewer system and pipelines, and the cracking of pavements and streets. In other localities landslides may be a threat (Parkes and Day 1975).

Although the civil engineer is usually dealing with depths much below levels of agricultural interest, most of the properties of the natural materials with which he works are the same. When individual houses are constructed, the scale is sufficiently small for properties in soil B horizons to influence foundations. This is frequently the case with the extensive hardpans in Eastern U.S.A. In the laying of pipelines, the analogy with ploughing is even closer. Soil conditions may influence the line of excavation and determine the amount of support for pipes when in position. With increased pressures in pipes and heavier traffic on roads above them, increased subsidence may create leaks. Most measures to limit or check such damage depend on a sophisticated knowledge of soil properties, as seen in the ventilation of soils in urban areas in western Europe where underground leakage from natural gas mains has become a major problem in the last decade. Corrosion rates of iron and steel piping depend partly on soil properties, especially those which influence soil air and water. Corrosion is particularly rapid in anaerobic environments where sulphate-reducing bacteria are most active. British gas, water and oil concerns have all encountered serious safety and economic problems associated with the corrosion of subsoil ferrous pipelines (Corcoran *et al.* 1977).

In tropical areas the main engineering concerns are roads as the arteries of urban growth. Lateritic soils and gravels are the common, naturally occurring material for road, airfield and earthdam construction and for structural foundations (Clare and Beaven 1962). They are gravel clays, with the clayey material binding the coarse aggregates. If undisturbed, lateritic soils may produce good subgrades which are permeable. However, weakly indurated gravels pulverise during compaction. If water is present, the soil may soften and loose strength by progressive increase in density, impermeability and plasticity. Possibly, it may begin to shrink and swell. Most roads in tropical countries range from dry-season tracks to roads with such gravelled surfaces, and once traffic exceeds 200 vehicles per day, corrugations and potholes develop.

(b) *Waste disposal*

Since the times of Ancient Greece, the soil has served as a medium for sewage disposal. This was the practice in Roman cities, too, and was taken up again in Britain at the beginning of the eighteenth century, with effluent irrigating meadows. Recently, as supplies of farmyard manure have dwindled, sewage sludge at rates of 25 tons/ha has been applied, some sewage works disposing of much of their sludge in this way (Berrow and Webber 1972). Currently in the U.S.A., 90 per cent of all waste disposal is to the soil and a view of the soil as a waste-treatment system develops urgently (Herschaft 1972). Soil removes BOD, viruses, and pathogenic bacteria from sewage effluent. Some knowledge of how soils and soil micro-organisms make an indispensable contribution to the city is therefore part of an understanding of urbanization. Soil micro-organisms oxidize a great variety of organic substances and synthesize new stable organic compounds. Their effectiveness on sewage relates to the huge surface area which a large volume of minute organisms presents to water percolating through the soil. A feature of continued suburban expansion in many parts of the world is the septic tank. In siting this system, the apparently simple percolation test is required. However, its results can be so variable that interpretation depends on some knowledge of soil properties, such as texture and porosity. Poorly drained soils will not absorb water at the required rate. In excessively free-draining soils, the effluent may not be adequately purified before it reaches groundwater (Bouma 1974).

The effective disposal of other urban and industrial wastes relies on a range of soil properties. Complex organic wastes include used motor oil which is often discarded on soil or used for some purposes that ultimately reach soil (Giddens 1976). Problems associated with the disposal of waste products from atomic reactors have led to many studies of the selective-sorption properties of clay minerals (Nishita et al. 1956). Where landfills are used, several alternating layers of rubbish and soil can be built up with barriers of undisturbed soil preventing lateral escape of waste. Porosity should be sufficiently fine to exclude insects yet large enough to allow air circulation to oxidize organic wastes and to prevent any build-up of toxic gas. Clay percentage must be limited to prevent the soil from cracking during dry weather, with the risk of desiccation fissures penetrating to the waste layers. The subsoil must be relatively impermeable at landfill sites. Landfills for municipal waste may not be safe for arsenic- and selenium-containing industrial wastes. For example, selenium has been reported to pollute groundwaters 5 km away from a dump on Long Island (Korte et al. 1976). It is therefore important to understand how such elements might be retained in the soil and their toxicity reduced.

It is ironic that animal excreta is now termed farmyard effluent, where not returned to the land mixed with bedding straw. Instead, semi-liquid slurries are spread untreated on the soil. Poultry manure at more than 10 m tonnes/year in the U.S.A. is returned to the soil from broilers,

65 per cent of which are situated in the south-east (Jackson *et al.* 1975). Waste water from agricultural processing plants also depends on soil filtration as from potato processing in Idaho (Smith 1976).

(c) *Landscaping and soils*

If plant growth is to cover reclaimed coal waste, one problem is the fixation of added phosphate in a form unavailable to plants. The large amounts of acids and sulphates formed do not encourage root development (Pulford and Duncan 1975) and the sulphides release sulphuric acid and metal sulphates on oxidation. The acidity of the drainage water brings iron, aluminium, and manganese into solution and to levels which may prove toxic to plants (Blevins *et al.* 1970). In the Peak District National Park, where 200,000 tons of solid material are discarded annually as slurry by the fluorspar industry, fertilizers are essential, as the low organic matter is expressed in very low nitrogen and phosphorus levels (Johnson 1976). Again, in reclaiming bauxite-mined lands in Jamaica, a high fertilizer application is recommended with a 30 cm spread of topsoil (Morgan 1974). The risk of mechanical instability in spoil tips is now too well known to require elaboration.

Large areas of soil materials in various stages of weathering are exposed where deep road cuts are made. The establishment of a protective vegetation cover to prevent soil erosion or slumping on the exposed soil materials is impeded by adverse chemical properties of the exposed material (Miller *et al.* 1976). Similarly, some soil and water conservation practices, such as land-levelling and parallel terracing, often involve removal of topsoil from part of the land surface. In the case of glacial till soils in the western Corn Belt, dense clay or clay loam subsoils exposed are infertile and unsuited physically from growing crops, and a sound fertilization programme including the addition of micronutrients is essential (Olson 1977).

Within suburban areas, soils knowledge is being increasingly used in landscaping. For example, in siting lakes in a suburb of Rotterdam, soil conditions were investigated to depths of 8–35 m below the ground surface, penetrating through clay and peat layers overlying a sandy subsoil (Haans and Westerveld 1970). For sports fields, specific admixtures of sand-soil-peat are prescribed so that the soil does not change structure when compacted (Bingaman and Kohnke 1970). Another urban problem is that impermeable soils will compact and puddle under excessive pedestrian traffic. School yards and playgrounds are particularly susceptible to this condition. A similar problem has been envisaged for the town-dweller seeking pleasure in the countryside, but confined to doing so from his car by the threat of soil on his shoes.

1.3 Soils and geographical method

An understanding of many aspects of both physical and human geography is enriched by some comprehension of the nature and significance of soil properties (Burnham 1973). Soils, however, offer more to the geographer than enlightenment about the environment or why man adopts certain agricultural or engineering practices. Being a virtually continuous but continually varying cover on the earth's landsurface, soil is one of the earth's most distinctively geographical phenomena. Intrinsically, therefore, soils are well-suited for exemplifying or practicing geographical method.

1.3.1 SOILS AND A REGIONAL APPROACH

(a) *Latitudinal variations*

Awareness of the degree of differences in soils in the major world regions has become increasingly highlighted by the limitations of concepts, methods, and techniques based on information acquired in mid-latitudes when applied to tropical soils or in semi-arid areas. For example, low organic content of arid-zone soils makes most of adsorption–desorption studies of little relevance as such studies usually relate to soils with relatively high organic contents. To offset this, the use of green manure is not necessarily effective. The low moisture content at the time when manure is ploughed under creates unfavourable conditions for decomposition.

Tropical soils are far too complex to fit readily into the framework of concepts based on mid-latitude soils (Gerasimov 1973). The degree and depth of leaching also adds to the geotechnical distinctiveness of tropical soils.

Fertilizing tropical soils raises distinctive problems (Engelstad and Russel 1975). There is little agreement on the method for determining the lime requirement of highly weathered tropical soils and appreciable losses of calcium by leaching have been reported (Mahilum *et al*. 1970). Whereas quantity of clay accounts for much of the phosphate retention in tropical soils, the relation is if anything reversed in British soils (Lopez-Hernandez and Burnham 1974). In such soils, phosphate deficiency is often so critical that plants hardly respond to any other nutrient than phosphorus and hence are unlikely to respond to rises in pH with liming.

Iron and aluminium hydroxides also influence engineering properties of tropical soils by coating clayey constituents and binding them into coarser aggregates. This suppresses the behavioural relationship between particle size and plasticity on which many mid-latitude soil engineering procedures depend. In high latitudes, the engineer's interest is the effect of extremes of freezing temperatures on engineering designs (fig. 1.16).

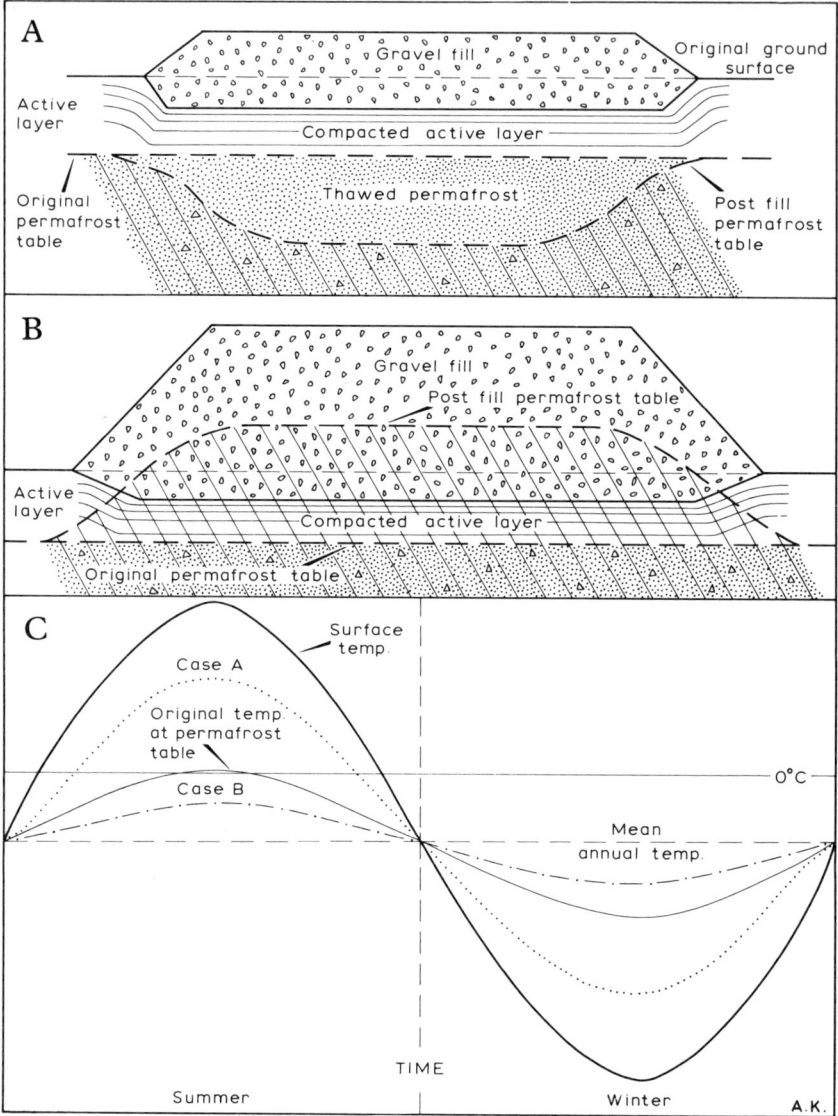

Figure 1.16 Effect of gravel fill on thermal regime of permafrost soils (Ferrians *et al.*, 1969). The removal of protective vegetation during road construction means that some insulating effect is lost, and if the ice thaws, the structure will be damaged.

A Too little fill—the insulating effects of the fill together with the compacted active layer is less than the insulating effect before construction. Permafrost thaws and the active layer increases in thickness.

B Insulating effects of fill and compacted layer are greater than original active layer. In this case, the permafrost table is raised because the amplitude of seasonal variation is smaller.

Agricultural interest in such soils is minimal due to the very slow biological activities in such climates. Nonetheless, there are soil groups specific to regions of continental climate and widespread permafrost (Sokolov and Targul'yan 1976).

(b) *Longitudinal variations*
At certain higher latitudes, the U.S.S.R. extends from west to east for more than 4500 km. Along this span, a variety of soils reflects the range of environmental conditions which pose to Soviet investigators a comparable range of problems and require specific agricultural and engineering methods for broadly similar soils. For instance, swamp soils on the Baltic become waterlogged mainly in spring or autumn, whereas similar soils of the Far East, with its monsoon climate, become waterlogged in summer. Even in the case of 'Mediterranean' countries, regional characteristics predominate over the general in soil properties related to rice-growing (Kyuma *et al.* 1974). Since several factors, variable from place to place, may influence the effects of a given saline water, many standards set up for water quality and use have only a local or approximate value. Similarly, along a given tropical latitude, soils have developed under

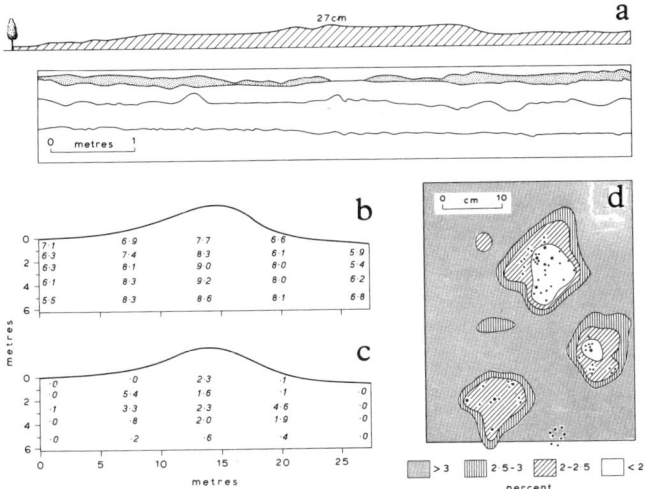

Figure 1.17 Local variability and micropatterns in soil properties.
a Greater development of leached horizon (stippled) on flanks of a low mound in the Semenkov Forest near Moscow (Rubtsova, 1967).
b and c. Distribution of pH and percentage carbonates below a termite mound in Rhodesia (Watson, 1962).
d Local drying of sandy soils by plant roots (Voronkov, 1967), at a depth of 150 cm on the floodplain terrace of the Borovka River.

conditions ranging from desert to rainforest, with highlands extending to the snowline. Thus, not all 'tropical' soils are strongly weathered and this variety complicates their understanding and optimum use.

(c) Micro-geographical variations (fig. 1.17)

It is generally known that the soil mantle is heterogeneous and demonstrates geographical patterns on broad scales. Less obviously, even on the smallest scales of study and utilization, soil remains intrinsically a geographical phenomenon. Micro-geographical variations in soil, termed soil heterogeneity, refer to changes in soil properties at short distances from some centimeters to several meters (Collins 1976). Such micro-geographical variations in soils have been most closely studied in forests. Stemflow water creates a wetter environment in soil contiguous to the stem than that prevailing farther from the stem. This results in an intensification of the podzolization process in soil nearest the stem (Gersper and Holowaychuk 1970, p.793).

Even on the smallest scales, therefore, the soil is not necessarily a uniform medium. It comprises a large number of micro-habitats with point-to-point variations in patterns of diffusion of moisture and solutes, and variations in microbiological and root activities. Micro-geographical knowledge is clearly needed about the precise extent of such changes in the form of large-scale detailed maps (fig. 1.17d).

(d) Soil mapping units as geographical regions

Inevitably there are no soil individuals as there are among plants and animals, and intermediate forms grade from one recognized soil type to another. However, the Dochuchayev school deduced that major soil types would be sufficiently individual to be mappable units. Subsequently, experience in soil mapping has benefited from increasingly objective, clearly defined criteria (Courtney and Webster 1973) and from precise methods of laboratory analysis. Areas can now be identified with little difference in soil properties within their boundaries (fig. 1.18). It is intriguing that the distinctively geographical interest in delineating regions (Whittlesey 1936) has faded from geography without careful evaluation of soils as the phenomenon most intrinsically and usefully regarded as geographical regions, recognizable, and now statistically definable as such, at all scales. Today, soil survey data is increasingly stored and processed in forms amenable to automated cartography (Ragg 1977). Four major methodological and theoretical problems and possibilities of defining and characterizing regions are exemplified in the classification and mapping of soil types.

First, the problem which most taxed the perplexed regional geographer of the Colonial Age, that of defining a boundary, is the everyday

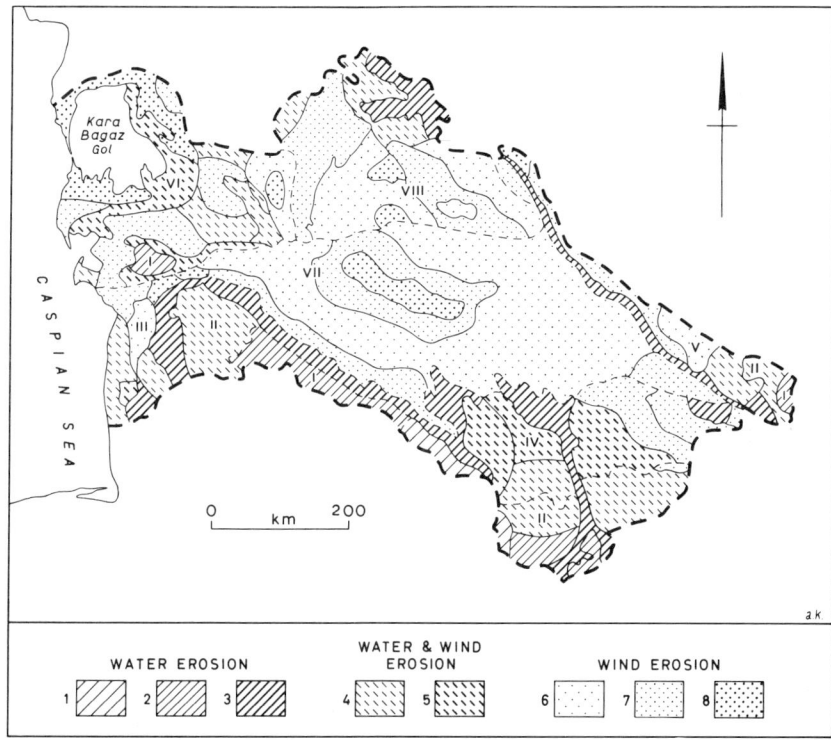

Figure 1.18 Regions of soil erosion in Turkmenia (Stepanov, 1966). The main control is the variability in origin of the parent materials. The degrees of erosion are: (1) slightly or unaffected, (2) moderate to severe, (3) downwash, (4) slight, (5) slight water erosion with moderate to severe wind erosion, (6) slight or unaffected, (7) moderate, (8) severe to very severe.

decision of the soil surveyor. Precision depends on the soil properties considered. On soils in eastern Kansas, sampled on a 10 m grid, pH and silt content change slowly, but sand percentages change sharply at the contact between adjacent soil types, by as much as 8 per cent within 20 m (Campbell 1977). In contrast, transition zones up to 90 km wide have been noted at the edges of areas of red earths in Australia.

Second, observations confirm whether the soil within a boundary is uniform. The density of observations needed to estimate homogeneity within a soil region is inversely proportional to the average distance between boundaries (Burrough *et al.* 1971). Also, the range of values for a given soil property tends to increase with the size of area sampled (Wilding *et al.* 1964). In north-west Ohio statistics summarized the variability of 48 selected soil properties for given mapping units. The most variable properties were depth to mottling, depth to fine-

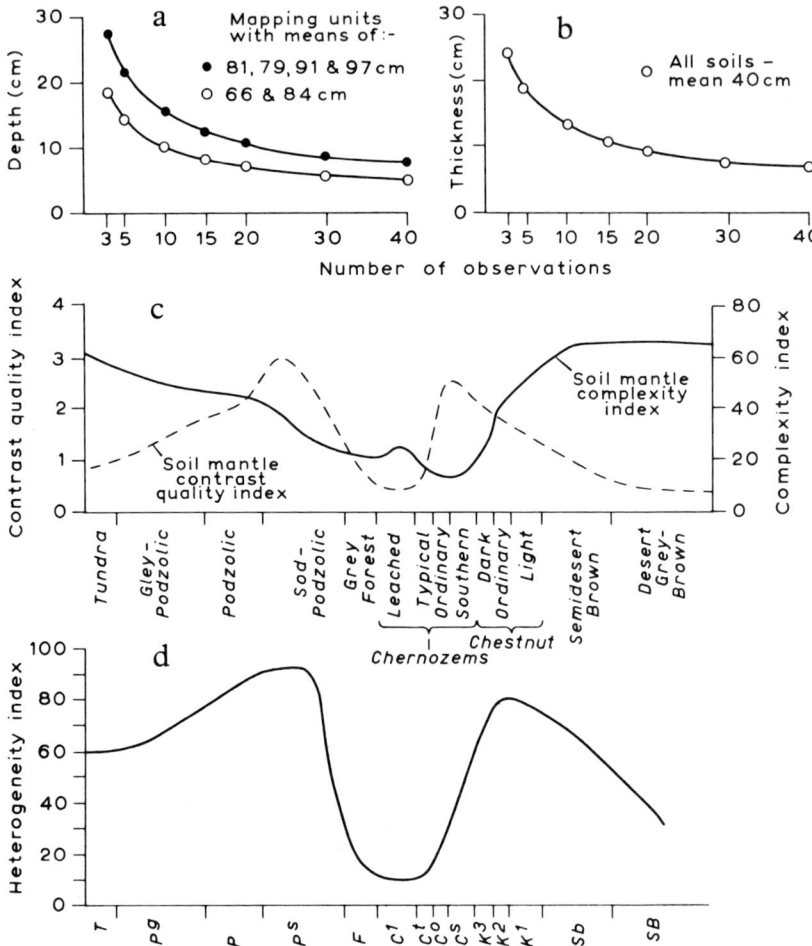

Figure 1.19 The influence on sampling and zonal distribution of soil heterogeneity.
a and b. The number of observations required to determine soil properties within certain limits of accuracy (McCormack and Wilding, 1969). The soils are coarse and medium-textured, with an illuvial, fine-textured B horizon, lying within the shorelines of late-glacial stages of Lake Erie. The properties are depth to subsoil discontinuity (a), and thickness of illuvial horizon (b).
c and d. Contrasts in quality, complexity, and heterogeneity of soils along the 55°E meridian in western U.S.S.R. (Fridland, 1967). The complexity index is based on the percentage of intrazonal soils in a mantle of a given zone. The contrast quality index is based on the range of soil quality within one zone, where best-quality soils are rated at 5 points down to 1 point for soils unfit for ploughing. The heterogeneity index is the product of these two.

textured discontinuity, horizon thickness, and texture. In contrast, soil colour hue and value, pH of all horizons, texture of the B2 horizon, and the size and shape of soil structures were the least variable. Depth of leaching curves suggest that 10–15 observations would define the mean depth of leaching within 10–20 cm for the mapping units used (fig. 1.19). Emphasis on establishing the degree of homogeneity within a mapping unit may conflict with the importance of establishing a boundary. In the Vale of White Horse, Berkshire, only 30 per cent of observations were made to verify either boundaries or their association with external features (Burrough et al. 1971), in contrast to the Netherlands where soil boundaries lack external expression and where 75 per cent of observations may be made to locate boundaries (Buringh et al. 1962).

Third, the validity of differences between mapping units may be examined. For any single soil property, the range of values in adjacent soil regions usually overlaps appreciably. The cumulative effect of several such differences, however, may add up to a distinctive difference overall. In Dodge Co., south-east Wisconsin, discriminant analysis measured the differences between the distribution of soil properties in drumlin and adjacent drumlin-free areas. The distinguishing criteria were all central to land-use planning and the definition of prime agricultural regions (Pavlik and Hole 1977).

Finally, the alternative premise, that soil regions do not in reality exist, may be considered. In the 1960s, soil classification schemes in many countries were revised or recast in new frames. In most cases, the precepts are those used in the U.S.S.R. and originate from the concept of 'natural regions' which dominated scientific views of the world at the turn of the century (Gersmehl 1977). From the belief that soil, vegetation and climate are arranged in distinct, discrete and discernible major regions, increasingly fine subdivisions have been made. Ultimately, local needs are met by local classification systems. In contrast, systematizers do not assume that the world is discretely regionalized. Soils can be viewed as a very complicated system of points made up of individual vertical columns of soil. This is both the approach and the problem of the Soil Survey staff of the U.S. Department of Agriculture, who adopted a new classification scheme in 1965. Soils are viewed as discrete individuals and are allotted to categories and at various levels in a hierarchy of categories, according to an intricate set of measurable limits. The basic principle is that soils should be classified according to characteristics and properties intrinsic to the soil itself rather than on environmental or genetic criteria. As these units are not devised for mapping, the system can prove difficult to use for soil mapping (Webster 1968). However, the system has been widely tested and is still popularly known as the 7th Approximation. The U.S.D.A. Soil Taxonomy is perhaps the most

40   Geography and soil properties

explicit description of the system. Its terminology is built up synthetically from Greek and Latin syllables and formative elements into a pedological Esperanto (see Appendix).

(e) *Spatial distribution of soil properties*

Despite the validity and usefulness of defining, classifying, and mapping soil regions or of identifying distinctive soil individuals, gradual and progressive changes are the inherent geographical expression of certain soil properties. Properties which are commonly continuous trends often relate to transported materials from a given source area, with gradients of change in wind-transported materials being particularly smooth

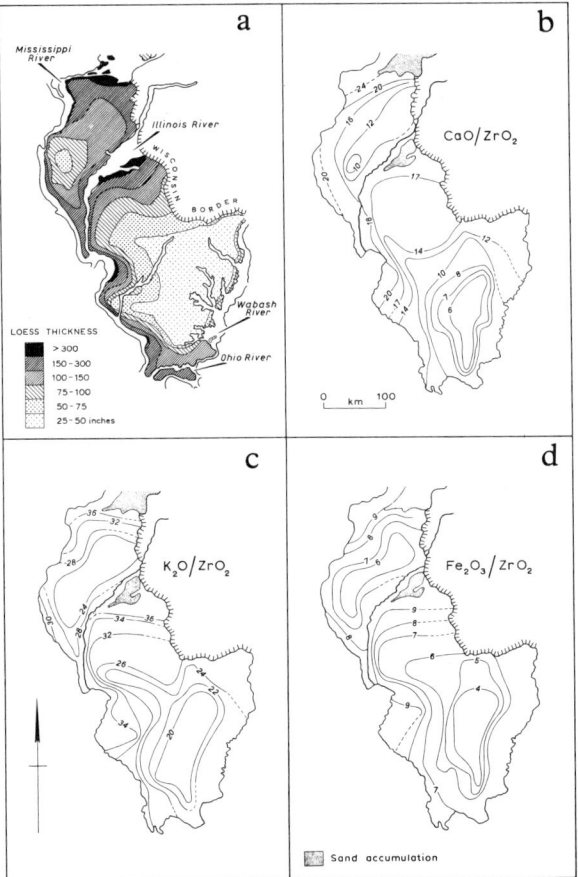

Figure 1.20   Downwind decreases in loess thicknesses and its influence on soil weathering characteristics (Jones and Beavers, 1966). The significance of the weathering ratios depends on the immobility and stability of zirconium, and shows that weathering is least in deep loess.

(fig. 1.20). Increasingly, source areas from which soil properties diffuse are 'hotspots' of human activity (Davies 1976), whether these be the over-grazed sites of nomads' camps where wind erosion starts, or the concentration of toxic compounds declining steadily with distance from a road (Minami and Araki 1975) or sewage disposal ponds (Lund *et al.* 1976). Thus, man repeats on a larger scale the point-patterns of activity of soil organisms, like the mole-hill field or the termitaria-speckled savanna.

1.4 Conclusions about soils and geography

From a sketch of impressions on the 'soils and geography' theme, several conclusions might be drawn. Two, in particular, appear worth outlining and incorporating into the following description of soil properties.

First, those who use the soil do not use the term soil in the same sense. The connotation varies according to the purposes to which the soil is put rather than there being any difference in the superficial materials used which can be defined concisely and uniquely. The main difference stems from the compartmentalization of knowledge and practice rather than lying in the natural body of the earth's uppermost mantle. Further, there are many soil properties, like texture or porosity, which do mean virtually the same to a diverse range of specialists. Therefore, it is suggested that the geographer, wishing to understand all man's activities, could find the approach of any one technical specialist too restrictive. It might be hoped that the geographer, despite his increasing urbanization, will always regard the agricultural use of soil his main interest, but increasingly required is some basic familiarity with soil properties beyond the context of soil as a medium for plant growth.

Secondly, anyone with an innate geographical sense of interest and enquiry will find the soil a particularly attractive subject for study. The soil reveals the scale of man's impact on the environment, it shows clearly how other environmental influences work in interaction and it is a medium ideally suited for the practice of the art and science of regional geography. The geographer, therefore, takes up very eagerly themes such as soil erosion or pesticide problems, or the influence of landsurface declivity of soils or sets off with auger or dividers to survey regional boundaries in the field or on the map. However, studies of man's impact on the environment, the influence of the environment on soils or the study of the distributions and patterns of soil geography will inevitably be confined to a very superficial level if these exercises are undertaken without a pre-existing knowledge of the nature of soil properties, assimilated from a basic systematic framework. Too readily and eagerly the geographer seizes on the very obviously geographical aspects of soils, but aided by an assumed knowledge of soil properties rather than in the light of systematically acquired familiarity with the nature of soil properties.

2 The mineral fraction of the soil

Many basic soil properties and the suitability of soil for plant growth are closely linked with the mineral fraction of the soil and with the size distribution of its particles. Its estimation in the field is a basic step in soil survey, and its detailed study is critical for many engineering purposes.

2.1 The classification of particle sizes

A variety of subdivisions are used to describe particle size, all being essentially empirically derived (fig. 2.1). The search continues for natural breaks in particle size scales, of interest to soil scientists, geologists (Tanner 1969) and engineers who are trying to find mutually acceptable particle-size classifications. Fortunately, two of the boundaries are widely agreed: 2 mm divides the coarser fraction of gravel, stones, and pebbles from coarse sand in nearly all schemes. In the revised British classification, which accords with engineers' practice in so far as size limits are based on M.I.T. engineers' size grades, materials containing more than 35 per cent stones by volume (50 per cent by weight) are classed as gravel. At the other end of the scale, the finest silt passes into the clay fraction at 0.002 mm in most classifications. To describe these small sizes neatly, the micron is used as a unit of measurement, and is represented with the Greek letter mu – μm.

The greatest difficulties arise at the demarcation between silt and fine sand. The usefulness of a division at 20 μm is supported by the observation that the greatest available water capacity in soils is associated with silts in the 5–20 μm range (Petersen *et al.* 1968). On the other hand, the International system, separating fine sands from silts at 20 μm has been found inconvenient in desert areas where particles of this size are abundant. The division at 50 microns in the United States Department of Agriculture (U.S.D.A.) system is not the most logical from the viewpoint of erodibility. The very fine sand (50–100 μm) particles seem to behave more like silt than like other sand particles (Wischmeier and Mannering 1969). There is, in fact, a rapid decrease in the effectiveness of wind transport between 60 and 100 μm.

Some of the difficulties introduced by arbitrary size limits are avoided by the use of cumulative frequency curves. As with particle-size scales, however, regional adaptations of the basic textural triangle are also required (fig. 2.1).

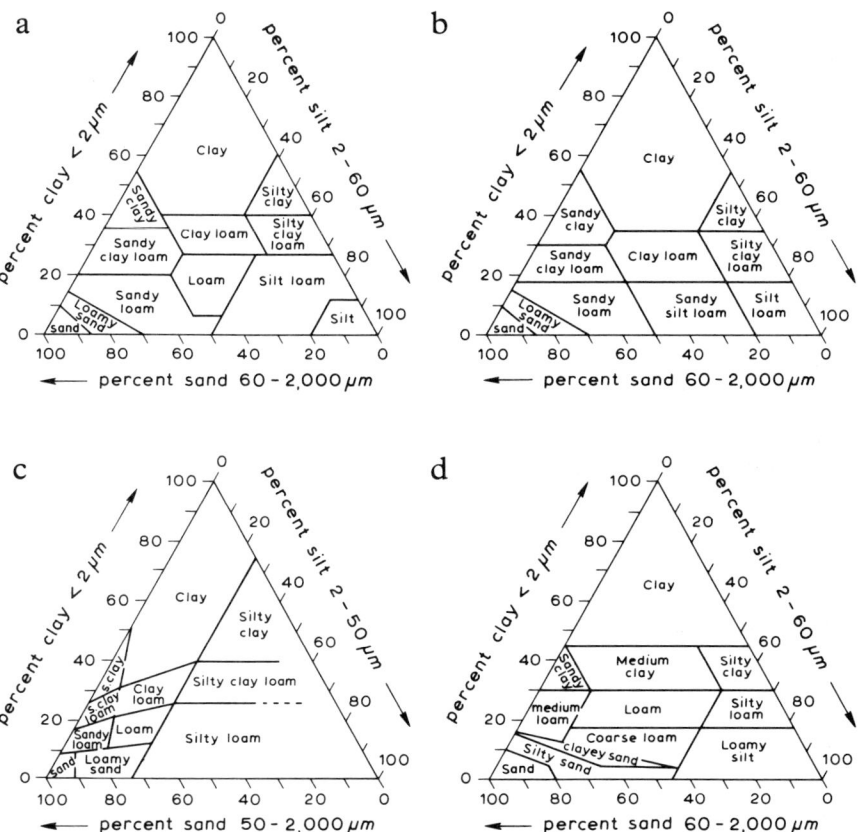

Figure 2.1 A range of textural triangles used to define limiting percentages of particle-size classes.
a Textural triangle as used in the U.S.A. since about 1947 (Whiteside *et al.*, 1967).
b Scheme now used by the Soil Survey of England and Wales (Avery, 1973).
c Soil textures defined for Australian use (Marshall, 1947).
d Textural classes used in soil mapping in East Germany (Lieberoth, 1968).

2.2 The coarse fraction

Detachment of fragments of weathered bedrock by freeze–thaw, windthrow, or by deep ploughing means that stones will make up some percentage of the volume of most soils weathered *in situ*. Equally, transported glacial, fluvioglacial, or colluvial materials contain appreciable proportions of stones, pebbles, and boulders (fig. 1.7a).

2.2.1 AMOUNTS OF COARSE MATERIALS IN SOILS

Downslope movement of coarse material explains why coarse materials are formed in soils developed on clay. Thus, the percentage weight of stones in soils from a wide range of parent materials in the Oxford area (Clarke and Beckett 1965) are representative of mid-latitude environments. The Canfield silt loam of north east Ohio includes an average of 4–9 per cent pebbles and stones (Miller *et al.* 1971). In humid temperate climates, soil fauna are very active in burying stones beneath the fine material which they transport to the soil surface. In contrast, in cold or very dry conditions, the land surface itself is stony and if fines occur they are found beneath the surface stones.

2.2.2 PREDOMINANTLY STONY SOILS

(a) *Skeletal soils*
A surface debris of stones and gravel occurs widely in higher latitudes. In spite of their thinness and other unfavourable properties, such skeletal soils produce timber on a moderate scale, as in the subarctic portion of Alaska (Wilde and Krause 1960). In mid-latitudes, particularly on limestones, coarse materials may bulk large at the ground surface. In Estonia, for instance, stoniness of underlying layers prevents deep ploughing over 30–40 per cent of the agricultural land. In the very stony north and west parts, 100 m^3/ha of stones are found on the land in its natural state. The cost of removing stones is an investment which is recovered in about 10 years (Kask 1965).

(b) *Transported coarse materials*
Stony soils are often an inevitable feature of fluvioglacial deposits. In Ontario, boulder spreads allow deeper penetration of tree roots and accumulations of fine mineral and organic fractions in fissures and crevices. These features markedly increase the rate of growth of jack pine and black spruce (Wilde *et al.* 1954). In Finland, some of the most stony parishes occur within the arc of the Salpausselka. Where stone gathering reaches a peak, 1000 cartloads may be removed in clearing half a hectare. The infertility of gravelly outwash soils can be offset by intensive applica-

tion of fertilizers and by irrigation to offset the small amounts of water retained in coarse, stony materials.

The way in which a surface lag develops on boulder clay is illustrated by the degree to which fragments of pre-Cambrian rock replace limestones in the surface soil horizons on drumlins in south-east Ontario (Rutherford 1971). Below escarpments or plateau edges, caprock fragments may be prominent on the ground surface downslope. Thus, coarse materials occur in surface soils in proximity to duricrusts. In most areas of stony tableland soils in north-west South Australia, the stones consist of quartzitic silcrete derived by the erosion on the duricrust plateau. Fragments are 0.5–15 cm in size, with occasional blocks of silcrete measuring 90 cm or more (Jessup 1960).

(c) *Stone pavements*

A striking feature of loamy, desert soils is their gravelly cover or desert pavement. Commonly, these are termed hammadas in the Middle East and they occupy many stretches of desert in Syria, Lebanon, and in the southern Negev. Stone pavements, however, are not restricted to extremely arid zones. In Solonetzic soils near the Barrier Range in western New South Wales, fine-textured soils, although they contain little quartz gravel and stone within their profiles, are completely covered by a gravel pavement, including stones up to 10 cm in length.

Scant grazing may be afforded by plants rooted in the soil beneath a stone pavement, as in the South Turkana desert in Kenya which is frequently used for grazing sheep. Clearly, a major role of stone pavements is to arrest further deflation in areas where the climate is too extreme for a continuous plant cover to become established.

2.2.3 SCIENTIFIC SIGNIFICANCE OF STONES IN SOILS

From the agricultural point of view, the problems of a profusion of stones in the soil is obvious and proverbial. However, coarse fragments in the soil give a visual indication of parent materials. The size, shape, degree of sorting, orientation and inclination of coarse materials in soils all reflect the nature of the transportation agents (fig. 1.4).

Stony or gravel soils are of particular scientific interest as favoured sites for habitation. These include the dry-point sites selected by man from earliest times onwards (fig. 1.9) to the soil fauna which everyday shelters beneath stones.

2.3 Sand

It is only in the sand-sized grade, with mineral particles less than 2 mm diameter, that 'fine earth' or soil in a strict sense is recognized. The lower

46 Geography and soil properties

size limit to sand is set at 50 μm in the USDA system and at 20 μm on the geometric scale of the International system. Sandy soils occur locally wherever sandstone outcrops are weathered. In addition, sandy soils are widespread in areas of Quaternary sedimentation (fig. 1.19). For example, one third of the Netherlands consists of Pleistocene sandy soils and similarly in the Lower Volga area sands cover a huge area.

2.3.1 MINERALOGY OF SAND PARTICLES

Mineralogy is a fundamental consideration in the understanding of soils (fig. 2.2). As plant growth depends critically on only a few macronutrients, it is easy to recognize if elements such as potassium, K, or magnesium,

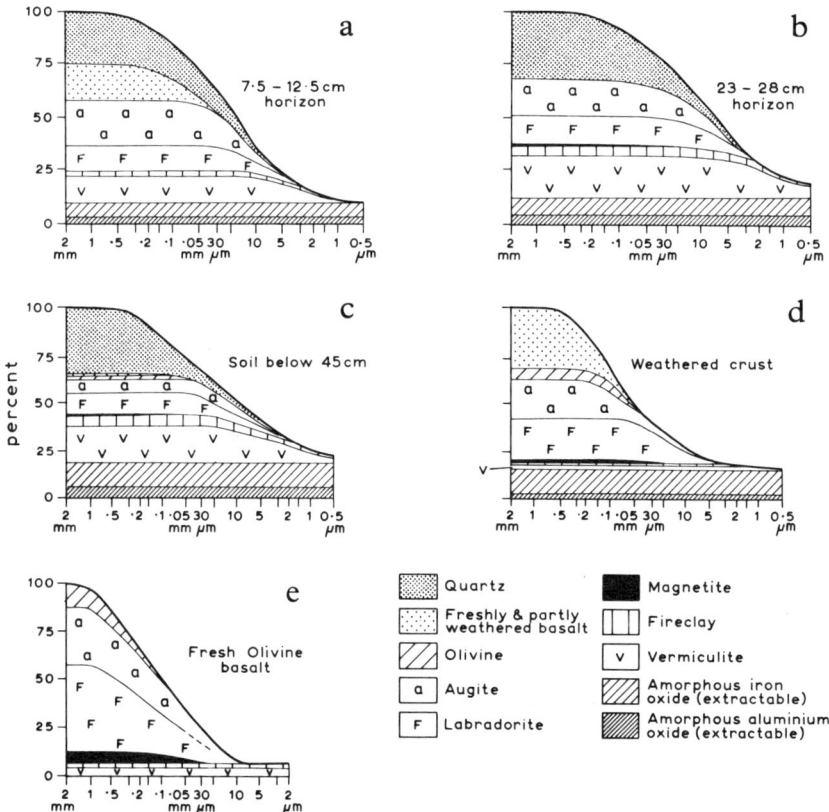

Figure 2.2 Physical and mineralogical breakdown of primary minerals in a fertile olivine basalt soil on the Antrim plateau, Northern Ireland (Smith, 1957). The quartz cannot be a product of *in situ* weathering of a basic rock. The wide range of quartz particle sizes may indicate an origin by glacial transport.

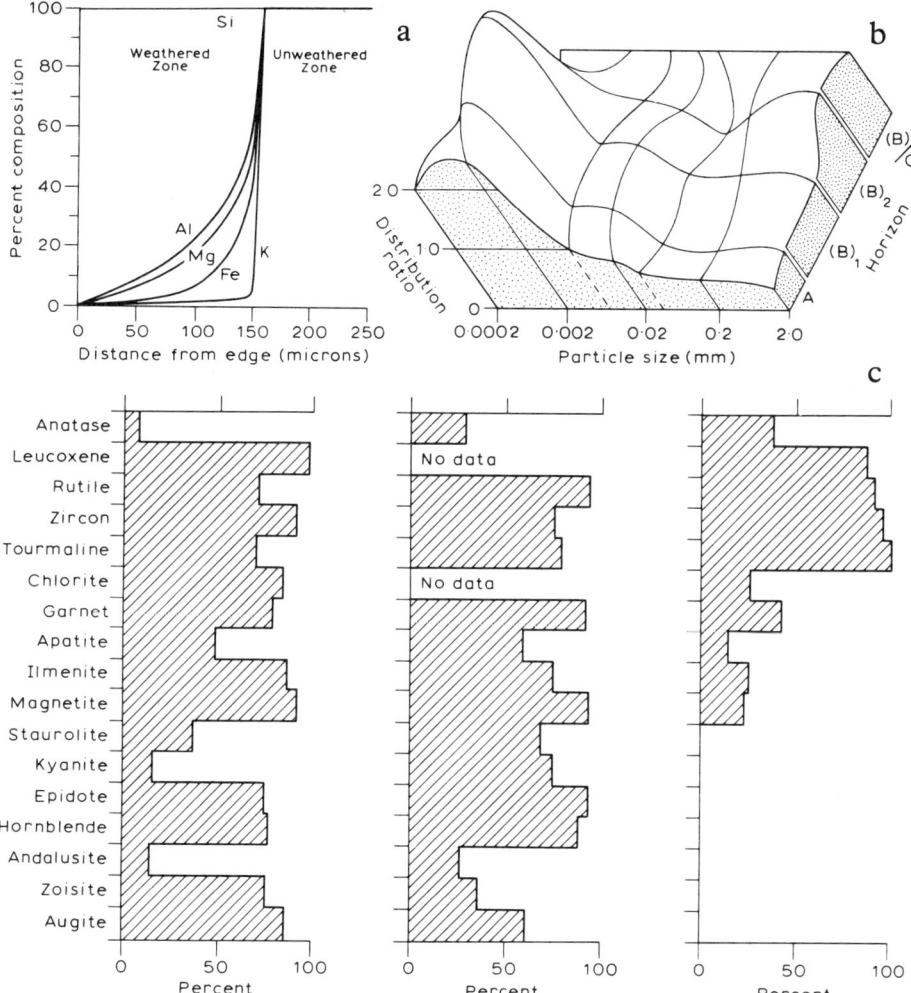

Figure 2.3 Influence of weathering on mineralogical characteristics of soils.
a Chemical composition of the weathered edge of a vermiculite flake, showing progressive release of cations (Sawhney and Voigt, 1969).
b High content of iron in coarser clay fraction and marked accumulation in the B_2 horizon of the Banbury clay loam (Storrier and Muir, 1962). The distribution ratio is obtained by dividing the amount of a particular material in the horizon under consideration with its level in the C or reference horizon. This analysis shows the form in which iron is weathered, translocated and deposited in a particle size.
c Frequency of occurrence of heavy minerals in Pleistocene deposits (left) compared with those of Carboniferous sandstones and grits (right) in North Wales, and with the world-wide pattern in Pleistocene sediments (Smithson, 1953).

Mg, might be released into the soil solution if a given mineral weathers easily (fig. 2.3a). In the scientific context, the mineralogical characteristics of even 1 or 2 per cent only of sands may, by their relative abundance and distribution, prove to be index minerals for studies of transportation and weathering processes.

(a) Light minerals

With a specific gravity of less than 2.70, the 'light' minerals such as quartz, feldspar, and mica commonly account for at least 95 per cent of the sandy fraction of soils.

Silica, SiO_2, usually occurring as quartz, is by far the most common of the many minerals which may occur as sand-sized grains in soils. For instance, in the sand and loamy sand soils along the Don River, the quartz content fluctuates between 90 and 100 per cent. As is evident from the formula for silicon dioxide, there are no basic cations to be released. This is in fact a secondary consideration, since quartz is virtually undecomposable. Thus, quartz becomes predominant in the sand fraction of soils and sediments (fig. 2.2).

Feldspars are a group of silicates of aluminum, Al, sodium, Na, and calcium, Ca. Owing to their greater susceptibility to subaerial weathering (fig. 2.2) feldpars are only common in sands derived from little-weathered source areas or parent materials. Thus, the alluvium of streams draining mountains created during geologically recent phases of orogenesis is typically rich in readily weathered minerals, such as feldspar. Floodplain deposits of the main rivers on the Indian sub-continent are relatively rich in feldspars, their sand fraction usually containing between 15 and 30 per cent feldspar.

Potassium feldspars, $KAlSi_3O_8$, are of two main types, orthoclase and microcline, which have the same composition but differ slightly in crystalline form. Potassium feldspars are the largest natural reserve of potassium, K, in many soils. From the K-feldspars to albite, $NaAlSi_3O_8$, which is sodium-rich, there is a continuous series of minerals with varying proportions of $KAlSi_3O_8$ and $NaAlSi_3O_8$, the so-called alkali feldspars. The sand fractions of soils developed on Tertiary formations in Bangladesh, for instance, contain 1–10 per cent feldspar, which is mainly alkali feldspar.

Plagioclase feldspar, compared with the alkali feldspars, forms a complete series from albite, $NaAlSi_3O_8$, to the most basic member, anorthite, $CaAl_2Si_2O_8$. In the sand fraction transported by the Nile floods, Ca-feldspar accounts for 26–28 per cent of the light fraction, K-feldspar accounts for a further 6–9 per cent, with quartz being less than half the total (fig. 2.4).

Feldspars, owing to crystallographic properties like their excellent cleavage or degrees of disorder within the framework of their crystal structure, are susceptible to physical breakdown. The effect of the grinding action within moving till, for instance, has been observed to be more pronounced on orthoclase than on microcline (Somarsiri and Huang 1971). In tropical soils, feldspar may be removed completely from surface horizons by weathering. In West Africa, even at depths of 2 m, all cleaved feldspar grains are much cracked, with clay formation proceeding along the cleavage planes (Nye 1955).

The mica group are the third main light minerals and are highly variable in composition and structurally complex. They are aluminosilicates with potassium, magnesium, iron, hydroxyl, and other ions. Micas account for between 5 and 30 per cent of the sand fraction in the floodplains of the main rivers in Bangladesh and about 3–4 per cent of the sand carried in the Nile floods. Muscovite, $K_2Al_4(Si_6Al_2)O_{20}(OH)_4$, is also known as white or potash mica. The more common is black mica or biotite, which is a magnesium–iron mica. Both types occur as flakes rather than grains, due to their easy, perfect basal cleavage. They are therefore, readily transported in water, but prolonged grinding action completely destroys micaceous structure, and transitional stages towards clay minerals are often observed (fig. 2.3a). Under weathering, biotite alters more readily than muscovite and it is therefore much less common in soils and recent sediments than muscovite. Nonetheless, in most Ganges deposits about 45–60 per cent of the mica grains are biotites. In the Brahmaputra and Tista deposits, biotite grains account for about 65–75 per cent of the total mica fraction (Huizing 1971).

(b) *Heavy minerals*

Heavy minerals have a specific gravity greater than 3.0 but, due to characteristically small amounts of ferromagnesian minerals in parent rocks, their proportion in soils does not often exceed 3 per cent. Much of the natural micro-element composition of soils in Quaternary deposits is governed largely by the heavy minerals in the sand fraction (fig. 2.3c). Also, because of the specific nature of their crystalline structure, many heavy minerals are broken down relatively easily. Thus, in the Euphrates fine sand deposits, the weight of heavy minerals in the 50–74 μm fraction varies between 11 and 20 per cent in the relatively unweathered desert soils, but represents 6 per cent or less of the recent alluvial soils (Al-Rawi and Sys 1967, p. 190).

Augite $(Ca,Mg,Fe)(Mg,Fe,Al)(Al,Si)_2O_6$ is one of the more significant heavy minerals found in the sand fraction of soils (fig. 2.3). For example, it is particularly characteristic among the unstable minerals of the soils developed on the Pleistocene deposits of North Wales (fig. 2.3). The considerable degree of weathering which has taken place, especially

50   Geography and soil properties

in many of the coarser interglacial sands, is shown by the deep indentations on the sand grains. Like biotite, augite is a slowly soluble source of magnesium.

Hornblende $(Ca,Na)_2(Mg,Fe,Al)_5(Al,Si)_8O_{22}(OH,F)_2$ also decomposes readily, the early stages leading to a chloritic product forming along the edges and cleavage traces of the grains. The Lower Volga area is an example of sandy soils which contain large amounts of hornblende, with a maximum of 19 per cent in the Astrakhan Oblast (Vakulin 1966, p. 1786). Hornblende may be a significant source of magnesium.

Apatite $Ca_5(PO_5,CO_3)_3(F,OH,Cl)$ is significant as the most common phosphorus-containing primary mineral in the soil. As it is slowly soluble and has a relatively high rate of weathering, calcium, fluorine, and chlorine may also be released together with phosphate ions when apatite is decomposed (Williams *et al.* 1969). Epidote $Ca_2(Fe,Al)_3Si_3O_{12}(OH)$ is another potential source of calcium, being the product of low-grade contact metamorphism of limestones (fig. 2.3). It is estimated that stretches of the marine Caspian sands may contain up to 41 per cent epidote. Finally, olivine Mg_2SiO_3 is noteworthy as one of the simplest of the magnesium silicates and as an easily weathered mineral (fig. 2.3).

2.3.2   UTILIZATION OF SANDY SOILS

In *Quiet flows the Don*, Sholokhov wrote 'A farm do you call it? ... we live from hand to mouth, and all our life is one long grind and struggle ... our land is sandy'. Wherever quartz is preponderant, sand is the sterile fraction of the soil, releasing no nutrients and retaining little water. Natural biological activity is also low in sandy soils, which accentuates their infertility. Quartz sands are unable to retain artificial fertilizers against the leaching action of heavy rains and crop and environment may be damaged if heavy applications are made. Sandy soils may be given over to pine plantations, other non-agricultural uses, or to golf courses near affluent suburbs. In certain economic circumstances, physical properties of sandy soils can be used to advantage. In market gardening enterprises, markets of large urban areas make economical many artificial improvements, controls on soil environmental conditions, and heavy fertilization. Loose, unbound sandy soils are easily worked by hand and for several crops, particularly asparagus, it is an advantage for them to be easily removed by hand. The low water-holding capacity of sands becomes an advantage, since dry, sandy soils warm up quickly in spring. Being loose and retaining little water, sands may pose special problems for the civil engineer (fig. 1.15).

Finally, sandy soils are not invariably inherently infertile. In regions where leaching is restricted and where major rivers gather sediment in recently uplifted mountain ranges, sand fractions may have considerable natural stores of nutrients to release on weathering. The names of such regions are household words to students of civilization.

2.3.3 SCIENTIFIC SIGNIFICANCE OF SANDY SOILS

In the study of the origin and evolution of soils and sediments, quartz sands are used as a weathering index mineral, owing to their persistence, abundance and wide distribution. Due to differential weathering, the quartz content of soils may increase in comparison with unweathered parent materials or alterable minerals in the profile. For example, in Iraq the quartz:feldspar ratio appears to differentiate between the desert soils and the alluvial soils as it ranges from 2 in the former to 1 or less in the latter (Al-Rawi et al. 1969).

Of the heavy minerals, the least destructible, like garnet and zircon (fig. 1.20), may indicate the relative degree of weathering in the horizons of a soil profile. In the Middle East, there is a relatively higher content of epidote and a lower content of green hornblende and augite in the Tigris levee deposits compared with those of the Euphrates. Since epidote is more resistant to chemical weathering than augite or hornblende, the Tigris sediments are more weathered or are derived from more weathered parent materials (Al-Rawi and Sys 1967). Heavy minerals, despite their usually small percentage, often provide the clearest evidence of the origin of soil horizons (fig. 2.3b). More commonly, the importance of the diagnostic value of specific heavy minerals is cited in identifications of source areas for transported parent materials (fig. 2.3c). The importance of garnet as an indicator of glacial drift has been established for some time.

2.4 Silt

Many soil properties grade imperceptibly from coarse sand to fine silt. In particular, soil mineral particles comminuted beyond any limit set for sand sizes remain preponderantly quartz. However, silts may consist of unstable minerals to a larger degree (fig. 1.20) than in coarser sand and, because of their smaller size and hence an enlarged surface area, they are a more readily available natural source of plant nutrients than sand-sized materials.

2.4.1 MINERALOGY OF SILT PARTICLES

The degree to which silts are a source of plant nutrients depends on median particle size within the silt range. For instance, the silt fraction of the Pegwell Bay brickearth contains 75–77 per cent quartz and 21–23 per cent

52 Geography and soil properties

feldspar. In the fine silt fraction, the feldspar content is similar, at 23–25 per cent, but the quartz fraction is reduced to 58–66 per cent and layer silicate minerals account for 8–16 per cent of the total (Weir *et al.* 1971). In Kansas, the Peoria loess is again slightly more than half quartz, with a quartz:feldspar ratio of about 4. Orthoclase is the prevalent feldspar with sodic plagioclase and microcline also present. One of the reasons for the contrast with the quartz:feldspar ratios of about 2 in the Sheringham or Pegwell Bay silts is the presence of volcanic ash shards in the Peoria loess which may range from a trace to as much as 10 per cent of the coarse silt fraction.

Source area is an important control on silt mineralogy. Where weathered from basic rocks, silts may be a source of natural soil fertility where partially weathered silt-sized fragments of ferromagnesian minerals such as augite or olivine are present. Thus, Herodotus described the Nile floodplain as 'black and crumbly as if it were the mud and alluvial deposit, brought down by the river from Ethiopia'. Significantly enough, the silt

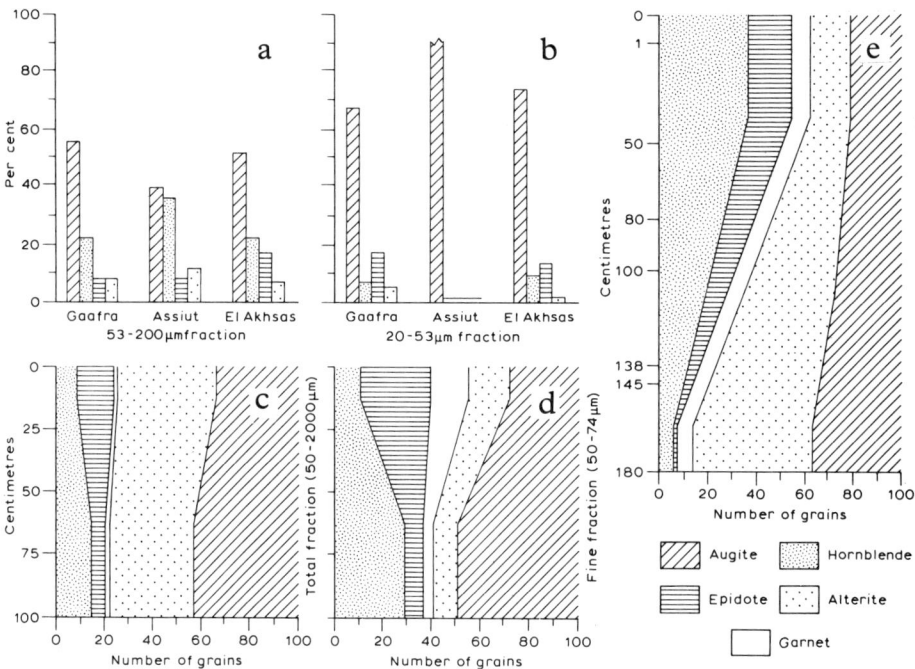

Figure 2.4 The basis of natural soil fertility in hearths of civilization.
a and b. Heavy minerals in the suspended sediment in the Nile, the source of the very fertile alluvial soils of Egypt (Nabhan *et al.*, 1969).
c–e Heavy minerals of desert soils on alluvial deposits of the Euphrates (Al-Rawi and Sys, 1967), on old alluvial terraces (c and d) and in recent basin soils (e).

carried in the Nile floods (fig. 2.4) is only 18 per cent quartz, with 20 per cent feldspar, 12 per cent hornblende, 6 per cent mica, 3 per cent augite, with 45 minerals in all being recorded, including 0.5 per cent apatite.

Subhumid and desert soils have high concentrations of calcium carbonate as individual particles, like those of the Mesopotamian plain which contain about 20–30 per cent carbonate on average (Altaie *et al.* 1969). They occur as either primary or secondary forms as calcite crystallites concentrated in the silt-size range. Similar sized crystallites of gypsum, hydrated calcium sulphate, $CaSO_4.2H_2O$, may also be present. A high percentage of calcite silts is a virtually inexhaustible source of calcium ions for soil solutions and the presence of gypsum is beneficial because it prevents alkalinity.

Calcareous grains of silt size pulverize readily if transported by water. For instance, only 1.2 per cent of suspended silt in Nile floods is calcium carbonate. A high content of lime silts is, therefore, distinctively a feature of wind-transported silts from a source area, by implication dry, and thus a non-leaching environment. Therefore, calcium carbonate silts are frequently significant in loessial deposits, even those derived from dry, former periglacial environments (fig. 1.9b).

2.4.2 UTILIZATION OF SILTY SOILS

Where a soil is predominantly silt with too little sand and clay to give it a loamy texture, the soil may have physical disadvantages, including impeded drainage as soil pores become too small to transmit water readily. There is not the advantage of a stable sub-soil or drainage fissures which clayey soils often possess. The absence of stable sub-soil can lead to compaction. As silt packs down much more after it is wetted, the planning of irrigation canals and controls has to allow particularly for subsidence effects of silts. In high latitudes problems of relict permafrost lenses in outwash sediments are linked with the occurrence of lenses containing a high percentage of silts. Sands and gravels are more permeable and are therefore more readily thawed by infiltrating water. They also have a lower ice content. However, the real heartbreak of silty soils is their susceptibility to wind erosion (fig. 1.10). Clearly, particles which arrived suspended in air can readily move on by the same process (fig. 1.20). The relationship between particle size and distance of transport and other related factors can be established precisely, or eloquently, as John Steinbeck has shown. Thus, he wrote in *The Grapes of Wrath*, 'In the roads where the teams moved, where the wheels milled the ground and the hooves of the horses beat the ground, the dirt crust broke and the dust formed. Every moving thing lifted the dust into the air; a walking man lifted a thin layer as high as his waist, and a wagon lifted the dust as high as the fence tops, and an automobile boiled a cloud behind it. The dust was long in settling back again'.

In contrast, silty soils may have a beneficial influence on the physical properties of soils, particularly in relation to soil water movement. The pore spaces defined by silt-sized particles often combine the hydrophysical properties of permitting relatively rapid water movement and hence rapid availability of both water and nutrients in solution to growing plants, with sufficient impedance to retain much larger quantities of available water than is found in sands. The first advantage is seen particularly in areas of intensive irrigation cultivation.

If a dominantly silty soil includes some clay and humus, the resultant texture and soil structure may be very suitable for a wide range of crops. A high calcium carbonate content is usually typical of loess (fig. 1.20b) and silty alluvium may have a rich and varied suite of weatherable, non-quartz minerals. It is not, therefore, surprising that Theophrastus (372–287 B.C.), writing around 300 B.C., noted the fertility of the Tigris alluvium, that the water was allowed to remain on the land as long as possible so that large amounts of silt might be deposited, or that river and lake silt have been applied laboriously for millenia to improve soils conditions in Egypt, India, China and elsewhere.

2.4.3 SCIENTIFIC INTEREST OF SILTY SOILS

The relatively high silt and low clay content seems to be typical of soils in alpine and arctic environments. For example, up to 90 per cent of the south-west Siberian lowland is mantled by silts. Therefore, one of the initial stages in breakdown may be as rock flour, produced by physical action within moving moraine. In addition, comminution of quartz grains to silt-size is most commonly attributed to freeze–thaw action (Sneddon *et al.* 1972, p. 109). A suggested lower limit below which little or no frost-splitting occurs is about 10 μm. The production of silt-sized particles in tropical environments may be a contrasted process, as clay and iron may be firmly aggregated into silt-sized particles.

Silts often pose a second problem relating to their mode of transportation from a source area. Silts are readily laid down as alluvial deposits on the floors of major valleys (fig. 1.21). More widely spread are wind-transported silts, often described by the German word, loess. This genetic term has become a source of much confusion where silt deposits, initially described as loess, have been re-interpreted subsequently as essentially water-borne deposits. The finest silts, predominantly in the 2–10 μm range, are more certainly recognized as wind-borne. They appear to originate in desert areas, perhaps the result of blasting off the corners of angular sand grains as they assume their distinctively rounded shape. They may be transported as aerosolic dusts taken up into the troposphere in periods of drought when the jet stream passes over arid areas, such as western and central Australia. Aerosolic dust remains airborne until removed from the troposphere by precipitation.

2.5 Clays

2.5.1 THE DISTINCTIVE CHARACTER OF CLAYS

To place the small sizes involved in clay mineralogy in perceptive, the Angström unit is adopted. Represented by the symbol Å, this unit equals 10^{-4} microns or one tenmillionth of a millimeter. Size at this scale is one of the fundamental considerations in appreciating the crystalline nature of clay minerals (Hendricks and Fry 1930) found in the soil.

(a) *Enlarged surface area*
The intrinsic nature of clay particles is linked with their small size simply because of the much enlarged surface area on such soil particles. Surface area is inversely proportional to particle diameter. In sandy soils it is the comparatively limited surface area which determines the relatively small storage capacity which coarser soils have for nutrients and for water. By contrast, the surface area of the tiny but innumerable clay particles is huge. In geometrical terms, a block 1 cm square has a surface area of 6 cm². If such a block were diced into cubes 1 μm in size, the total surface area would become 6000 cm². In real clays, the total surface area is much larger. For example, in soils developed on a Gault clay in Sussex, the total surface area of the clay is 200 m²/g.

(b) *Ionization of elements*
A second fundamental characteristic of clays is that the elements in clay minerals tend to ionize. They exist not only as atoms, with the positive charge in the nucleus exactly balanced by the negative charge of its orbiting electrons but, having lost or gained one or more electrons, as ions such as Si^{4+} or Al^{3+}. Where the atom loses one or more electrons, it becomes a positively charged ion, termed a cation. Where the atom has gained one or more electrons, it acquires a negative charge and is termed an anion. Although the only abundant anion in soil clay minerals is oxygen, O^{2-}, it accounts for 60 per cent of all atoms and ions in the earth's crust. Acquiring an electric charge is the vital characteristic of ions, but also equally significant to explaining soil behaviour is that the effective size of the element is changed during the course of ionization. Where the atom has lost one or more electrons, the remaining electrons in the cation are pulled closer to the nucleus and the ionic radius is correspondingly less than the atomic radius of the element. Conversely, the

56 Geography and soil properties

ionic radius of the anion, having gained one or more electrons, is larger than the atomic radius of the same element. Clay mineral crystals are held together by the attraction between neighbouring positive and negative charges on the ions and by multiple-charged ions sharing their charge between two adjacent layers in the crystal structure. This interlacing of electrical forces between adjacent ions and layers of ions is commonly interrupted at the sides and edges of clay minerals. This unbalanced condition, which is overall a net negative charge on clay mineral particles, is the basis for some of the physicochemical reactions of their surfaces. Thus the distinctive character of clays is not simply the very large surface area of countless, tiny particles, but that this surface carries an electronegative charge as well.

(c) Basic units in clay mineralogy

In the building up of the basic units in clay mineralogy, in their modification by weathering, or in their changes following the addition of artificial fertilizers, size of the individual elements is a critical constraint. Size limits which elements can occupy 'holes' of a given size in the basic crystal framework. Their size also determines which of elements present in clay and soil solution might replace each other in a given space in the crystal framework. Such frameworks are silicate sheets, made up of a basic unit of oxygen and silicon. This basic unit is visualized as a tetrahedron with four oxygen ions at each corner of a pyramid with a triangular base which encloses one silicon ion (fig. 2.5a-d). Silicon is one of the very few cations which can be incorporated inside such a tetrahedron, since the compact oxygen ions, with their large ionic radius of 1.4 Å, are spaced only 2.6 Å apart, whereas the ionic radius of silicon of 0.42 Å is very small. Therefore, the arrangement of the silicon–oxygen tetrahedron, referred to as the basic silicate unit, is readily understood. However, such tetrahedra of one Si^{4+} and four O^{2-} ions are not electrically neutral and therefore they do not exist as isolated units. In fact, each of the three oxygens at the base of the tetrahedron are shared by two silicons of adjacent units (fig. 2.5e). This sharing forms a silicate sheet, with the tetrahedral units packed in such a way that the upper surface of the silicate sheet has hexagonal 'holes' (fig. 2.5f).

Aluminium has an intermediate-size ionic radius of 0.51 Å. It is significant that Al^{3+} is sufficiently small to fit inside an oxygen tetrahedron. However, it more commonly occurs in the less crowded octahedral coordination with Al^{3+} equidistant from six oxygens, idealized as a six-pointed shape with eight sides (fig. 2.5h). Each oxygen is shared by two aluminium ions and thus sheets of aluminosilicate are formed by the joining of aluminium–oxygen octahedra by shared oxygen ions (fig. 2.5i,j) in the same manner as silicon–oxygen tetrahedra combine to

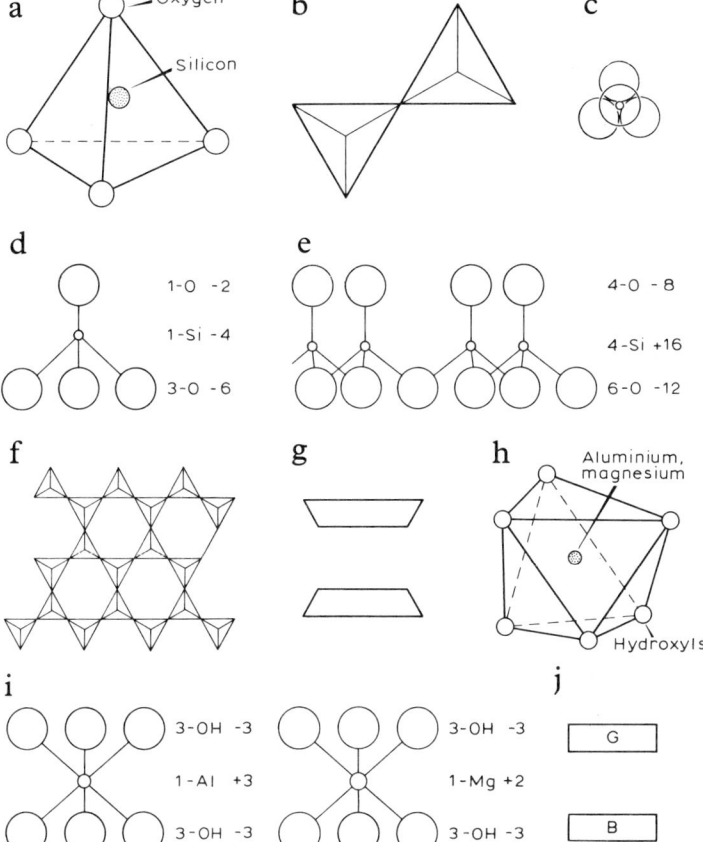

Figure 2.5 Schematic diagrams of silicate units.
a–d Silica tetrahedra, showing (a) oblique view, (b) plan view of double tetrahedra, (c) plan view showing relative sizes and spacings of atoms, and (d) edge view representation of atomic structure.
e–g Silicate sheets, showing (e) edge view representation of atomic structure, (f) plan view, and (g) simplified symbol used in portraying the stacking of silicate sheets in clay minerals.
h–j Aluminium or magnesium octahedral units, showing (h) oblique view, (i) edge view representation of atomic structures, and (j) simplified symbols used in portraying clay minerals, gibbsite sheet (G) and brucite sheet (B).

from silicate sheets. However, only two-thirds of the possible octahedral centres are occupied by aluminium.

Clay particles are always hydrated, being surrounded by layers of water molecules termed adsorbed water. Hydrogen is equally abundant as aluminium but the space it occupies is negligibly small. H^+, as a lone proton, is held so closely to an oxygen anion that the ionic radius of the hydroxyl anion, OH^-, is the same as O^{2-}. In consequence, the hydroxyl anion can occur at points in mineral structures where oxygen is positioned. This affects the balance of electrical charges and if the sheets of octahedra enclosing Al^{3+} consist entirely of OH^- ions instead of O^{2-}, the sheet is electrically neutral. This arrangement is that of the stable mineral, aluminium hydroxide, $Al_2(OH)_6$. This formula is commonly simplified to $Al(OH)_3$ and the mineral termed gibbsite, a basic unit in clay mineralogy (fig. 2.5j).

Magnesium is the main element, other than aluminium, which can be incorporated within the octahedron formed by six oxygen or hydroxyl ions. This arrangement is again a function of medium-sized ionic radius, that of Mg^{2+} being 0.66 Å. If magnesium is present in place of aluminium, it also combines into a stable mineral if the octahedral sheets consist of hydroxyl ions in place of oxygen ions. This magnesium-rich mineral, $Mg_3(OH)_6$ is termed brucite and is commonly expressed simply as $Mg(OH)_2$. Such a brucite sheet is another basic unit in clay mineralogy (fig. 2.5i,j).

The binding together of either a gibbsite or a brucite sheet with one or two silicate sheets is the basic structure of the clay minerals found in the soil. Clay minerals are often referred to as layer silicates because they are 2, 3, or 4 sheets bound together in layers. A silicate sheet and a gibbsite sheet may be bonded together by shared oxygen ions, the apex oxygen ions of the tetrahedra in the silicate sheet replacing some of the hydroxyl ions in the octahedral sheet. Such clay crystals of a silicate sheet bonded to a gibbsite sheet are called two-sheet or 1:1 group of clay minerals. Most of the other clay minerals occurring in soils are three-sheet or the so-called 2:1 layer silicates in which either a gibbsite sheet or a brucite sheet is bound to two silicate sheets sandwiched on either side.

(d) *Expansion of 2:1 layer clays*

Expansion in three-layer clay minerals is an interaction between two forces. On the one hand, liquid molecules adsorbed on to the oxygen surface of the silicate layer require space between the layers. Through their adsorption forces they act as a lever to prise open the clay layers. The strength of this adsorption force determines the effectiveness of water molecules in separating the silicate layers. The tendency to prise the silicate layers apart is resisted, however, by the bonding forces exerted by whatever charged interlayer material is present. As this balance of forces is affected by particle size, no simple correlation can be assumed between charge density and expansion properties and the structural cations present significantly influence the unit cell dimensions.

2.5.2 AGRICULTURAL SIGNIFICANCE OF CLAYS

In addition to the retention in clay soils of water in a form available for plant growth, the retention by clay of plant nutrients is equally vital. The clay minerals have the unique property of sorbing both cations and anions. Such exchange reactions can store nutrients for periods of maximum plant growth. Thus, in rubber plantations in Malaya, a fairly close relationship was found to exist between phosphate retention in the soil and its percentage clay, and similar relations between phosphate and clay have been observed in Florida. Similar patterns are observed for micronutrients and clay. In addition, the soil organic matter, being a vastly more dynamic soil component, is stabilized within the physical and chemical framework of clays. Thus clay is necessary in a soil for the holding of nitrogen which in its natural state is incorporated in organic matter. Held within a clay structure, organic matter is not accessible to soil micro-organisms, and its rapid decomposition is arrested. Other important soil biochemical processes have also been noted to increase with greater clay contents of the soil, such as increased activities of enzymes hydrolysing starch.

Clays lend coherence to a soil which supports plant roots and limits water and wind erosion. In fact, the greater resistance of soil to erosion is usually very highly correlated with increased clay percentages. Owing to sorption by clay, toxic compounds may be maintained at lower levels in the soil solution. For instance, arsenic toxicity in plants decreases as the clay content of soils increases.

In short, the overall effect of clays on agriculture is profound. If complicated in detail, it is often a relationship between clay percentage and yield which can be expressed as a simple correlation. For example, it is generally accepted that optimum conditions for growing rice in tropical Asia are only met in heavy clay soils. Similarly, some of the highly productive areas of Egypt include examples of soils with a high clay content. In arid, alluvial clay plains in the Sudan, clay contents ranging from 40–70 per cent are positively correlated with relative cotton yields. Within these clay ranges,

$$y = -49.3 + 2.57x$$

where y is the per cent relative yield and x is the per cent clay content, gives highly accurate predictions. A relative yield of 100 per cent corresponds with a clay content of 58 per cent (Finck and Ochtman 1961).

In contrast, there are also instances where per cent clay does not necessarily correlate with agriculturally advantageous properties. For instance, Finnish claylands do not provide satisfactory permanent grass for grazing. In Alaskan podzols, clay content correlations with moisture are weak, with $r = 0.37$. Correlations between clay and organic carbon, $r = 0.21$ or with extractable iron, $r = 0.15$, are both insignificant. Clays may fix plant nutrients and water in forms unavailable to plants and they may also impede root penetration or air and water movements. Thus, the agricultural desirability of tilling clay soils is also much conditioned by the risk of excessive wetness, the susceptibility of structural crumbs to slaking, and the risk of moisture deficiency. The wetness hazard is also a limit on the timing and weight of vehicles crossing the soil.

2.5.3 RANGE OF INTEREST IN CLAYS

The technical and scientific interest in soils now spans a wide range of problems, such as the use of clays as catalysts for organic reactions in petroleum refining, as carriers for pesticides, or as a disposal medium to decontaminate radioactive wastes.

There is an overlap of interest in clays between soil science and geology because soils were the main source of the clay fraction of modern alluvium, marine deposits and sedimentary rocks. Clays are the most abundant minerals in sedimentary rocks, accounting for as much as 40 per cent of the total. Clay may reflect the history of the sediment, the characteristics of its source area, the distinctiveness of physicochemical conditions in its depositional environment and perhaps post-depositional changes (Linares and Huertas 1971).

When clays are heated to 400–700° C, they lose their hydroxyl water, and the loss may disrupt or alter the mineral structure, with new crystalline phases developing at higher temperatures. Clays are the most abundant raw material of the ceramic industry and the modifications of their physical and chemical properties on firing is intensively studied.

In engineering practice, most of the problem soils are clays. Clay soils with the same particle-size distribution may differ in their hydrophysical properties, depending on their mineralogical composition. This affects properties like plasticity, shear strength, and consolidation. Larger differences in engineering behaviour are explained by the character of the water held in the soil, which is dependent on the nature of the clay-mineral surfaces.

2.5.4 PROBLEMS IN IDENTIFYING AND INTERPRETING CLAY MINERALS IN SOILS

Prior to the 1920s there were no adequate analytical techniques for determining the nature of particles as small as clays. The true nature of clay minerals was not revealed until 1930 with Pauling's X-ray diffraction studies on the crystal structure of micas. Later, other sophisticated techniques like differential thermal analysis and electron microscopy observa-

The mineral fraction of the soil 61

tions (Bramao et al. 1952) have assisted a wide range of specialists to identify clay mineral characteristics and types. In soils, however, clay minerals are complicated mixtures and their interpretation requires caution.

(a) *Complexity of clay minerals in soils*

Pure monomineralic clays rarely exist in soils (Schultz et al. 1971). Clay itself is commonly only a fraction of the fine earth sample and those sizes below 2 µm include pulverized particles of primary minerals like quartz, mica, and feldspar. Conversely, clay minerals may commonly occur in the fractions coarser than 2 µm, especially in tropical soils (Merwe and Weber 1965).

Clay minerals are not stable entities in soils. They change continually, on contact with water, some minerals being dissolved and others re-precipitated. Apart from such temporal changes, clay minerals may change with depth in the profile. Thus, nearer the surface of tills of Wisconsin age, the more hydration of clay minerals has occurred (Droste 1956). Similarly, but on a microscopic scale, weathering attacks the outer surface of clay minerals first which may, therefore, have a different composition on the edges compared with the inner parts of the crystal. It is therefore difficult to identify and to separate the relative abundance of the clay minerals in complex mixtures like soil clays, especially when the constituents are poorly crystallized.

(b) *Interpretation of soil clays*

The origin of soil clays may be linked with the parent materials. Clearly, clay in soils developed on igneous and metamorphic rocks must be primary. Conversely, soils like the Houston black clay undoubtedly inherited clay minerals for the sedimentary rock parent material (Van Houten 1953).

Chemical weathering superimposed on diagenetically altered parent material may produce diverse soil clays. Synthesis of clay minerals may even involve a complexing of Al^{3+} with organic acids. Drainage waters with a distinctive chemical composition may direct clay mineral changes taking place as might traces of added volcanic ash. Expressions of distinctive, local environments illustrate why the interpretation of clays as indicators of environmental conditions in a broad spatial or temporal sense requires care.

2.5.5 CLAY MINERAL GROUPS

(a) *Kaolinite*

Kaolinite is found in at least small amounts in almost all soils (fig. 2.6). Conversely, it is because the clays of African soils are largely of a kaolinitic

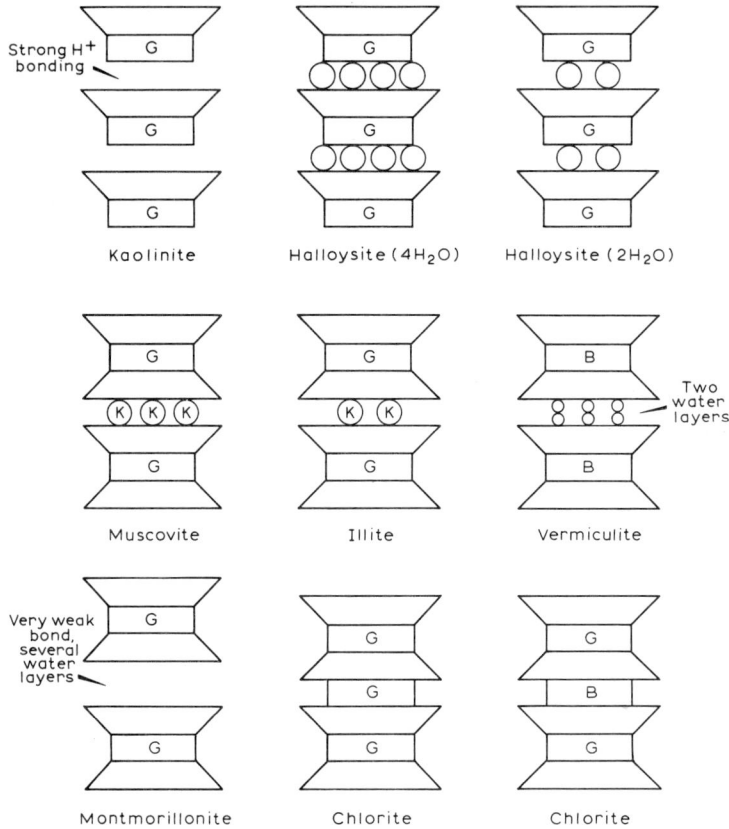

Figure 2.6 Clay mineral crystal structures, with simplified symbols representing the stacking of tetrahedral and octahedral sheets, as illustrated in fig. 2.5.

nature that their fertility is relatively low. Chemically, kaolinite is the most resistant of the clays common in soils, both in acid and in alkaline soil waters. Feldspars constitute the chief source of kaolinite (Ross and Kerr 1931). Physically, it is the clay type most frequently encountered by the engineer. Kaolinite deviates very little from its structural formula $Al_4Si_4O_{10}(OH)_8$. The oxide formula for kaolinite is $4H_2O \cdot 2Al_2O_3 \cdot 4SiO_2$. It consists of a silicate sheet and a gibbsite sheet, the two being bonded together by shared oxygen atoms. It occurs as relatively large platelets, 0.3–3 µm diameter and 0.05–0.2 µm thick, often being hexagonal in outline. The hydrogen of the hydroxyl ions in the gibbsite sheet of one layer is also weakly attracted to the oxygen in the adjacent sheet of the next layer. This is termed hydrogen bonding and is sufficiently strong to link layers together fairly rigidly. This distance between layers, termed basal spacing, is only 7.2 Å. The combined effect of crystal lattice rigidity and of small basal spacing prevents water from entering between the layers. In consequence, kaolinite has little or no tendency to shrink when dry or swell when wetted, and extra ions cannot enter between the layers. These distinctive characteristics are all related to the small electric charge on the surfaces of kaolinite layers, since isomorphous substitution in kaolinite is uncommon. Possibly, aluminium replaces silicon in the silicate sheet only once for every 400 silicon ions. The only electric charge in the lattice is restricted to the broken bonds at the edges of the layers. Assuming a thickness of 0.05–2 µm, calculated edge areas lie between 0.4 m^2/g for well-crystallized kaolinite to 4 m^2/g for poorly crystallized varieties.

An acid environment favours the formation of kaolinite and it is stable under such conditions. The genesis of kaolinite depends on a strongly leached environment and the removal of Ca^{2+}, Mg^{2+}, Na^+, K^+, and Fe^+. SiO_2 is also possibly removed and H^+ added. A high concentration of Al is expressed in a wide Al:Si ratio and oxidation of Fe^{2+} to Fe_2O_3 is characteristic. As its formation depends on a leaching regime in the soil profile to remove bases and silica, kaolinitic minerals are found in large quantities in the clay fractions of coarse-textured soils. Thus, they predominate in the clay fraction of podzols. Also, as the removal of bases and particularly silica takes time, kaolinite belongs to the penultimate stage of weathering transformations. It is predominant in ancient soils like lateritic and Red Earth soils, where it often occurs together with final weathering products like gibbsite and goethite.

(b) *Halloysite*

Halloysite is also one of the kaolin group of clay minerals and has essentially the same composition as kaolinite (fig. 2.6). It is less well structured than kaolinite, with an irregular stacking arrangement of the silicate sheets. Isomorphous substitution is more prominent, however, since aluminum substitutes for silicon once for about every 100 silicon ions. The most significant difference between halloysite and kaolinite is a layer of water molecules between the layers of the halloysite crystal. There are two degrees of hydration, usually distinguished as halloysite ($2H_2O$) and halloysite ($4H_2O$) (Marshall 1964). Basal spacings range from 7.2 Å in dehydrated forms to 9.6–10 Å in hydrated materials. The presence of water introduces a strain into the inter-layer spaces which result in halloysite curling up into distinctive tubular particles about 250 Å diameter. Partial dehydration occurs at relatively low temperatures, especially in dry conditions. In natural soils, mixtures of halloysite and kaolinite are common, but perhaps in inverse proportions. In a climatic range of soils in California, kaolinite and halloysite are negatively correlated.

In some cases the hydrated form, halloysite, appears to have formed first and then been replaced by kaolinite (Eswaran and Heng 1976).

(c) *Illite*

The hydromicas, of which illite and vermiculite are the two main groups which are clearly recognized, are intermediate products of weathering and may even be little-modified primary mica minerals inherited from decomposing rocks. Illite (fig. 2.6) is structurally and chemically similar to muscovite, but contains more water and less potassium than the rock-forming mineral. The clay develops as a result of the hydronium ion (H_3O^+) replacing potassium. Potassium averages approximately 60 per cent of that found in muscovite or biotite, but a K_2O content of at least 6–7 per cent is assigned as a critical value in defining the clay mineral. Illite layers are about 0.1–2 µm diameter, with thicknesses about one-tenth of the diameter. They are 2:1 silicates, with a gibbsite sheet sandwiched between two silicate sheets. The layer surfaces have a high negative electrical charge due to numerous and varied isomorphous substitutions. One in seven if not one in four of the tetrahedral positions in illite are occupied by Al^{3+} in place of Si^{4+}. In turn, magnesium and iron may replace aluminium in the octahedral sheet. The negative charge is balanced by K^+ ions held between adjacent layers. The potassium ions fit into hexagonal holes in the silicate sheets, bonding six oxygens from one silicate sheet to the adjacent six oxygens of the silicate sheet of the next layer. Thus adjacent layers of illite are held firmly in a rigid lattice. The basal spacing of approximately 10 Å is too small for water to enter between the layers and therefore illite does not expand when wet. The stability of the illite lattice is responsible for fixation of potassium ions which may enter the soil solution, including those added as fertilizer. It is difficult for other cations to displace K^+ held in the non-expanding illite lattice. Also, in the span of a growing season, such potassium ions are not available to plants. However, they gradually become available as they are released by weathering processes, but illites pose problems where a balanced supply of nutrients has to be added to maintain crop growth. Problems in the study of illite include distinguishing between that formed at low temperatures and that known to form at high temperatures (Hower and Mowatt 1966). Also illite commonly occurs in association with other clay minerals.

(d) *Vermiculite*

Vermiculite is the second main clay group of the hydromicas. It is similar to illite, the main difference arising from its formation by the leaching of potassium from biotite. Since it is related to biotite, vermiculite belongs in chemical composition to the magnesium aluminosilicates. Exchangeable bases such as Ca^{2+} or Mg^+ occupy interlayer positions as do the potassium ions in biotite. Thus vermiculite is essentially a magnesium or magnesium + calcium mica. The octahedral sheet, sandwiched between two silicate sheets in the 2:1 arrangement, is brucite (fig. 2.6). Part of the negative charge on the layer surfaces is due to a considerable amount of aluminium replacing silicon in the tetrehedral sheets, as in illite, but the characteristics of the interlayer space between adjacent layers is different and distinctive. Two layers of water molecules occupy the interlayer space and the lattice as a whole is partially expanding. Compared with the basal spacing of 10.1 Å in biotite, that of vermiculite can expand to 14–15 Å. Owing to isomorphous substitution of magnesium by aluminium and iron, there are plenty of Mg^{2+} ions between layers to offset the considerable negative charge in the silicate sheets of the vermiculite layers. Although not commonly abundant, vermiculite is an important clay mineral, noteworthy because the Mg^{2+} ions in the two water layers can be exchanged with other available plant nutrients. As a weathering product of biotite, it occurs as an accessory mineral in many clay soils.

In general, the hydromicas are found in most soils developed on basic (fig. 2.3) and intermediate rocks apart from Red Earths and lateritic soils in the humid tropics where hydromicas are completely decomposed. They tend to be readily dissolved in strongly acid conditions and are therefore most common in slightly alkaline soils, the presence of vermiculite depending also on a high magnesium content.

(e) *Montmorillonite*
The montmorillonite group clays, often referred to as smectites in geological literature, have a wide range of chemical composition and striking physicochemical characteristics. Montmorillonite is the predominant 2:1 silicate clay mineral, forming from micas, illite, vermiculite, chlorites, and non-crystalline substances including volcanic glass. In contrast to kaolinite, silica is relatively high in comparison with Al^{3+} and H^+. Instead, montmorillonite is like illite insofar as the basic crystal consists of a gibbsite sheet sandwiched between two silicate sheets (fig. 2.6). However, isomorphous substitution occurs mainly in the gibbsite sheet and whereas K^+ links adjacent layers together rigidly in illite, K^+ is easily displaced from montmorillonite. In consequence, the interlayer space is occupied by a range of readily exchangeable cations and by water and it has a low ratio of K^+ to $(Na^+ + Ca^{2+})$ compared with illite. Therefore, whilst a general equation for montmorillonite is $OH_{12}Al_4Si_6O_{16} \cdot nH_2O$, this clay mineral may also occur in either sodium, magnesium, or calcium-bearing varieties.

Clay particle size is notably small in montmorillonite. A coarse fraction exists, which is about 0.1–2.0 μm diameter but only about one-hundredth as thick. Also, some particles are so small that they are only a few silicate sheets in thickness. Thus the surface area of such narrow crystals may exceed 600 m²/g, including both external and internal surfaces, and a surface area of even 0.6 ha/g has been calculated.

The surface charge in montmorillonite is only partially due to isomorphous substitution of Al^{3+} for Si^{4+} in the silicate layers. Far more significant is the substitution within the sandwiched gibbsite layer of Mg^{2+} for Al^{3+}, and Mg^{2+} is essential for the formation of montmorillonite (Ross and Hendricks 1945). About 80 per cent of the exchange capacity is due to substitution within the structure. It seems that a charge in the inner octahedral layer would be less capable of holding the silicate layers together. Also the hydroxyl ions which remain in the octahedral sheet are enclosed by the tetrahedral sheet so there is no hydrogen bonding to hold layers together either. The only force holding layers together in montmorillonite is simply the attractive force which like materials exert on each other when in close proximity, the very weak van der Waals attractive force. This is also due to electric interaction but does not involve two net charges and is not strong enough to prevent water from entering between the layers. Montmorillonite is therefore highly hydrous and the basal spacing varies substantially according to water content. The basal spacing ranges from 9.6 Å when all water is removed to 21.4 Å or more, depending on how much water is present. Such spaces cannot only accommodate all plant nutrients and release them readily, but even organic molecules such as proteins can enter into the clay mineral lattice. The expanding lattice may accommodate substantial increases in clay volume after the wetting of dry montmorillonite clay. After swelling, the volume may be twice if not three times that of the dry volume. Such physical properties are profoundly affected by which ions are predominant in the interlayer space. Sodium montmorillonite, in particular, forms a colloidal gel that may change to a fluid if subjected merely to an abrupt jar. On the other hand, aluminium and iron hydroxides are commonly present in the interlayer position of soil montmorillonite. This is not readily exchangeable and is a considerable complication in calculating crystal structure and properties.

Pure montmorillonite is rarely found. In particular, illite is often mixed if not inter-layered with montmorillonite. These clays are usually found where there is slowly draining, base-rich soil water. As silica and hydrated iron and aluminium oxides are also incorporated into such clays, they also tend to be a characteristic of warmer environments. Montmorillonite is, therefore, typically prevalent in the clay fraction of warm, arid zones. In cooler, humid climates they constitute only a small fraction of soil clays, owing to rapid leaching and other unfavourable factors. Thus, subaerial weathering of montmorillonite has created kaolinite several feet thick over parts of central Florida, following leaching of SiO_2, Fe_2O_3, MgO, and K_2O (Altschuler et al. 1963). The environmental controls on montmorillonite can be expressed quite precisely. For instance, montmorillonite content of soils has been observed to be inversely correlated with the percentage of highly resistant minerals.

(f) *Chlorite*
Chlorite occurs as a complex clay mineral, probably related to the dispersion of large-grained primary chlorites released in weathering. The repeating layers of the chlorite minerals are composed of a gibbsite sheet sandwiched between two silicate sheets. These 2:1 micaceous layers are held together, however, by brucite or gibbsite sheets (fig. 2.6) and are therefore sometimes described as 2:2 layer silicates. The lattice is partially expanding. Such chlorite-like minerals with aluminium hydroxide in the interlayer spaces may represent a stage in the de-alumination of clay minerals. Inter-layer aluminium may contribute to the low cation exchange, toxicity of aluminium and the fixation of phosphate commonly encountered in some soils. They inhibit the expansion of montmorillonite from 14 Å. The formation of aluminium interlayers is essentially chloritization of the expanding type 2:1 layer silicates and such clays might be described as chloritized montmorillonite or chloritized vermiculite (Sawhney 1958). Chlorites are, in fact, often inter-layered with montmorillonite. Chlorite is difficult to distinguish from vermiculite in X-ray analysis and its significance may have been underestimated in consequence.

(g) *Allophane*
Allophane, as a general term, describes any amorphous substance which may be present in the fine-clay fraction in which the constituents are not specifically identifiable. More specifically, it is the principal mineral in the clay fraction of little-weathered soils derived from andesitic and rhyolitic volcanic ash showers. Although the instantaneous cooling inhibits the development of an identifiable structure, many important mineral nutrients may be released on weathering (Fieldes 1955).

2.5.6 NON-CRYSTALLINE COMPONENTS OF CLAY-SIZE

In both high and low latitudes, soils contain particles smaller than 2 µm which have properties different from those of the crystalline aluminosilicate clays. In alpine and polar areas 'clay'-sized particles may merely be the chemically unweathered, pulverized fragments of parent materials and minerals (Milestone and Wilson 1977). In dry areas of the U.S.S.R., where negative temperatures prevail for 8–9 months of the year, 40–80 per cent of clay-sized fragments are unweathered mica fragments.

In a wide range of soils, the hydrous oxides of Fe and Al, the weathering products of crystalline minerals, are present as amorphous, clay-sized materials which lack the order which distinguishes clay crystals. These oxide-hydroxide compounds of Fe and Al are termed sesquioxides and are represented as a group by the general formula R_2O_3, and are now often qualified by the term 'active' to emphasize their importance in soil processes. Their significance is much greater than the relatively small amounts present might suggest, particularly in tropical soils. The bulk of red tropical soils are layer silicates, partially veneered with coats of iron and aluminium oxides which may be monomolecular but are more commonly continuous and several layers thick and stable. Amorphous sesquioxides are important in physicochemical reactions relating to nutrient exchanges and environmental protection (Huang et al. 1977). The sesquioxides also influence the engineering properties of tropical soils by coating clayey constituents and binding them into coarser aggregates. This granular structure accounts for the initially desirable engineering properties of some undisturbed laterite soils. Until recently, amorphous sesquioxides were the unmentionable stains on laboratory specimens. They were removed from soil samples to 'clean up' clays to provide 'better' data for comparison with model clay systems. As a result, their fundamental role in many physical and chemical reactions in the soil was underestimated.

It is axiomatic that the widely contrasting characteristics of soil clays affect man's utilization of soil in innumerable ways. It is also clear, particularly in the contrasting properties of the two main clay minerals, that the distinctiveness of certain clays is localized in certain environments or is representative of broadly developed regional traits. Perhaps the intrinsically geographical nature of soil clays is epitomized by the derivation of the names of the three main contrasting soil clay mineral groups. Montmorillon is a small town in southern France, the term 'kaolin' is derived from Kaoling, a mountain ridge in Kiangsi, China, where the clay was quarried centuries ago, and illite hails from Illinois.

3 Soil organic matter

3.1 Supply of organic matter to the soil

3.1.1 INTRODUCTION

Soil organic matter contributes substantially to the physical, chemical, and biological processes in soils, and is expressed in several distinctive soil characteristics. By absorbing solar rays, green plants and living organisms transform their energy into the energy of chemical bonds of organic compounds (Mustafayev *et al.* 1976). The difference in the functions in the soil of the main fractions of organic matter influence fundamentally most phases of crop or floral development.

Soil organic matter releases up to 98 per cent nitrogen, 5 to 60 per cent phosphorus, and between 10 and 80 per cent of sulphur found in soils. In addition, a large part of the total soil reserves of vital microelements, such as boron and molybdenum, are organically combined. Amounts of such nutrients in the soil are, therefore, related to the total quantity of organic matter, but their availability is linked to its rate of decomposition. Apart from being a potential source of certain nutrients, organic matter influences soil physical properties, such as structure, water-holding capacity and resistance to erosion.

There are three broad categories of soil organic matter. First, there is the coarse fraction of dead or dying fragments which remains or even accumulates in the uppermost soil horizons (fig. 3.1). Humus, as a pool of organic complexes and resistant residues, is the second organic fraction. Thirdly, there is the biomass of active or resting living soil animals and plant roots which constitutes the pulse and pumps of many chemical elements and physical functions in soils.

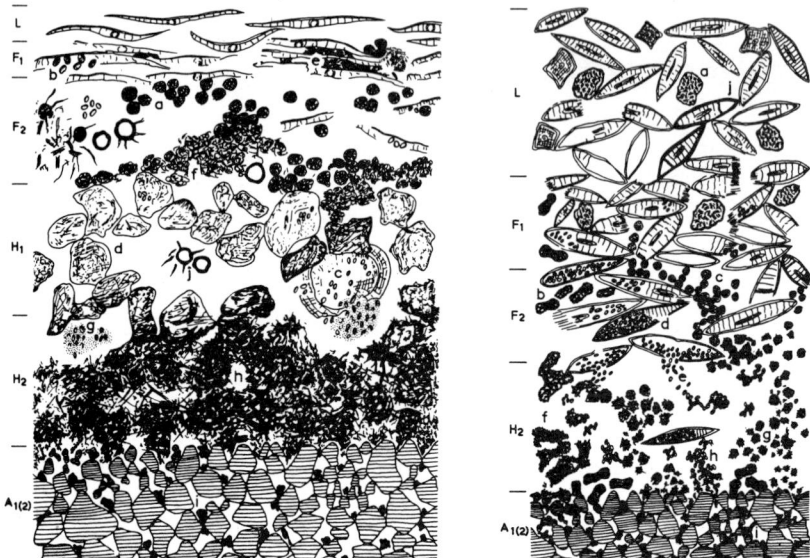

Figure 3.1 Development of O (organic) horizons beneath woodland (Bal, 1970), on sandy soils near Arnhem, with different sources of litter.
a Profile beneath red oak *Quercus borealis* Michx. Indicated are the excrements of the fungus gnats' (Mycetophilidae) larvae (a), *Nothrus silvestris* which is a beetle mite (Oribatei) (b), *N. silvestris* and *Rhysotritia minima*, another oribatid, after exogenous and endogenous root glutting (c), the earthworm *Dendrobaena rubida* (d), and enchytraeid worms and *Onychiurus*, a soft, sluggish Collembola (e).
  Also indicated are ageing excrements of Mycetophilidae larvae (f), Oribatei (g), and *Dendrobaena rubida* (h). Faecal parts are illuviated from the H-horizon into the A_{12} horizon (i), and there are root remnants filled by fungi with black humic substances (j).
b Profile beneath Douglas fir *Pseudotsuga manziesii* Mirb. Indicated are the excrements of the crane fly (Tipulidae) larvae (a), the *Adela* caterpillars of leaf-mining lepidopteran insects (b), and Mycetophilidae larvae (c). Needles are mined by the beetle mites (Phthiracaridae) and filled with excrements (d). Ageing excrement adheres to the outside of needles (e). Other ageing excrements are indicated for *Adela* larvae (f), Mycetophilidae larvae (g), and Phthiracaridae. Faecal parts are illuviated from the H-horizon into the A_{12} horizon (i), and black humic substances fill needles (j). *Adela* sp. cut leaf pieces and hold them together by spinning silk.

The subdivisions of the O (organic) profile include the L (litter) layer, which is essentially undecomposed except for leaching of soluble constituents and discolouration. The F (fermentation) layer is made up of partially decomposed litter, which passes into the well-decomposed H (humus) layer, which is low in mineral matter. The A-horizon is the mixed mineral-organic surface layer. Identifiable changes within the subdivisions are designated by consecutive numerical subscripts.

3.1.2 BIOMASS AND LITTERFALL

Although the upper limit of the soil body is apparently defined physically by the earth's surface, soilforming processes start in the tree tops. In mid-latitudes, approximately half the total radiation is photosynthetically available. The photosynthetic efficiency, estimated from the annual rate of dry matter production, is 1–2 per cent or more of this, and the calorific values of plant compounds are commonly in the 4000-5500 cal/g range of oven-dry matter. In addition to the volume of dead vegetation about to fall to the ground, considerable quantities of soluble organic compounds may be washed from the canopies. For instance, in oak woodland in Grizedale Forest in north Lancashire, this supply of energy in soluble form can be up to 86 kg/ha/month (Carlisle 1965). Soil materials start to form in niches where branches meet, in bark and in fallen trunks. As the ground is approached, the composition of the fauna increasingly resembles that found, at least part of the time, within the soil. Even certain species of earthworm climb trees.

Litterfall or leaf-fall is a particularly significant quantity. For instance, the rate of litterfall is generally a comparatively small proportion of the biomass of arctic environments. In contrast, the volume of tropical rainforest can exist on a near-sterile soil because of the scale of litterfall combined with the rapidity with which mineralization releases the volume of nutrients to the growing vegetation. Leaf and litter production are comparatively short-term stores of energy. Wood production is long-term energy storage since it resists decay and is unsuitable for grazing.

In cool environments large volumes of litter often indicate the resistance of organic materials to the slow decomposition processes rather than an abundance of re-cycling materials readily incorporated into the soil. In polar areas, where organic matter is largely supplied by lichens, algae and diatoms, the major vegetation component, the lichens, are extremely slow suppliers of organic matter, owing to their slow growth rates. Nonetheless, mosses growing in damp, sheltered niches can build up appreciable amounts of peat beneath their clumps. From alpine and tundra environments, reported biomass volumes range from 1 to 30 tonnes/ha, although a biomass of 5 tonnes/ha, as calculated for the northern taiga on the Kola Peninsula, is typical. Here half the biomass may accumulate as litter, with a litterfall of 0.4 tonnes/ha/year (Rudneva *et al.* 1966). The same figure has been recorded for Scottish heaths growing on sandy dune soils on an exposed coast but this is a very low figure compared with most British heaths. In the northern Pennines, at 550 m O.D. on the Moor House Nature Reserve, annual litterfall was 3 tonnes/ha and the biomass 13 tonnes (Forrest 1971). In the drier environment of the Forest-steppe of European U.S.S.R., leaf-fall is about 0.5–0.7 tonnes/ha/year.

Of a biomass of 15–30 tonnes, roots account for 4–5 tonnes and litter 1–1.5 tonnes/ha (Rublin and Dolotov 1967). Red pine woodland in the Adirondack Mountain region adds 2–3.8 tonnes/ha/year of dry-weight pine needles to the plantation floor (Stutzbach *et al.* 1972). In the southern hemisphere, *Eucalyptus* forests in Victoria, Australia, shed 6.9–8.1 tonnes/ha annually, and biomasses of over a thousand tonnes/ha have been recorded. In the dry subtropics biomass figures decline to around a median figure of 50 tonnes/ha with litterfall in the 2–4 tonnes/ha/year range. Owing to the high temperatures, plant litter is rapidly decomposed and the litter accumulation is not usually more than 5 cm thick. In more moist sub-tropical areas, quantities of forest litter of 4–9 tonnes/ha/year have been estimated. Measurements of litter production in African and American equatorial rain forests have given results ranging between 10.2 and 15.3 tonnes (Bray and Gorham 1964). For instance, a litterfall of 11.2 tonnes/ha/year has been calculated for a mature secondary forest in Ghana (Nye 1961) and litterfalls of up to 14.4 tonnes/ha/year have been measured in lowland forest in Malaya.

Litterfall illustrates rates of semi-natural renewal of fresh organic matter at the soil's surface and many studies are summarized by Bray and Gorham (1964). Apart from its organic bulk, the 1 or 2 per cent ash content of litterfall is a vital return of chemical elements to the soil. The effectiveness of vegetation in recycling nutrients can be estimated from the ratio of the amount of an element in the litter accumulation to that in the annual leaf-fall. If the ratio is less than 1, that element is being leached more rapidly into the mineral horizons of the soil. The significance of the litter accumulation as a reservoir of major nutrients can be illustrated by observations of the forest floor of a lodgepole pine stand located at over 2500 m altitude in the Front Range of Colorado (Moir and Grier 1969). Of a total forest floor weight of 25–40 tonnes/ha, total nitrogen in the litter ranged from 330 to 540 kg/ha, a value similar to N accumulations in organic matter beneath pine stands in northern Europe and beneath pine forests in south-east U.S.A. Phosphorus amounted to 22–50 kg/ha, potassium 70–170 kg/ha, and calcium was between 100 and 250 kg/ha.

3.1.3 OTHER SOURCES OF ORGANIC MATTER

Roots are an important source of organic materials in soils, and virtually the only source for many arable soils at the present-day. The dry matter of roots and short stubble of grass and cereal crops adds about 2–4.5 tonnes/ha/year at least. For example, in Sod-Podzolic loam soils on the Sloboda State Farm near Moscow, awnless brome grass produced about 15 tonnes/ha of roots, Meadow fescue produced about 10 tonnes/ha, and timothy produced about 8 tonnes/ha (Shcherbakov 1970). The quality

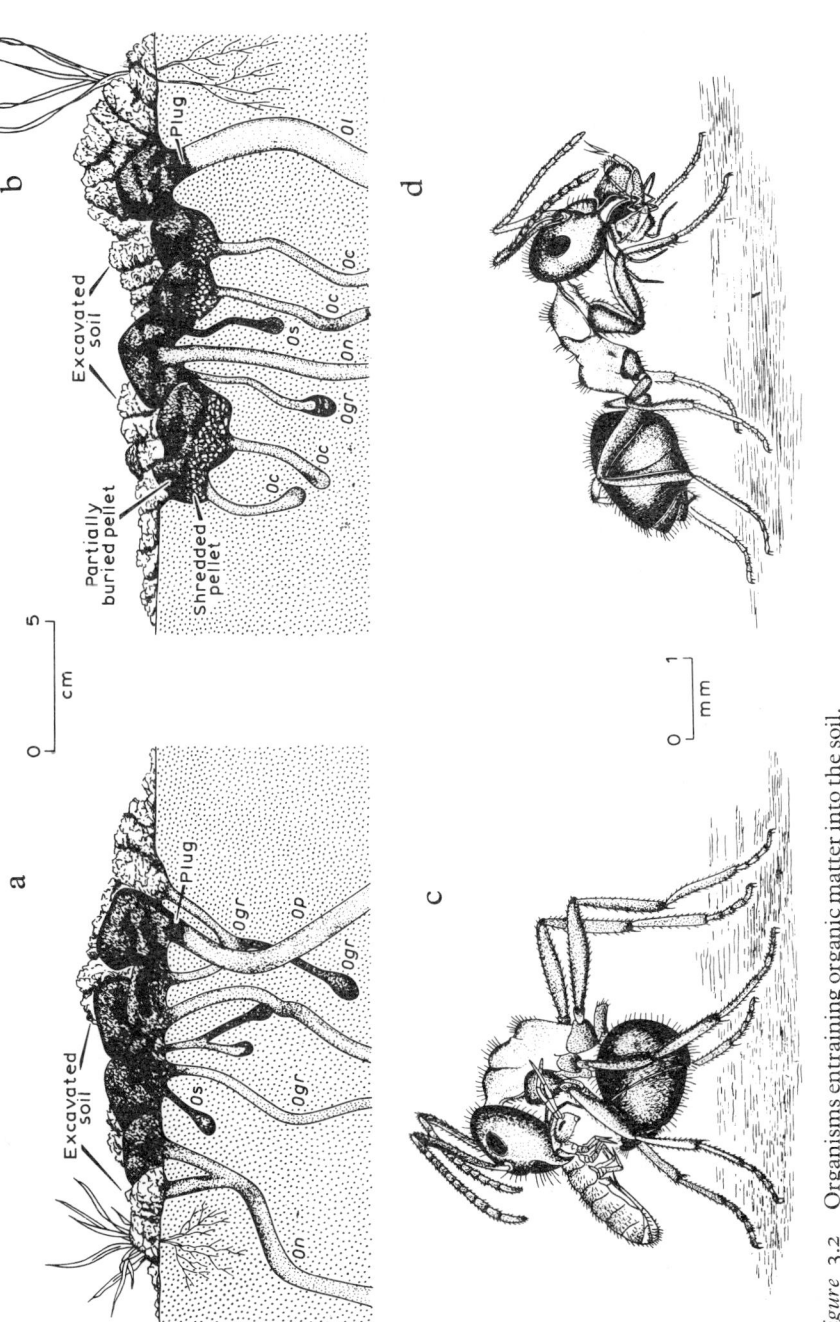

Figure 3.2 Organisms entraining organic matter into the soil.
a and b. Pellets of wombat excrement (a), being buried by the activity of the dung beetle (Bornemissza, 1971).
c and d. Worker ant paralyzing larger prey with a spray of formic acid, with folded 'tail' of Collembola *Isotoma virdis* prey evident (c) and worker ant *Lasius emarginatus* carrying young Collembola (*Bourletiella hortensis*) between its mandibles (d) (Vannier, 1971). Nine out of 16 of these ants were to be seen carrying Collembola in preference to any other form of organic matter

of the residues varies according to the crop. About 2.5 times more nitrogen is supplied to the soil by the residues of forage peas than by residues of silage corn, reflected in a 0.9–1.4 tonnes/ha increase in winter wheat yield (Gerkiyal 1974). In uncultivated areas, the return of dead organisms to the soil is a significant source of organic matter, as is the excrement of living animals (fig. 3.2).

Farmyard manure, abbreviated to FYM for convenience, has been a major source of organic matter for arable soils. On pastures excreta from grazing livestock is an important source of organic matter. The composition of excreta varies according to the quality and composition of the food supply as much as it reflects the type, age and state of health of the livestock. Typically, however, about 75 per cent of the nitrogen in the plant materials consumed is re-cycled, and proportions of phosphorus and potassium returned as excreta may be 5–10 per cent more than this. The actual amounts of nutrients added to the soil depends on the amounts of extra feedstuff ingested. For instance, each additional 10 tonnes/ha of feed, may increase the available phosphorus in the topsoil by 17 kg/ha (Benacchio *et al.* 1970). Nitrogen returns are commonly between 180 and 320 kg/ha/year.

In certain cultural areas human excreta, termed night soil to avoid embarrassment, is an important source of organic matter in arable soils, with application rates of 3–6 tonnes/ha being estimated for the most intensively worked agricultural lands. In other areas, where re-cycling is less customary, the feasibility and acceptability of utilizing sewage sludge as a fertilizer on soils is being increasingly evaluated (Sommers, 1977).

3.1.4 AMOUNTS OF ORGANIC MATTER IN SOILS

Despite the addition, subtraction, and re-cycling of organic material being continuous and varied processes, the percentage of organic matter in a given soil is usually contained within narrow and distinctive limits. For instance, most of the soils of Uruguay are classified as one or other type of 'Praderas' (Prairie) because of 3–5 per cent organic matter maintained in the soil by the grassland vegetation.

Polar desert soils may have up to 6 per cent organic matter in the topsoil, as the very slow accumulation rates may locally exceed the even slower decomposition rates. However, excluding peaty soils, the percentage organic matter in a soil is a guide to its store of nutrients and values are often highest under forest. In the U.S.S.R., for instance, Dark Grey Forest soils have organic matter contents between 5.2 and 10.5 per cent. Chernozems and related grassland soils are consistently above 4 per cent, as in the Dneiper River area where values of 4–6 per cent occur on the right bank, increasing to 5–9 per cent on the left bank. In deserts, bare soils have a predictably low amount of organic matter. A figure of 0.85 per cent at 15 cm depth has been reported from the Sonoran desert. From tropical areas there are reports of a mean organic matter content of 3.75 per cent in Hawaiian soils, but the comparable 3.54 per cent in humid areas in Puerto Rico declines to 1.84 per cent in more arid localities. In the Tarai region of Uttar Pradesh, the organic matter content is moderately high, ranging from 2 to 3.5 per cent in the $Å_p$ horizons (Desphande et al. 1971). In the Kenya Highlands values average over 4 per cent but fall to 2.5–4 per cent where wheat cultivation is continuous and crop residues are burnt (Birch and Friend 1956). For the savanna of West Africa a mean percentage of 1.36 organic matter has been suggested.

Organic matter content in soils declines rapidly in the profile below the distinctively dark organic horizon to which quoted values usually refer. However, where comparisons include A horizons which are unusually deep, such as Chernozems in which organic matter may still be as much as 1 per cent at 1 m depths, estimation of soil organic matter based on total volume is worthwhile. In the United States soils typically have a total bulk of organic matter in the 25–85 tonnes/ha range for the top 30 cm of the profiles. A mean figure for East Africa is 60 tonnes/ha. In Central and south-east Europe, the total amounts of organic matter decrease from 170–310 tonnes/ha in Chernozems to 120–280 tonnes/ha in Brown Earths and to 85–190 tonnes/ha in drier areas (Bedrna and Mičian 1967).

3.1.5 SOIL TEXTURE AND AMOUNTS OF ORGANIC MATTER IN SOILS

Particle size of the inorganic fraction is a source of significant local variations. Sandy fractions, owing to their relative infertility, reduce productivity and with large pores favouring free air circulation within the soil, oxidation rates are accelerated. Significant correlations have therefore been established between percentage sand and organic matter, with $r = -0.42$ ($N = 33$) for soils in Israel and $r = -0.87$ ($N = 55$) for freely drained soils in Iceland.

The relative importance of the silt fraction, compared with that of clay, varies. In Nigerian forest soils around Ibadan by far the largest part of the organic matter occurs in the coarse-silt fraction, with only 1.6 per cent in the sand (Bates 1960). Organic matter content in Compact Chernozems in Moldavia is 2.5 times higher in the silt fraction than in the clay and is also associated with fine silt in Chestnut soils. The importance of clay is seen in the Ukraine where clay loams and clay Chernozems have 4.7–6.5 per cent organic matter compared with 3.9–4.9 per cent in medium loams and 3.0–3.9 per cent in sandy loams (Krupskiy *et al.* 1970). In the Kenya Highlands, a 10 per cent increase in clay content appears to favour an increase in organic matter by 0.3 per cent (Birch and Friend 1956). However, the relationship between clay and organic matter percentage is not simple outside a certain range. In Vertisols in West Africa the organic fraction starts to decline with increases in the clay content beyond 35 per cent (Jones 1973).

3.2 Breakdown of organic matter in the soil by macrofauna

Depending on its feeding habits, the soil macrofauna includes phytophagous animals which feed on living plant materials and saprophagous forms which feed on dead plant material (fig. 3.3). Carnivores include moles, many beetles and centipedes. Directly or indirectly, therefore, the combined effect of the soil macrofauna is to initiate the physical breakdown of organic materials and its subsequent biochemical decomposition (Dickinson and Pugh 1974). Most animals initiate the comminution process by mastication. The digestive apparatus of the earthworm, however, is distinctive, being equipped with a gizzardlike organ which facilitates the grinding up of coarser fragments of organic matter (Burges and Raw 1967). The worm *Dendrobaena rubida* has been observed to reduce organic material to particles less than 15 μm by 15 μm and to sizes as small as 4 μm by 4 μm (fig. 3a). The excrement also includes leaf fragments of more than 400 μm in size and whole excrements of small arthropods.

Ants feed on both plant and animal substances. Leafcutter ants, like the Texas leaf-cutting ant *(Alta texana)*, cut pieces of leaves which they transport to their nests for food. Fragments are also masticated to produce a compost on which a special type of fungus is grown.

Termites (Lee and Wood 1971) are primarily vegetarians, of which the wood-feeders, fungus growers, and the humus feeders are the three main kinds. Some species of termites eat their way through logs and fallen trees, leaving only a thin outer layer. All species have a rich intestinal flora of bacteria. The wood-feeding termites have a rich fauna of protozoa which release cellulase in the gut, the enzyme which breaks down the cellulose into products assimilable by termites. The digestive processes of termites are so efficient and the utilization of ingested food material so complete that they make little contribution to the liberation of energy sources and nutrients to the soil. Certain termites block the cycling of organic matter to the extent that it may be immobilized in mounds for more than 50 years in areas in West Africa. In savannah woodlands of Australia, decomposition of organic matter may take 60 years. The extreme example is the black-mound termite's nest, in which nothing is wasted.

Soil organic matter 77

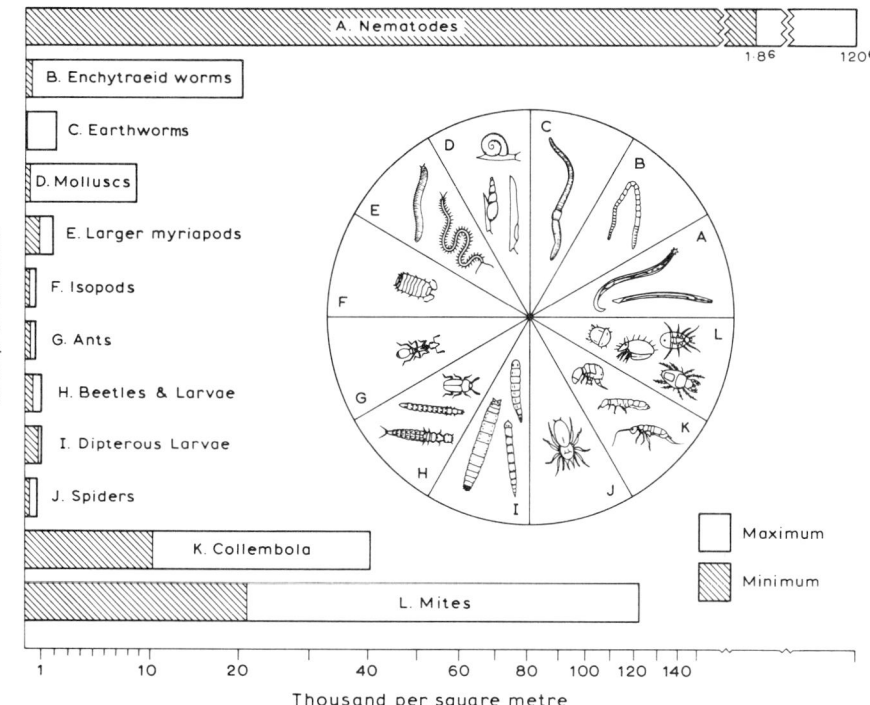

Figure 3.3   The groups and numbers of smaller organisms in the soil of a European grassland (Baker and Snyder, 1965). Spiders and centipedes are carnivorous and ants are predominantly carnivorous. Beetles, fly maggots, mites and nematodes range widely in their diet according to species. The remaining groups feed largely on decaying organic matter, with many Collembola being closely associated with fungi and other micro-organisms.

Dead bodies are devoured and excrement is eaten again and again until every particle of organic matter is extracted. In such cases the removal of organic matter from the soil is permanent.

Arthropoda especially Springtails (Collembola) are involved in the mastication of organic fragments (fig. 3.3). They inhabit a wide range of damp soils, leaf mould, and organic mats that accumulate at the base of plants. Most species subsist on fungi, spores and pollen grains in addition to plant detritus (Murphy 1953). Springtails reduce leaves to skeletons and breakdown the faeces of larger arthropods. The mites (Acari) are found in all situations where vegetation occurs, particularly amongst leaf litter (fig. 3.3). Mites are a diverse group, including carnivores, phytophages, saprophages and parasites. They reduce twigs, roots and leaf ribs to particles as small as 25 µm. They excrete particles about 50 μm diameter, of irregular shape (fig. 3.1). Decomposing grassland herbage is also readily consumed (Curry 1969).

The variety and numbers of beetles is similarly large and their feeding habits diverse. Some species operate in the leaf litter, whereas others burrow to 10 cm or more into the soil. Larvae are chiefly active on the underside of leaves, eating holes and voiding excrements about 250–600 μm diameter. Molluscs are also important in the processes of comminution of organic fragments as their digestive fluids contain powerful cellulases.

The biochemical decomposition of the smallest particles is also begun, particularly where earthworms are active. The chemical composition of worm casts, compared with the surface soil of an arable field, usually shows greater proportions of available nutrients. About 6 per cent of non-available nitrogen ingested by worms was found to be excreted in forms available to plants. Worm casts may also be less acid than the ingested food. Some genera of earthworms contain calciferous glands and convert ingested calcium into calcium carbonate even in the absence of a food source containing carbonate.

Rates of decomposition are greatly accelerated in tropical areas (Nye 1955). In some instances, crops may be damaged, as by nematode activity which reduces herbage production in pastures (Yeates 1977).

3.3 Biochemical compounds in soil organic matter

As decomposition of organic matter in the soil releases individual chemical compounds, a relative accumulation of the more resistant ones is built up (Whitehead *et al.* 1975). In addition, new organic substances are synthesized by micro-organisms (Dubach and Mehta 1963). This complicates an elucidation of the decomposition routes. Unlike the command role in the determination of specific structures in living matter played by DNA (deoxyribonucleic acid), there is no template to determine a specific, unambiguous structure of organic matter synthesized in the soil. Indeed, the end product, humus, is usually defined as that portion of soil organic matter which is completely amorphous and has no cellular structure like that of plants, animals, and micro-organisms.

3.3.1 THE NON-METAL ELEMENTS

As with the composition of soil clays, in organic cells the non-metals oxygen (65 per cent) and hydrogen (10 per cent) are preponderant. A major difference is the presence of carbon (18 per cent) as well as the vital 3 per cent nitrogen. Phosphorus and sulphur are also present in some compounds. These elements are held together in compounds

Soil organic matter 79

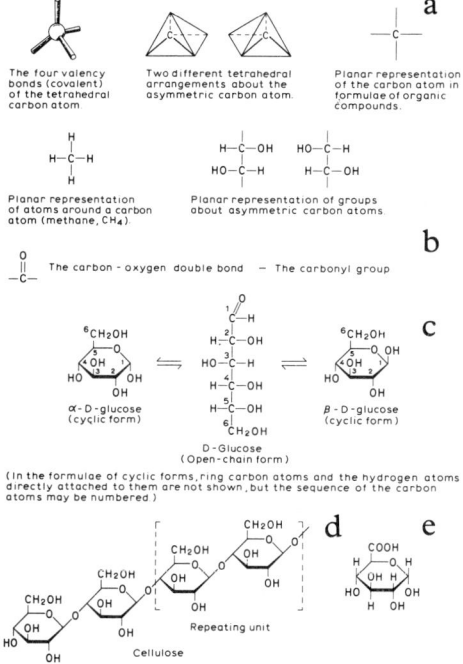

Figure 3.4 Representations of carbon in organic compounds and formulae for saccharides.
a Perspective view of the tetrahedral distribution of valencies of carbon and convenient planar representations of this 3-dimensional arrangement in organic compounds.
b The carbonyl group.
c The cyclic form of monosaccharides, showing the two crystalline forms of glucose, designated α and β.
d Linkage responsible for the linear nature of a polysaccharide.
e Uronic acid, a six-carbon sugar acid in which the carbon-6 has been oxidized to the highly reactive carboxylic acid group (− COOH).

by bonds formed by the sharing of electrons. In the covalent bond, for instance, two electrons are shared between two atoms, each atom providing one electron to be shared. The utilization of all possible types of electron bonds allows for great versatility in the molecular structure of organic compounds.

Carbon is the key element in organic materials and may account for 45–55 per cent of organic residues in the soil. It also unites with oxygen and is lost as carbon dioxide. Carbon is quadrivalent and all four of its valency linkages are regarded as being equal in strength and in every way alike. The four valency bonds (covalent) are distributed in space so that the four atoms or groups of atoms attached to the carbon atom are at a maximum distance from each other. Thus the four bonds are imagined to be distributed tetrahedrally in space, with the carbon atom at the centre of the imaginary tetrahedron (fig. 3.4a). A carbon atom to which four different atoms or groups of atoms are attached is said to be asymmetric. Asymmetry greatly increases the number of organic compounds as the four different groups at the corners of the imaginary tetrahedron may exist in two different spatial configurations, one the mirror image of the other.

Oxygen makes up nearly all the bulk of water molecules and it complexes with other elements to form many important ions and organic compounds. Molecular oxygen ensures destruction by oxidation and plays an important role as a dehydrating and carbonizing agent in the combined reactions of destruction and synthesis of organic materials. The oxidation is largely due to microbial activity, although purely chemical oxidation may also be involved.

Hydrogen is important in balancing charges without producing any increase in volume. It bonds with carbon and nitrogen in organic compounds. Nitrogen plays a key role in soil organic matter as it is the natural store for this vital plant nutrient. A wide array of nitrogenous organic substances occurs in intimate association in the soil.

3.3.2 CARBOHYDRATES

Many of these 'hydrates of carbon' have the general formula $C_x(H_2O)_y$ with others containing some nitrogen and sulphur as well. Carbohydrates are the main constituents of plants (Oades 1967), making up 60–90 per cent of their dry mass. In soils, 7–25 per cent of total carbon in podzols is carbohydrate, 12–23 per cent in Chernozems and 20–29 per cent in Red Earths (Orlov and Sadovnikova 1975).

Cellulose is the most abundant carbohydrate, accounting for over half the carbon found in plants. The basic carbohydrate unit which cannot be broken down by hydrolysis into smaller molecules is the monosaccharide or simple sugar (fig. 3.4c). This may be an open-chain structure containing the important chemical grouping, the carbon-oxygen double bond, termed the carbonyl group (fig. 3.4b). As this is a site of unsaturation, the electrons of the carbon–oxygen double bond are affected by approaching compounds more than electrons in single bonds. Because of the geometry of the simple sugar or glucose molecule, one result is that the carbonyl group can react with the hydroxyl group on carbon-5 to form a six-membered ring containing one oxygen and five carbon atoms (fig. 3.4c). Polysaccharides are made up of monosaccharides joined by carbon–oxygen–carbon linkages. These oxygen bridges are described by the general term glycosidic linkages, with the more specific term, glucosidic linkage, referring to the linking of glucose units as the monosaccharide (fig. 3.4d).

In litter layers, polysaccharides are present as components of plant and animal tissues. Plant cell walls are almost exclusively structured by polysaccharides, particularly cellulose which accounts for about 40–45 per cent of dry wood. In its simplest form, cellulose consists of glucose molecules, bridged by C–O–C (glucosidic) linkages (fig. 3.4d). Glucose molecules, 300–3000 in number, form long threads or polymers which are about 100–300 Å wide and about half as thick. The almost exclusively linear pattern of the cellulose chains facilitates close packing of the polymers, so that the fibre resists pulling apart (Gupta and Sowden 1964). Cellulose is totally insoluble in water and is not digestible by mammals. The digestion of cellulose and other polysaccharides is achieved by the enzymes which most micro-organisms can produce. Also, an important soil-forming process starts in the stomachs of cattle and other ruminants, where cellulose digestion depends on enzymes produced by the bacteria which live symbiotically in their digestive tracts.

Chitin is structurally similar to cellulose but it also incorporates the nitrogen-containing amino groups. Up to half the cell wall of fungi is made up of chitin which is also found in insects, other arthropods and worms. All hard, tough and strong parts of insects and crustacea, such as shell, exoskeleton, wings or claws, are made up of this nitrogen-containing polysaccharide. Hemicellulose is another polysaccharide which accounts for 15–30 per cent of woody materials. This is a large heterogeneous polymer, chemically unrelated to cellulose. It comprises a mixture of monosaccharide units, a common feature of which is the presence of acidic groups. Dead organic materials may also contain some of the polysaccharides which were stored as an energy reserve. For instance, seeds and grains may contain up to 80 per cent starch. This large polysaccharide molecule, made up of several thousand glucose units, is also stored in stems and roots. In higher animals the comparable metabolic reserve is glycogen which consists of branching chains of glucose molecules.

In the soil, polysaccharides like starch or hemicellulose are rapidly re-cycled (Cheshire *et al.* 1969). Even the chemically resistant polymers of cellulose are not stable. Thus cellulose and starch are rapidly hydrolysed in the soil to glucose which is taken up by living organisms and are not present in the decomposition end-product, humus. However, some 5–20 per cent of soil organic matter is in the form of polysaccharide (Swincer *et al.* 1969). It is important in the weathering of rocks and minerals, and is highly effective in binding or aggregating soil particles. These substances are also particularly active in exchange reactions with nutrients in the soil solution. In their carboxyl groups (–COOH), the hydrogen ion tends to ionize away, leaving a negatively charged site. Further, the importance of soil polysaccharides in the chelation of metals and in the flocculation of clay particles has been widely demonstrated.

Soil polysaccharides are clearly different from plant polysaccharides like cellulose or starch (Lowe 1968). They are complex mixtures with a wide range of molecular size and shape. In addition to a range of monosaccharides, they incorporate non-carbohydrate components like 3–7 per cent amino sugars, amino acids, and several sugar acids termed uronic acids (fig. 3.4e). These sugars common in soil polysaccharides are not abundant in higher plants. All these differences between plant polysaccharides and soil polysaccharides are attributable to the fact that there are two continuous and simultaneous micro-biological processes in operation on organic compounds in the soil. Not only are micro-organisms digesting and degrading plant and animal polysaccharides but they are also re-synthesizing new carbon compounds from a range of sources, including the polysaccharides undergoing decomposition. It seems that soil polysaccharides are more the product of microbiological metabolism, particularly the products synthesized for cell walls, rather than the direct derivatives from decomposing organisms. Their resistance may be due to interaction and association with polyvalent metallic ions and with stable organic fractions (Cheshire *et al.* 1974). They may also be held within soil aggregates in niches too small for micro-organisms to penetrate.

3.3.3 PROTEINS AND AMINO ACIDS

Proteins themselves are large molecules, the fundamental constituents of protoplasm and thus found universally in the cells of living organisms. The majority of proteins contain 50–55 per cent carbon, 20–25 per cent oxygen, 15–20 per cent nitrogen and 6.5–7.5 per cent hydrogen. Some proteins also contain phosphorus and nearly all contain some sulphur. Proteins are readily decomposed in soil as rather slight changes can easily disrupt their stabilizing forces (fig. 3.5a). On hydrolysis they release amino acids, the primary units of which proteins are constructed. These units are relatively small molecules containing amino ($-NH_3^+$) and carboxylic acid ($-COO^-$) groups around the carbon atom, hydrogen, and a side chain, R. The symbol R is used whenever the organic chemist wishes to represent a general class of compounds. For instance, in the formula $R \cdot CH(NH_2).CO_2H$, the radical R may include carboxyl, amino, amido and hydroxyl groups, or it may be an aromatic or heterocyclic ring. The ionizable groups in amino acids (fig. 3.5b) are either proton donors, acting as weak acids, or as weak bases (proton acceptors), depending on the relative acidity or alkalinity of the medium. For example, the

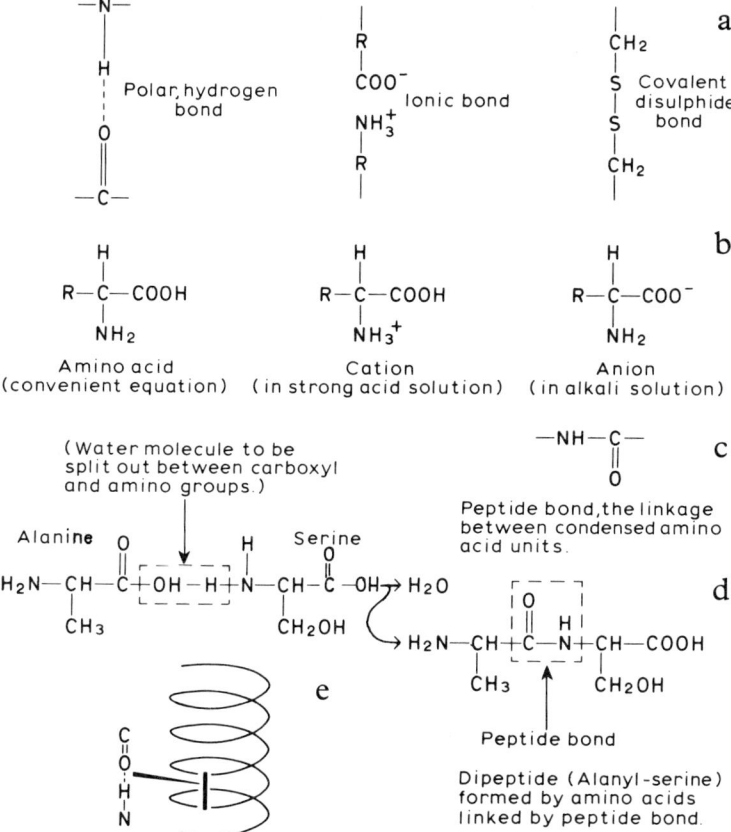

Figure 3.5 Linkages, constituents, and structures in proteins.
a Side-chain interactions in protein which are readily split in soil decomposition processes.
b Amino acid, conveniently represented as a non-ionized form (left), is usually ionized in solution. If the hydrogen ion concentration of the solution is great, amino groups accept protons to give $-NH_3^+$ (centre). Conversely, if the hydrogen ion concentration is low, protons are lost from the carboxyl groups which are then left with a net negative charge, $-COO^-$ (right). The fourth valence is satisfied by either a hydrogen ion or by a wide range of organic radicals, represented by R.
c Amino acids can link together by a peptide bond, since each possesses both an amino group and a carboxyl group.
d and e. Condensation, if dehydration continues, produces a polypeptide composed of a string of many amino acids. Commonly this chain may exist in a helical form, with the loops linked by hydrogen bonds (e). If the helix, in turn, is folded, the proteins are globular and act as enzymes.

undissociated acid RCOOH is a proton donor and the charged carboxylate anion is a proton acceptor, since dissociation of the hydrogen of the carboxyl group leaves a negatively charged site, whilst the addition of hydrogen to the amino group creates a positively charged site. Thus, according to certain conditions, an amino acid can behave as either a cation or as an anion. In strongly acid soils the amino acid is present largely as a cation, and in strongly alkaline soils largely as an anion (fig. 3.5b).

Since some 20–50 per cent of the organically bound nitrogen in the soil is in the form of amino acids, and since humic acids release about one-tenth of their weight as amino acids when subjected to acid hydrolysis, their character is of special significance. As the first products of protein decomposition, amino acids may exist as free acids in the soil. However, they are detectable only as traces, probably because they are such appropriate food for soil micro-organisms. Their persistence in soils results from their adsorption on to other soil constituents or their combination with more resistant decomposition products of lignin, tannin and other quinones or polyphenolic compounds. Polypeptides are formed by long chains of amino acids linked by the peptide bond (fig. 3.5c). The peptide (−CO.NH−) linkage is a covalent bond formed by the elimination of a water molecule between the carboxyl group of one amino acid and the amino group of an adjacent amino acid (fig. 3.5d). This elimination of water is termed condensation, a reaction fundamental to the re-synthesis of organic compounds in soils. Polypeptides, unlike polysaccharides, are coiled in the form of a helix, but this configuration is maintained by weak interactions only (fig. 3.5e) which are all readily broken during decomposition.

Enzymes are highly distinctive particles of protein which play a key role in soil processes. An enzyme is a catalyst, produced by living cells, in the presence of which certain chemical changes are accelerated. Enzymes are liberated during plant root metabolism. Numerous micro-organisms produce and excrete extra-cellular enzymes. Although these catalysts are colloids manufactured in the protoplasm of living cells, they often require a non-protein portion and the presence in the soil of certain metals including iron, magnesium, copper, zinc, manganese or cobalt is often necessary for their activity. After the death of cells some intracellular enzymes may persist in the soil for a certain period in an active state. Enzymes are tolerably resistant to denaturation in a soil environment and measurable activity has been found in 8700–9500-year-old permafrost samples.

By their catalytic activity, enzymes direct and control complex metabolic patterns in living organisms and decaying organic compounds in the soil. It seems that the soil microflora as a whole produces the complete required array of enzymes for decomposing litter and enzymes which readily diffuse into living plant roots and permeate plant cell walls (Kiss *et al.* 1975). Possibly the reacting compounds are brought into close contact by being adsorbed on the colloidal enzyme.

A given enzyme can act only on one substance. On the other hand, the structure of the substrate molecule which is acted on may be common to many enzymes. Several act on the $-CO-NH-$ links and many oxidize the $-CHOH-$ link to $-CO-$.

Enzymes in the soil are primarily involved with the metabolism of carbohydrates, compounds containing phosphorus, and in catalysing oxidation–reduction processes. They are classified, and terminology built up, according to the reactions which they accelerate, by adding the suffix '-ase' to the substrate on which they act.

Hydrolases are a very large group of enzymes catalysing hydrolysis:

$$AB + H_2O \longrightarrow AH + BOH.$$

They are subdivided according to the chemistry of the substrate which they cause to be hydrolysed; for example, amidases, or deaminases, effect the hydrolysis of amides. The metabolism of nitrogen compounds is closely associated with peptide hydrolases which hydrolyse peptide bonds. Esterases include a subgroup with a specificity for hydrolysing esters of phosphoric acid. In the carbon cycle, hydrolases which act on glycosides. The glycoside hydrolases include invertase, which is present in all soils, and particularly in peaty marshes. Invertase hydrolyses sucrose in glucose and fructose. Similarly, amylases catalyse the hydrolysis of starch. Urease greatly increases the conversion of organic nitrogen to inorganic nitrogen by the hydrolysis of urea to ammonia and carbon dioxide:

$$CO(NH_2)_2 + H_2O \longrightarrow 2NH_3 + CO_2.$$

Urease is found in higher plants and a large number of bacteria and fungi.

An important role in life in the soil is played by the enzymes of the oxidoreductase group as they provide energy for living cells. The dehydrogenases are part of the respiratory system of any cell, tissue, or organism. They catalyse the oxidation of substrates by implementing the removal of hydrogen. These enzymes effect the transfer of hydrogen from hydrogen-containing compounds to another hydrogen carrier:

$$XH_2 + A \longrightarrow X + AH_2.$$

Aerobic dehydrogenases utilize molecular oxygen directly as a hydrogen acceptor. Oxidases are widely distributed in most soil types, catalysing oxidations of various organic compounds.

The soil environment may exert a significant control on enzymatic activity, with particle size being particularly significant. Enzyme activity is suppressed following their adsorption on to soil particles. This action greatly increases as particles decrease to colloidal sizes. Chernozems have a most powerful suppressing effect where extracellular enzymes are adsorbed on clays. In montmorillonite clays, the enzymes may be trapped and their activity suppressed within the crystal lattice.

3.3.4 LIGNIN

Lignin is predominant in the contributions to the litter layer of many soils. It is present in finely divided form as the vascular tissue of fallen leaves and in decaying rootlets. In living plants, lignin is distinctive among the non-saccharide components of cell walls. It makes up some 15 to 35 per cent of the supporting tissues of higher plants and in trees accounts for some 17–25 per cent of hardwoods and 25–35 per cent of softwoods. Although it contains carbon, hydrogen, and oxygen as do carbohydrates like cellulose, the proportion of these elements makes lignin chemically quite different from cellulose. In fact, the exact chemical composition of lignin is indefinable. Lignin is not chemically combined with cellulose either. It is, however, intimately interwoven around cellulose fibres as the characteristic cementing constituent of woody tissue, and it has proved impossible to isolate lignin from wood in a form identical with that in its natural state.

In the plant, the lignin building units appear to originate with the aromatization of carbohydrates or with the conversion of living cells into compounds containing benzene-type rings. Aromatic hydrocarbons of this type were originally called aromatic because many of them have fragrant or spicy odours. Now the adjective describes a particular type of chemical structure, of which benzene (C_6H_6) is the simplest member (fig. 3.6a). The benzene ring, commonly found in animals and especially in plants, is a prominent compound in soil organic matter in general and in lignin in particular. The benzene ring can be represented as a compact hexagon with six carbon atoms each spaced 1.39 Å apart and linked by three carbon–carbon single bonds, alternating with three carbon–carbon double bonds. In reality, it seems that these six bonds between the carbon atoms of the benzene ring are a hybrid of both single and double bonds, a characteristic described as resonance. Thus, the chemical stability of the benzene ring is explained even though the molecular formula C_6H_6 indicates that there is only one hydrogen atom for each carbon atom in the molecule, and special conventions are followed in diagrams of aromatic hydrocarbons (fig. 3.6b).

Vital components of soil organic matter are formed with the removal

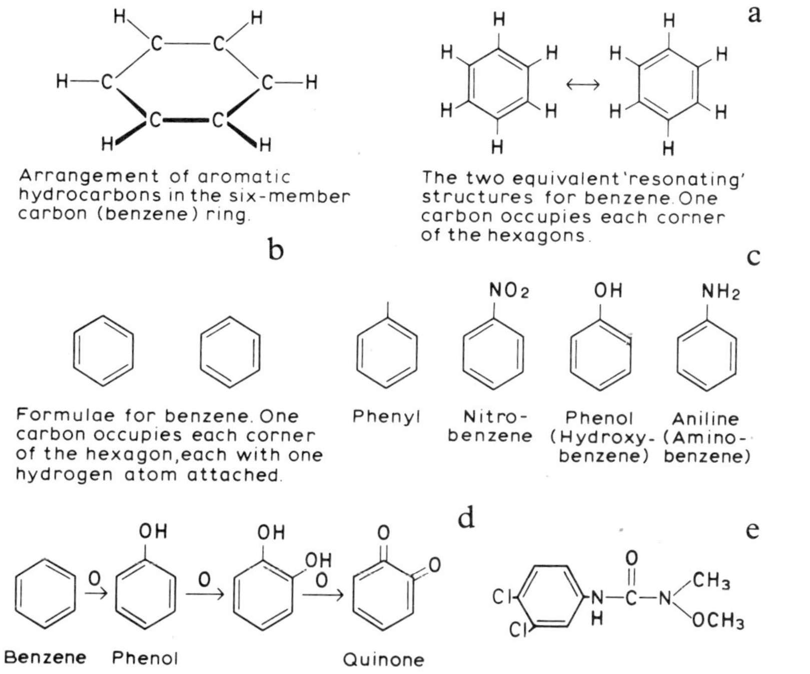

Soil organic matter 87

a

Arrangement of aromatic hydrocarbons in the six-member carbon (benzene) ring.

The two equivalent 'resonating' structures for benzene. One carbon occupies each corner of the hexagons.

b

Formulae for benzene. One carbon occupies each corner of the hexagon, each with one hydrogen atom attached.

c

Phenyl   Nitro-   Phenol   Aniline
         benzene (Hydroxy- (Amino-
                   benzene) benzene)

d

Benzene   Phenol   Quinone

e

Figure 3.6 Aromatic hydrocarbons, the cornerstones of the more resistant soil organic fractions.
a Schematic perspective (left) and planar (right) views of the carbon-carbon double bonds of the benzene ring, with the position of carbon and hydrogen atoms indicated.
b Hexagons as convenient formulae for benzene, with presence of carbon and hydrogen atoms assumed in position.
c Aromatic hydrocarbons resulting from the removal of a hydrogen atom (phenyl) and additions and substituents on the benzene ring.
d Oxidation of benzene to phenol and to quinone which results in condensation and polymerization.
e Structure of linuron, a herbicide which combines with clay and organic matter through the oxygen of its carbonyl group.

of a hydrogen atom from the benzene ring, as a phenyl is then formed. If NO_2, OH, or NH_2 are attracted to this site, nitrobenzene, phenol (hydroxybenzene), and aniline (aminobenzene) are formed. Many of the C atoms in the lignin molecule are connected to OH radicals (phenolic groups) in which H behaves like the carboxyl groups of organic acids. More complicated aromatic compounds are built up if there is more than one substituent attached to the benzene ring (fig. 3.7c). Lignin is a polyaromatic molecule arranged in a complicated and irregular network of phenolic residues which are linked by short hydrocarbon chains

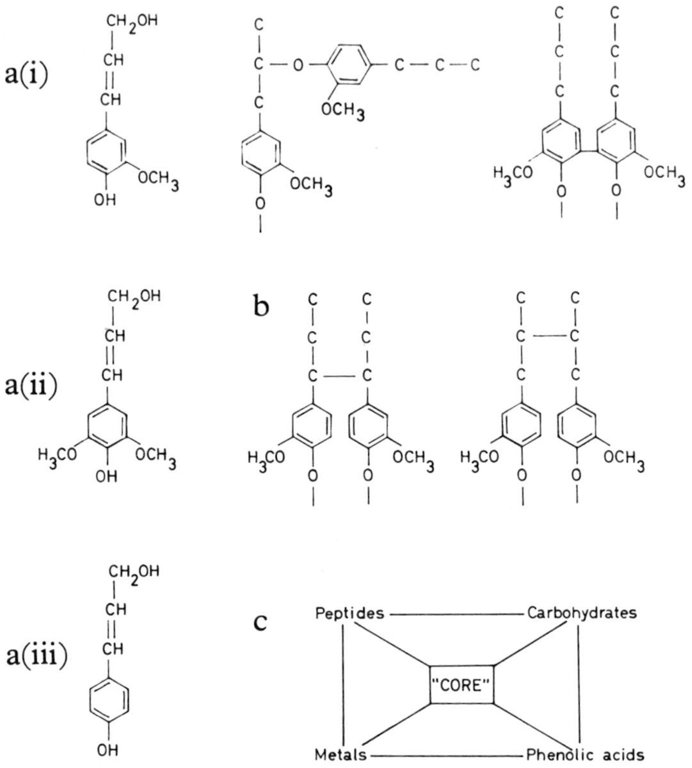

Figure 3.7 Building blocks and bonds characteristic of lignin.
a Building blocks of lignin found in different plants (Flaig, 1964). The building unit of coniferous lignin is almost entirely coniferyl alcohol (i). Lignin from broad-leaved trees is derived from both coniferyl and sinapyl alcohol (ii). Both occur in the lignin of grasses which also include *p*-coumaryl alcohol as a monomer (iii).
b Principal types of bond characteristic of lignin (Manskaya and Kodina, 1968). For simplicity, and since lignin structure changes continually during decomposition, the atoms or double bonds satisfying the spare valencies of the carbon atoms in the side chains are not shown. About one-third of the structural links of lignin are joined by ether bonds and two-thirds by carbon-carbon bonds.
c Diagrammatic representation of a humic acid (Haworth, 1971). The order of attachment of the groups is uncertain.

(fig. 3.7a,b). Lignin yields only slowly to chemical and biological attack probably because about two-thirds of the structural links in lignin are carbon bonds which are only 1.4 Å in width and difficult to cleave. As it slowly breaks down into simpler aromatic substances, it releases carbon and energy for microbiological metabolism.

3.4 Humus

3.4.1 HUMIFICATION

The process of humification includes the decomposition of the initial organic materials to the intermediate and final products of mineralization. Some preliminary crushing of highly orientated polymers is particularly important, to increase their susceptibility to hydrolysis. Decomposition is often initiated by fungi and carried on by lower invertebrates and bacteria.

The formation of humic substances in well-aerated soils involves the microbiological cleavage of linkage groups in lignin, especially the enzymatic fission of the benzene ring. Oxidation is catalysed by phenolase enzymes produced by lignin-decomposing fungi. Phenol, or carbolic acid (C_6H_5OH) is the first member of a series of hydroxyl compounds in which

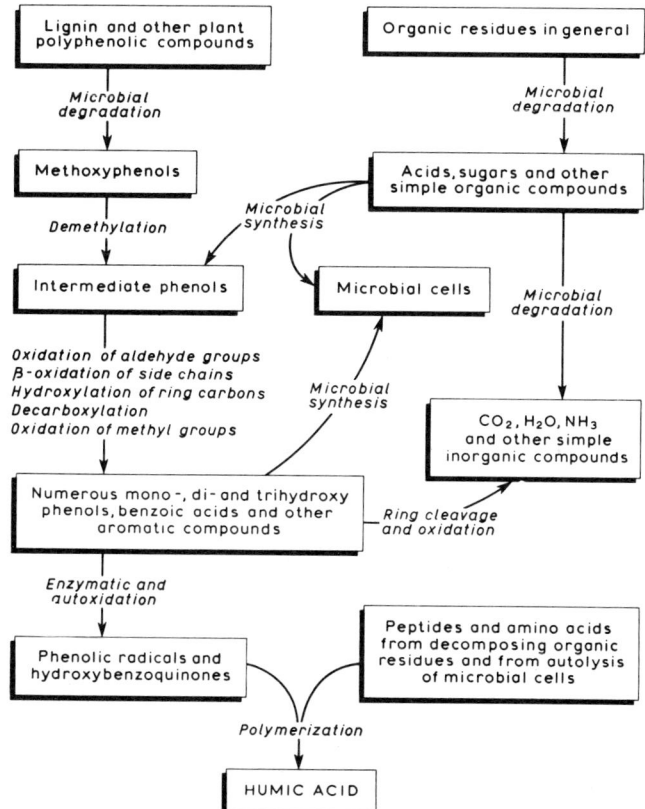

Figure 3.8 Schematic representation of the complexity of humic acid synthesis (Martin and Haider, 1971).

one to four hydroxyl (–OH) groups are attached directly to a benzene ring nucleus. Thus, a series of reactions takes place. The oxidation of monohydric to dihydric phenols may be followed by further oxidation to quinones, in which two hydrogen atoms on the benzene ring are replaced by oxygen (fig. 3.6d). Atmospheric oxygen is required, since humification involves the oxidative breakdown of lignins as a fundamental stage. Ultimately, some metabolic products are degraded to simple inorganic substances like water, carbon dioxide, and ammonia.

Humification also describes the re-synthesis of simple compounds in the soil. Quinones, for instance, are highly active radicals which react with other phenol units, peptides and amino acids and link up into large molecules of humic acid. Thus, the soil humus compounds which persist or are re-synthesized incorporate polycyclic aromatic ring structures in their core formed during the condensation of aromatic compounds especially polyphenols and the nitrogen-containing amino acids and peptide compounds (fig. 3.7c). In the side chains, carboxyls, phenolic and alcoholic hydroxyls, and carbonyls are the main oxygen-containing functional groups. Thus, many phenolic units resemble those in lignin, and possibly a larger part of humus is made up of complex phenolic polymers (fig. 3.8).

3.4.2 HUMIC ACIDS, FULVIC ACIDS, AND HUMINS

Because the make-up of humus is so heterogeneous and so easily changed, any method of fractionating it is somewhat arbitrary. However, recent work shows that there is a qualitative similarity between humic and fulvic acids isolated by conventional means and that extracted by water (Linehan 1977). Also, electron microscope examination of fulvic acids show sponge-like aggregates 20,000–30,000 Å long, punctured by voids 200–1000 Å in diameter (Schnitzer and Kodama 1975). Together, the humic and fulvic acid molecules make up 50–80 per cent of soil humus. They are a group of substances with a common molecular structure, with inter-related transitional forms, and are not individual chemical compounds. Their solubility in mineral acids is used as a basis for their differentiation. Thus, humic acid is a highly unspecific term denoting that fraction of soil humus which is soluble and extracted by alkaline solution and precipitated at pH 2 by acid. It is brown if not almost black in colour. The organic material which does not precipitate on acidification is called fulvic acid. Fulvic acids consist essentially of humic acid-type molecules, but with a relatively smaller molecular weight range and with polysaccharides.

A difference between humic acids and fulvic acids is seen in their elementary composition. The mean percentages of carbon, oxygen, hydrogen, and nitrogen in humic acids of Russian soils is about 61, 31, 3.7 and 4.7, respectively. Comparable figures for fulvic acids are 46, 48, 3.5 and 2.5 per cent (Kononova 1961).

All the humic substances have an aromatic nucleus containing one or more ring compounds. Open-chain, or aliphatic, compounds make up the peripheral structure. The major components, apart from polysaccharides, are thought to be polymers formed from the recombination of units such as quinones, amino acids, and aldehydes, compounds of varied complexity and without ordered sequence. Since most of these compounds have reactive side groups such as $-COOH$, $-NH_2$ and $-OHH$, linkages between them probably create finally a complex, three-dimensional structure (Whitehead and Tinsley 1963). Aromaticity is most marked in humic acids whereas the aliphatic, or open-chain, structure is most marked in fulvic acids. The reciprocal relationship between the proportion of aromatic structures, which have hydrophobic properties, and the aliphatic structures which are hydrophilic, accounts for the diversity in the properties of humic compounds. In particular, it affects their interactions with electrolytes and their degree of dispersion. The close relationship between humic acid content and quantity of water-stable aggregates is illustrated particularly by Chernozems (Ponomareva and Plotnikova 1975). Complex formation of metal ions with humic acids contributes to their role in enriching the soil with many trace elements. Peripheral open-chains, being hydrophilic, tend to disperse readily. Fulvic acids, therefore, may remain in solution after complex formation. This favours the downward leaching of elements in the soil profile.

The relative amounts of humic and fulvic acid have an important bearing on immobilization and translocation of many soil constituents. For instance, the humus of the Taiga soils, subject to prolonged freezing, is typically of the fulvate type, with the HA/FA ratio being less than unity. According to M. Kononova, HA/FA ratios for podzols are about 0.9 to 1.1, but areas with values between 0.57 and 0.81 have been studied. In Meadow Chestnut soils there is a tendency for humic acids to accumulate because of greater supply of plant residues entering the soil and the more favourable conditions for humus formation. The humus content of Ordinary Chernozems in the Ukrainian steppe has a high humic acid component, with HA/FA ratios of 2.0 to 4.7. In drier areas the proportion of fulvic acids increases. In desert-steppe soils in the Caspian Lowlands humic acids have a weakly condensed aromatic nucleus, fulvic acids are predominant, and HA/FA ratios are, in consequence, less than unity (Titova 1968). The ratios are narrower still in all types of Sierozems and in Solonchaks and Solonetzes. Owing to dryness, HA/FA ratios may be 0.6 to 0.8 in upper horizons and as low as 0.3 to 0.4 in lower horizons (Durasov and Marchenko 1967).

Soil types overlap one another with respect to their HA/FA ratios. There is, however, a broad trend which may indicate an increase in carbon from podzols of wet, cool regions to the more strongly condensed and polymerized humic compounds of the Chernozems. In general, a predominantly fulvate group can be recognized with HA/FA ratios of less than 0.75, distinguished from a predominantly humate group with values over 1.30. At a ratio of 0.95 the overlap is minimal.

Sources of more localized differences include texture. For example, in the Ukraine where coarse clay and clay loam Chernozems have HA/FA ratios of 3.4 to 4.3, sandy loam and loam Chernozems have ratios of 2.0 to 2.7 (Godlin and Son'ko 1970). The availability of certain minerals may also be significant, as it seems that both nitrogen and phosphorus are immobilized more readily by humic than by fulvic acids (Tan *et al.* 1972). Significant correlations exist between soil nitrogen content and HA/FA ratios. A similar relationship has also been noted for organic phosphorus. Within the humic acid group three fractions are recognized by soil scientists in the U.S.S.R. They differentiate between the first (free) and the second (bound with calcium) fractions, the latter being a prominent feature of Chernozems. The 'third' fraction of the humic acids is bound with silicate forms of Fe_2O_3.

In the fractionating of soil humus, a third major component is recognized. This is the residue which dissolves in neither the alkaline nor the acid solutions, termed humins. These are large molecular substances which are similar to humic acids. Formerly, this insoluble fraction from which it was impossible to isolate any specific fraction was termed 'lignoprotein' by S. A. Waksman and his co-workers, following their deductions from indirect evidence.

In field soils it has been suggested that humic acids become humins when the soil dries or freezes, or follows when humic acids are irreversibly adsorbed by soil clay minerals. The content of humins in Sod-Podzolic soils is very low and their distribution in the soil profile is inversely proportional to fulvic acids. In contrast, there are greater accumulations of humins in soils in dry steppes and deserts, where continental conditions result in frequent repetitions of drying and freezing processes.

3.4.3 CARBON–NITROGEN RATIO AND OTHER RATIOS

The elemental composition of soil organic matter changes as it decomposes. Micro-organisms use much of the carbon as a source of energy which is converted to carbon dioxide. Water is the other simple compound which is produced and lost. Most of the nitrogen, however, becomes part of the tissue of micro-organisms. The relative accumulation of nitrogen in relation to carbon in decomposing soil organic matter evidently describes its degree of humification and indicates the availability of nitrogen. The carbon–nitrogen or C:N ratio is, therefore, a valuable index for describing soil organic matter. Nitrogen-rich accumulations provide more suitable food sources for soil fauna than organic matter with a high C:N ratio. Each plant has a specific requirement for nitrogen but trees, in general, grow well with C:N ratios up to 35.

The main characteristic of C:N ratios is the close similarity maintained in soils widely separated in latitude. In Iceland, a large majority of soils from freely drained sites have C:N ratios between 11 and 17 with a mean value around 15. The C:N ratio of British arable soils is typically around 10. A range between 8 and 12 is reported for representative soils in the U.S.S.R. (Aseyeva and Velikzhanina 1966). In areas of shifting cultivation in Africa, Reddish Brown Earths, in which micro-organisms were active, have C:N ratios between 8 and 12, with values between 14 and 17 in Red Yellow Podzolic soils (Nye and Greenland 1960). Low C:N ratios typify arid and semi-arid areas, possibly related to the activity of nitrogen-fixing blue-green algae which are typically present in surface soil crusts. In Dark Chestnut solonetzic soils in the Caucasus, C:N ratios averaged 8.9 falling to 7.0 in the Volga and Western Kazakhstan provinces.

The C:N ratio is influenced by particle-size composition in the mineral fraction of soils, being wider in sandy loam and sandy soils than in clay loams. The wider C:N ratio in coarse-textured soils results from lower nitrogen contents in these soils owing to their greater permeability and, therefore, susceptibility to leaching. Conversely, an increase in nitrogen in relation to carbon as particle size decreases reflects the combination and retention of a higher proportion of nitrogen-rich materials in the clay fraction.

The source of soil organic matter is significant. Generally, herbaceous litter in grasslands has a relatively low C:N ratio since it contains more nitrogen than woody species or even crop residues. In woodlands the leaf litter of deciduous species has a lower C:N ratio than that of conifers and decomposes relatively rapidly. Needles of conifers have a high C:N ratio, and a high polyphenol content, properties which explain their resistance to decomposition. Most common crop residues have C:N ratios of more than 25–30 which micro-organisms can decompose rapidly even though the nitrogen content is less than that required for their cell synthesis. The nitrogen required is taken from that already available in the existing soil organic matter, but at the expense of a supply to the growing crop.

Any conditions which favour decomposition usually lower the C:N ratio, and the degree of aeration in a soil is a major control on the C:N ratio. Thus soils in which anaerobic conditions reduce or even suppress biological activity, such as water-logged moorland soils, C:N ratios may be as wide as the raw undecomposed litter. C:N values close to 100 have been recorded in high moor peats in the U.S.S.R.

Despite its utility as an index of fertility, the C:N ratio does not indicate the degree of humification of soil organic matter. As a relatively transient element, the quantity of nitrogen depends mainly on factors other than the state of decomposition of soil organic matter. However, the relative proportions of the other three main constituents, oxygen, hydrogen, and carbon can be usefully compared. All hydrogen values decrease slightly with increasing humification whereas oxygen values change very little with progressive decomposition. Therefore, as hydrogen content decreases more rapidly than oxygen, all O:H values increase slowly with time. This reflects a change in oxidation as well as dehydration in the humic compound formed. Thus the ratio of oxygen content to hydrogen increases from the wet and cold environments towards warmer, drier grassland and steppe environments, attributable to an increase in hydroxyl groups and to the greater state of oxidation in the more humified molecules. Values of the O:H ratio typically widen from 6 to about 12.

Whilst oxygen values change very little with progressive humification and hydrogen values decrease slightly, the carbon content of humic acids increases. The increase in carbon over time is attributable to preferential loss, either by attack by micro-organisms or by leaching, of the easily decomposed organic substrates which contain smaller proportions of carbon in their molecules. With the loss of such substances, like sugars and hemicelluloses, substances like lignin, with comparatively larger amounts of carbon, accumulate. Thus it can be assumed that a C:H ratio will correlate with the degree of condensation of increasingly humified organic matter. For example, a narrow C:H ratio of humic acids in Taiga soils in central Siberia indicates the relatively low degree of condensation of the aromatic carbon nucleus and the high content of side radicals in the molecule, and podzols in European U.S.S.R. are similar. In contrast, wide ratios of Chernozems, approaching 10, indicate a high degree of condensation of the aromatic nucleus (Bel'chikova 1966). C:H ratios confirm the tendencies shown by O:H ratios but appear to be a better index of the degree of humification.

3.4.4 RATES OF HUMIFICATION

The rate at which humification proceeds is fundamental to most aspects of plant nutrition as this regulates the cycling of vital nutrient elements. Regional climate is a decisive control on rates of humification. In tropical latitudes, heat and humidity favour an explosive microbiological mineralization of organic matter. Green manures, for instance, are decomposed within a month and no residues accumulate in the soil. Annual decomposition rates approach the total volume of organic reserves in humid tropical areas. For example, in the rain forest in Java, the decomposition of organic matter occurs at a rate of 8.1 tonnes/ha/year. In the humid temperate climate about 2 per cent of the total organic matter store is decomposed annually. The major components of organic tissues vary greatly in their decomposition rates. At one extreme, substrates like glucose decompose almost completely within a month. Cellulose, hemicellulose and proteins are fairly readily decomposed by enzymatic activity into their hexose and pentose sugar building stones and others such as uronic acids, peptides and amino acids. Thus, of labelled carbon in ryegrass root and tops, 33 per cent remained after a year and about 19 per cent persisted for 4 years (Jenkinson 1965). Lignin stands at the other extreme in decomposition rates and may retain distinguishing chemical structures for several years at least.

Humification rate is affected by land use. In humid temperate environments, under arable cultivation, plant debris from a crop decomposes and mineralizes almost completely before the next crop matures. In woodlands, and even in long ley pastures, organic matter often accumulates in relatively stable compounds. Deciduous leaf litter has been observed to disappear completely within 14 months, whereas fir needles in an F_1 horizon have been noted to lose only a third of their dry matter weight after 2 years and to reveal no anatomical change.

A final consideration is the rate at which humus itself is broken down. Some of the new substances synthesized during humification resist decomposition and form complexes with clays or metal ions. Estimates of rates at which humus itself is broken down demonstrate why this substance has earned the title of the frugal custodian of soil nitrogen. For a grey wooded podzol, for instance, a half-life for humus of 360 years was calculated whereas that for a grassland Chernozem was 2200 years.

3.4.5 ACCUMULATIONS OF HUMUS

In soil organic horizons, the persistence of specific types of source material as coarser fragments reflects their proportion of the more resistant constituents (fig. 3.9). In fact, forest soils in mid-latitudes give rise to two distinctive types of humus formation. A distinction is drawn between mull and mor groups.

The mull humus formation occurs on moderately well-drained soils beneath mixed or deciduous woodland. Few plant residues persist and most organic residues are decomposed to colloidal size and intimately combined with the clay in the profile. A large proportion of the organic carbon is concentrated in the humic horizon, and the organic matter

96 Geography and soil properties

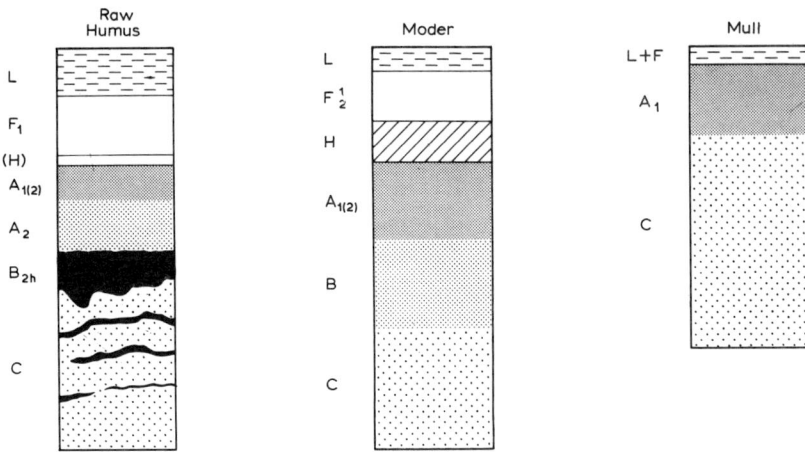

Figure 3.9 The three main soil profile expressions of accumulation of organic matter in soil profiles in mid-latitudes (Bal, 1970). The subscript 'h' indicates illuvial accumulation of organic matter in the B horizon of podzols developing beneath raw humus.

contains a large amount of nitrogen compared with more soils. There are quantities of calcium present and the pH range is normally about 4.5–8.0. Films of calcium carbonate protect organic matter from microbial degradation (Duchaufour 1976).

Mor is characteristic of well-drained acid habitats, with pH in the 3–6.5 range. Accumulating plant residues form layers of varying stages of comminution, separated from the underlying mineral soil by a sharp boundary. Mor is typically developed on heathland and beneath coniferous woodland. The raw-humus accumulations can be divided into a litter layer on the surface, passing down to the F layer where decomposition is taking place but where the external structures of litter are still evident. The F layer overlies the structureless brown mor of the H layer. In this raw humic material there is little available calcium, and nitrates and ammonium nitrogen are present only in very small quantities.

The effect of a different type of litter fall is compounded by the role of soil animals which predominate. Deciduous litter favours earthworms which flourish in near neutral soils and thoroughly mix the organic matter with the mineral soil. Typical mor is inhabited by mites and springtails which are not capable of comminuting all the organic debris produced by the woodland or heath and a vast store of plant nutrients remains on the forest floor. The differences between mull and mor are sufficient to persist long after the woodland has been felled. Intergrades between mull and mor are sometimes distinguished and moder-type humus, with excrements of soil fauna prominent but limited to a discrete organic horizon, may form in drier sites where earthworms are absent.

3.5 Peat

Decomposition of soil organic matter is largely an oxidation process. In waterlogged soils, where the circulation of air may become greatly restricted and the oxygen store depleted, the dead plant residues which are being continually produced by the vegetation, decompose less rapidly and peat begins to accumulate. The rate of decomposition of cellulose and hemicellulose may be particularly slow in the case of Sphagnum plants, which are often among the principal peat formers. Sphagnum grows most abundantly in acid sites where nutrients like calcium and phosphorus are scarce, and where cellulose-decomposing organisms are rare. The surface of peatlands is classed as soil in part because the accumulation of organic matter as peat is simply the extreme case of a major soil-forming process. Equally, profile development is usually discernible, with the top, biologically active horizon showing initial stages of decomposition of organic matter.

At least 1 per cent of the total land surface of the earth is peat-covered (Taylor 1964), being most extensive in the cool and moist land masses in the northern hemisphere, especially to the north of 60° N. The most extensive area is on western Siberia where the area surrounding the Ob and Irtysh rivers is almost entirely covered by peat, as is the lowland south of the Hudson Bay. About a third of Finland is covered by peat and about a fifth of the Boreal zone in Sweden lies under peat. Peat is also found in waterlogged areas in tropical zones, covering one-fifth of Sumatra and extending along the coasts of Malaya, Borneo, and the south coast of New Guinea.

3.5.1 DEFINITIONS OF PEAT

Fibre content and bulk density are two measures commonly used to estimate the degree of decomposition of peat soils (Farnham and Finney 1965). Depth is an important criterion. In Britain, a soil is defined as peat if it has an O horizon at least 40 cm thick, starting at the surface or at less than 30 cm depth. Subtypes of organic soils, or Histosols as they are termed in the U.S.D.A. Soil Taxonomy, are defined in terms of fibre content and the base status. Three main types are definable. In the raw, fibrous or 'fibric' (Latin *fibra*, fibre-containing) horizons, the nature of the original plant material is largely recognizable in well-preserved plant remains. Two-thirds of the fragments should exceed 0.1 mm in size. At the other extreme, 'sapric' (Greek *sapros*, rotted) horizons are highly decomposed, with less than one-third being fibres larger than 0.1 mm in size. Sapric Histosols have high bulk densities and occur commonly at the surfaces of most drained and cultivated organic soils. Thirdly, an intermediate type is recognized as the 'mesic' (Greek *meso*, intermediate) horizon in which identification of fibre is still possible using a microscope but in which the fibres are physically broken up and partially humified.

3.5.2 COMPOSITION OF PEAT

Peat is made up of an irregular sequence of plant species remains which reflect past environmental changes (Bascomb et al. 1977). The main expression of the damp conditions in which peat accumulates is that, instead of oxygen ions being added to substances, hydrogen ions are attached instead and humic substances modified accordingly. Considerably more than 2 per cent of the total organic matter may be in the form of fats, waxes, and resins, and about half of peat may consist of insoluble humic acids. This demonstrates the remarkable preservation of the structure of plant aromatic substances under conditions of peat formation. Nitrates accumulate in peat as one of the end products of the decomposition of organic matter, commonly to about 1.5–3 per cent. The C:N ratio decreases with increasing degree of decomposition. The distribution of available nutrients in peats shows that the soil-forming processes in the upper horizons are accompanied by the accumulation of several nutrients, indicating that there is a gradual increase in fertility, but that the accumulation of nutrients by natural means is slow. Boreal peatlands are often only weakly minerotrophic, particularly within the large Archean Shield areas of Canada and Fennoscandinavia. The ash content depends on the ash of the original plants (Kondô 1974) and therefore, on the minerals in the waters flowing through the bogs. Dust in rainfall may be significant, especially near sources of volcanic ash. Mineral deficiencies are common. In the peat soils which are widespread in the Ukraine, for instance, deficiencies in potassium and copper are marked.

3.5.3 SIGNIFICANCE OF ORGANIC SOILS

Many areas of improved and well-managed organic soils produce excellent yields of market garden crops like celery, onions, and lettuce. Such soils are invaluable if the peatlands lie close to large markets, like the Chat and Ashton-under-Lyne mosses near Manchester. Grassland yields from peat soils may be 2–3 tonnes/ha higher than from sandy or clayey soils, owing to the higher protein and nitrogen yield (Levin and Shoham 1972). In cool environments, a substantial proportion of the total agricultural production may come from organic soils. Peaty soils provide valuable meadowlands in the north of East Germany, and, in general, such areas provide very promising sources of additional cropland, cultivated hay fields and pastures (Okruszko 1975). Even in tropical areas, peatlands are viewed as a main source of increased agricultural area for rice-growing. Reclaiming organic soils for growing rice has posed several difficult problems for engineers, agronomists and pathologists. There are very acid peats in which iron is reduced to toxic compounds, as in Sierra Leone, Nigeria, and Indonesia. Other organic soils are slightly acid to neutral peats in which iron is in short supply, as in Japan or in the peaty soils of the Everglades in Florida. Physical problems include the submerged logs in newly reclaimed peat which must be removed as they appear at the surface, as in rubber plantation areas in Malaya. Shrinkage of organic soils after drainage is inevitable as oxidative decomposition proceeds. In the low moor peat soils in the older polders of the western Netherlands, oxidation owing to shallow drainage over a period of 8–10 centuries, has lowered the soil surface 1–2 m below sea-level (Schothorst 1977). In the deep peat soils of the East Anglian Fens, the Holme Fen Post in Huntingdonshire has been lowered 380 cm since 1852 (Richardson and Smith 1977).

Water retention, porosity, hydraulic conductivity and water yield vary greatly from one peat type to another and within a given area (Boelter 1965). In irrigation schemes the vertical and lateral seepage in peat can be so great that it is difficult to keep land inundated.

There is always an initial high cost to engineering works built on or across peatlands. Continued maintenance operations are also required for constructions on this type of soil. Several methods have been used to increase the stability of peat foundations. Shearing resistance, which determines stability, is very low in peat and impossible to determine accurately. Also consolidation strength is very unpredictable. Differences in peats and their effects on road and railways construction are related to permeability and rates of water movement through organic soils. The local variability of peat is a great handicap, both to successful construction and to laboratory investigation.

Peat may be removed and used for a variety of purposes. In some economies, peat is widely used as a fuel, raw material for the chemical industry, bedding for cattle, and for organic fertilizers. The use of peat for bedding and preparation of composts is partly based on the capacity of peat to absorb actively liquids and gases. In several areas, peat or peat compost has been applied to podzols in efforts to raise their productivity.

Finally, peatlands play a vital role in studies of late-Quaternary history. Owing to their slow and often steady development, and the preservation of old fragments of wood (fig. 16b), peat horizons are invaluable sources for materials suitable for radiocarbon dating (Martel and Paul 1974) from which local and world wide chronologies of events during and before the last glactation have been built up. Similarly, traces of the annual pollen rain have been preserved in peats, and samples from borings through peat horizons reveal variations in frequency of pollen types which can be linked with climate or geomorphological changes in the past. In the higher peat horizons, pollen records man's early activities in forest clearance and early agricultural practices.

4 Soil structure and porosity

4.1 Soil structure

4.1.1 DEFINITIONS, ORIGIN AND DESCRIPTION OF SOIL STRUCTURE

Soil structure refers to the binding of primary particles into aggregates or 'peds', with larger aggregates consisting of a group of smaller ones (fig. 4.1a). Although individual sand-sized particles may be held together by clay-sized substances (fig. 4.1b), particularly in tropical areas, sandy soils do not usually have well-defined structure. Spatial arrangement into a structural pattern, whether it be of primary particles or of aggregates, is a distinctive feature of soil structure and has been termed 'fabric' (Brewer and Sleeman 1960). 'Consistency' is a term to describe the nature of the contact between particles or aggregates and the nature and degree of bonding that exists. Other terms include 'tilth', which is considered to be the structural state resulting from cultivation of a soil. 'Mellowness', in contrast to 'rawness', describes the crumbly surface tilth of a well-farmed soil. Aggregates or peds of the B-horizon, usually the result of physicochemical processes, are referred to simply as 'structures'. Slaking is the disintegration in water of soil crumbs into discrete fragments.

Laboratory study using a petrological microscope has followed from Kubiena's development of a method for producing thin-sections of soils and the appearance in 1938 of his volume *Micropedology*, and has advanced to the contemporary use of the scanning electron microscope (Lynn and Grossman 1970). Microfeatures observable in thin-section include *glaebules*, accretions which differ in fabric from the enclosing soil matrix (Brewer and Sleeman 1964). *Pedotubules*, either organic or

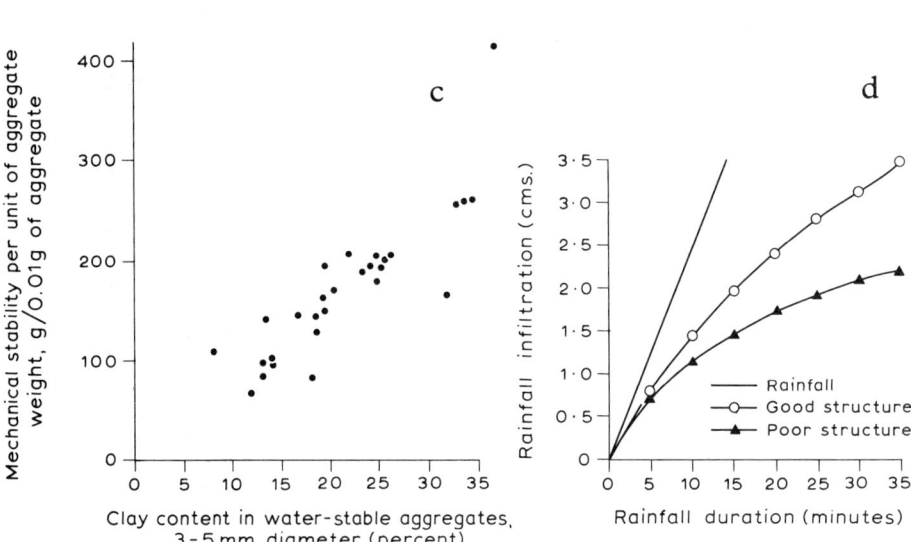

Figure 4.1 Nature and significance of soil crumb structure.
a and b. Simplified model of a soil crumb (a), and possible arrangements of domains, organic matter, and quartz particles in a soil crumb (b) (Emerson, 1959). PVA is polyvinyl alcohol, a soil conditioner. A clay domain is a group of clay crystals sufficiently orientated and close together to act as a single unit. The types of bond are quartz-organic matter-quartz (A), quartz-organic matter-clay (B), clay-organic matter-clay (C), with face-face (C_1), edge-face (C_2), and edge-edge (C_3), and clay-clay, edge-face (D).
c Effect of clay content on mechanical stability of soil aggregates (Kuznetsova, 1967). Soil samples were taken from a variety of soil-climatic zones. Mechanical stability is the value of breaking load per unit of aggregate weight (0.01 g) which is not an absolute value.
d Time variation of mean total infiltration into an undisturbed lateritic soil (Rose, 1962). The soil is the agriculturally important Kikuyu Red Clay Loam in Kenya.

Soil structure and porosity 103

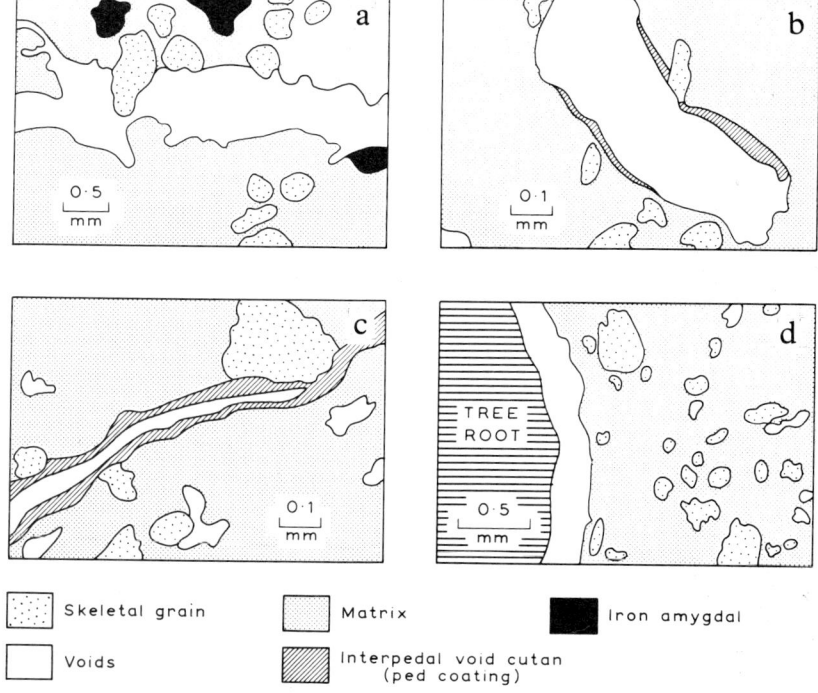

Figure 4.2 Micromorphology of pores and ped coatings.
a-c Voids and ped coatings in a Manitoba podzol profile (Pettepiece and Zwarich, 1970). Sinuous voids without coating in eluvial A horizon (a), development of cutans in the B horizon (b), thicker cutans and smaller pores in the lower B horizon, due to fewer physical disturbances (c).
d Size and shape of pore created at root-soil interface in Miami silt loam (Blevins et al., 1970).

sesquioxidic, have the form of voids filled by soil material and have an approximately tubular form (fig. 4.2) (Brewer and Sleeman 1963). *Plasma* is the mobile active part of soil material, mineral or organic, of colloidal size and relatively soluble. Plasma, together with the mineral skeleton and voids, is one of the three basic components of what is termed the *s-matrix*.

Most soils, provided they contain sufficient clay, will shrink on drying. Plant roots may penetrate such openings and create new pores in their immediate proximity. By their drying action, plant roots contribute to the creation of the environment needed for their own growth. As the area adjacent to the root zone is one of increased physical forces, as well as chemical and biological activity, clay skins or *argillans* are characteristically not commonly developed along the pores containing active roots

104   Geography and soil properties

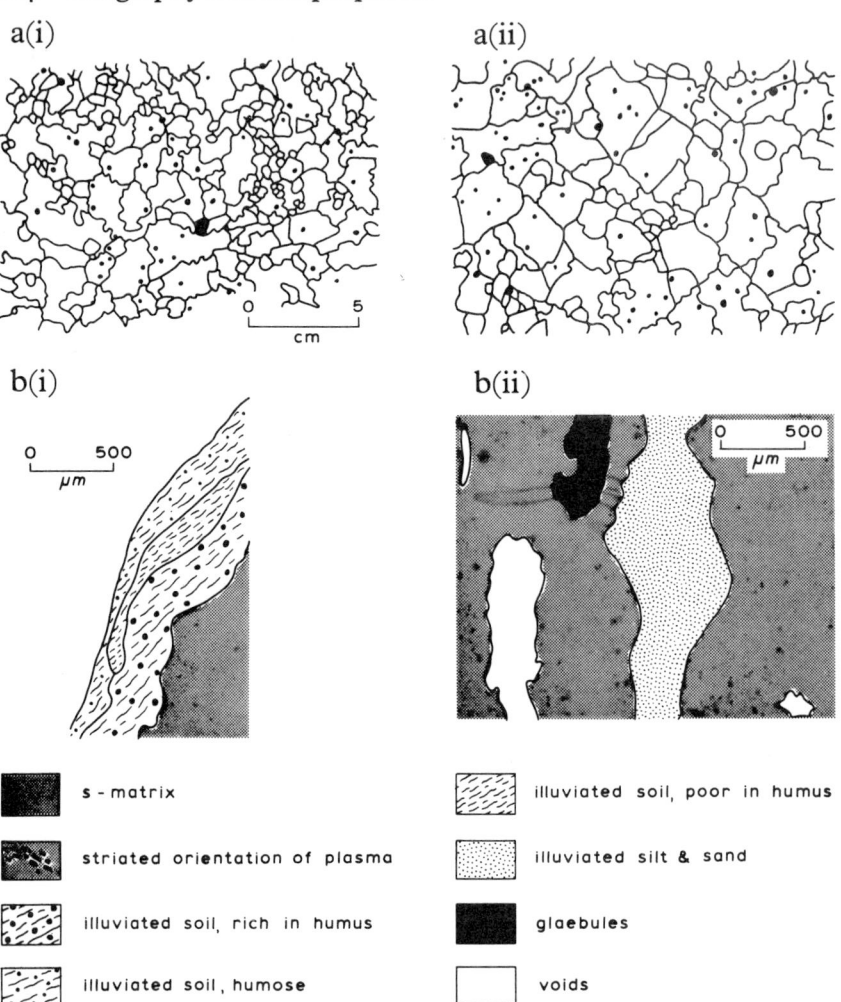

Figure 4.3   Pattern and detail of ped boundaries modified by land use.
a Contrast between uncultivated (left) and cultivated (right) silt loams (Bouma and Hole, 1971), the Tama silt loam of the Great Lake region. Moderate erosion has brought coarser peds closer to the surface.
b Ped surface coated with surface soil (left) which has been saturated and liquified, and which may completely fill dry cracks (right) (Jongerius, 1970).

(fig. 4.2). Porous aggregates 4–5 mm diameter are common in grassland soils, developed in part by root hairs (fig. 4.3a). These occur at less than 3 mm intervals behind the growing points of root hairs entering desiccation cracks. Since well-managed grass usually extends its new root

Soil structure and porosity 105

growth every season, it creates an intimate, ramifying network of cracks throughout the volume of a soil which contains some clay (Russell 1971).

4.1.2 SIGNIFICANCE AND NATURE OF AGGREGATION

The importance of aggregation is due mainly to its control on porosity in the soil and the type of surface it presents to precipitation. The correlation between stability of soil aggregates and other physical properties is usually high, particularly with those properties related to erodibility and the infiltration and movement of water (Emerson 1967). The maintenance of water-stable aggregates is essential if splash of particles is to be minimized. Infiltration is increased if the soil surface is in large clods, due partly to surface retention and partly to increased porosity (fig. 4.1c). The amount of energy required to initiate runoff is a function of the size of clods in a tilled soil and large clods are equally important in delaying the time when runoff begins. The rapid decrease in infiltration resulting from low structural stability and consequent surface sealing may be largely offset by greater cloddiness after tillage (Moldenhauer et al. 1967).

Aggregation is at the root of many other aspects of soil fertility and desirable attributes of agricultural soils (Allison 1968). Favourable aggregation provides a fine seedbed through which water and air will move freely and in which an initial root system can develop rapidly. Micro-aggregation is responsible for fine-textured soils accumulating more organic matter than coarse-textured soils under the same environmental conditions. Yield differences which cannot by accounted for by lack of nutrients may be primarily related to the structural state of a soil (Low 1973).

The significance of the cementing agent is linked with the tendency for aggregation to decrease with continued cultivation, which deforms soils and thus alters the structure (fig. 4.3). Soils under 100-year pasture are nearly four times as stable as those which have been under cultivation since 1860 (Low 1973). Soils with a natural porosity of 36–48 per cent have, in extreme cases, fallen as low as 6 per cent porosity.

Organic content is often correlated with degree of aggregation (fig. 4.4), although it has little effect on soil structure if undecomposed. First, the cells and filaments of organisms have certain mechanical binding effects in the early stages of organic matter decomposition. Fungal hyphae, particularly those with a 'woolly' surface, could be responsible for about one-third of the stable aggregates in Rothamsted soils. More specifically, adherence of soil particles to mucilage-covered hyphae rather than physical entanglement is a major aggregating mechanism by which filamentous micro-organisms stabilize soil aggregates (Aspiras et al. 1971).

106 Geography and soil properties

Secondly, microbial synthesis products have cementation effects, with bacterial slime being particularly important in aggregating soil particles, owing to the presence of polysaccharides and amino sugars. A number of studies show good correlations between microbial gums or polysaccharides and aggregation (Harris *et al.* 1966). Relationships between soil polysaccharides and aggregation are, however, more complex than was formerly envisaged, and in certain soils (Hamblin and Greenland 1977), such as old grassland or forest soils, they do not appear to play any significant role. However, the phenolic substances, infiltrating as leachates from decomposing plant remains or even from living tissues, could be very active in soil aggregation. Organic matter may also contribute indirectly to aggregate stability by preventing swelling, slowing the wetting process, and by decreasing wettability by modifying the attractive forces of clays for water molecules. Substances like fats and resins arrest slaking by rendering aggregates waterproof.

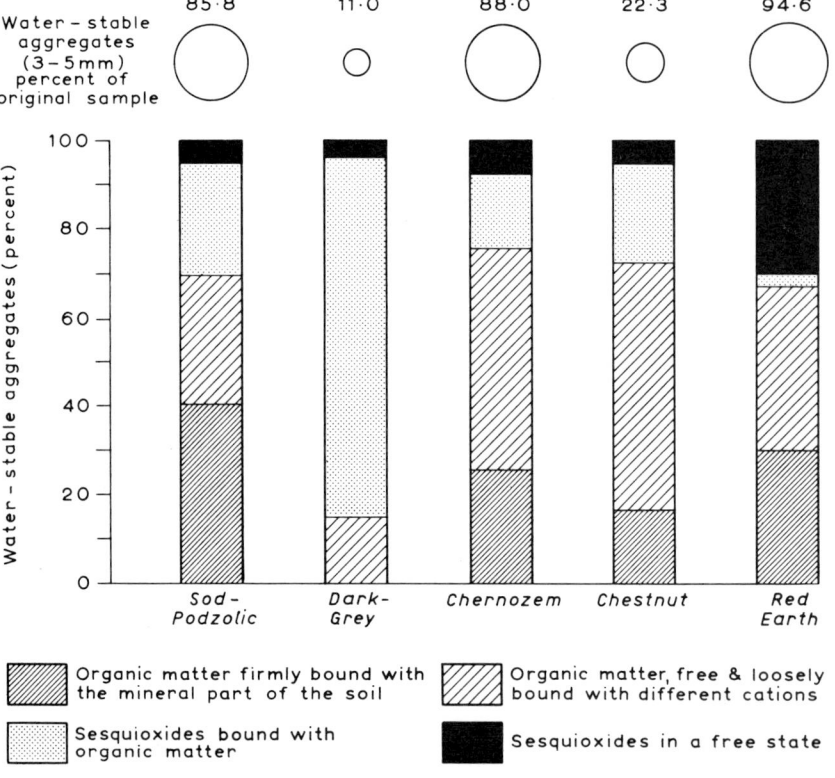

Figure 4.4 Participation of various soil constituents in maintaining water-stable aggregates (Kuznetsova, 1966).

Clay content is a major factor (fig. 4.1b) in determining aggregate stability. In fact, a minimum of approximately 16 per cent clay for the development of structural aggregates has been suggested and almost linear relationships established, with correlations as high as $r = 0.86$, $N = 37$ (Kuznetsova 1967). Intense aggregation is promoted by the flocculation of colloids, the cementing of particles by colloid films and their adhesion under the influence of van der Waals forces, residual valencies and hydrogen bonds.

Polyvalent metals such as calcium, magnesium or iron may link organic and inorganic soil particles. The linkages clay–polyvalent metal–clay and those of the organo-metallic complex–polyvalent metal–organo-metallic complex contribute to aggregation in most soils (fig. 4.5) (Edwards and Bremner 1967) and thus aggregation will be influenced by the exchangeable bases present in the soil. First, cementation may be due to the precipitation of a hydrated iron gel and its irreversible dehydration. High positive correlations between free iron oxide and degree of aggregation were first observed in lateritic soils and later in other soils. Secondly, iron in solution can prevent deflocculation. Iron actively affects the physicochemical properties and structural state of finely dispersed silicates. Iron films deposited on particle surfaces change the adsorption capacity of minerals and bring about the aggregation of the clay material. Thirdly, iron can form organic-mineral compounds of humic acids with free sesquioxides and its binding effect may be observed only in the presence of organic matter (fig. 4.4). Although the structural stability of iron-rich soils has often been attributed to free iron oxides, free aluminium oxides may be more important bonding agents in old, highly weathered soils. The apparent correlation between iron and aggregation may really reflect a still closer connection with free aluminium oxides (Deshpande *et al.* 1968).

4.1.3 CREATION AND CHARACTERISTICS OF PEDS

A number of processes tend either to accentuate aggregation at certain points, or to create planes of division. Physicochemical processes involved in creating crumbs include the flocculation of colloids. Alternate wetting and drying episodes may create discrete crumbs, perhaps due to orientation induced in clay and humus particles. Also, the shrinking of capillaries on drying, by bringing soil particles closer together, increases the atomic and molecular forces of attraction between them. Channels or cracks are formed throughout the soil, adding definition to the surfaces of discrete crumbs. Deeper horizons in such clayey soils often show slickensiding, indicating strong compression as the peds expand during moistening. Freezing removes pure water from suspensions and a soil may be

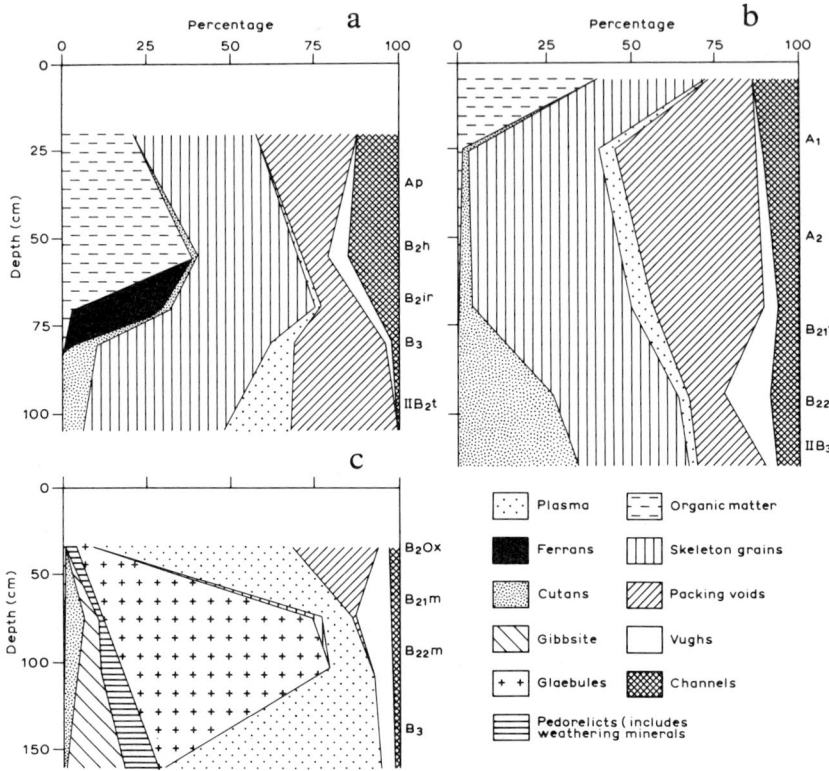

Figure 4.5 Micromorphology of soil structure in three Soil Orders of the U.S.D.A. Soil Taxonomy (Eswaran, 1968).
a Spodosol (Podzol) from Belgium, showing predominance of organic matter in B_2h horizon, with ferrans (iron coatings) on sand grains in the B_2ir horizon.
b Alfisol (grey-brown podzol, degraded Chernozem, non-calcareous brown earth) from Belgium, showing illuviation of clay only.
c Oxisol (lateritic soil) from Madagascar, the predominant feature being the maximum amount of glaebules present in the B_{21} and B_{22} horizons, which are indurated.

The suffixes to the horizon indices signify destruction by ploughing or other forms of cultivation (p), humified, well-decomposed organic matter (h), illuvial accumulation of iron (ir or fe), illuvial accumulation of clay (t), residual accumulation of sesquioxides (ox), strong cementation (m), and Roman numerals indicate a different, contrasting soil layer at a pedological discontinuity in the profile.

compressed between the zones of ice formation in the soil. These seasonal effects of frost action or wetting and drying in creating soil crumbs have been appreciated for many years.

Biological processes are involved, particularly in forming larger secondary crumbs. Again, processes either add to the coherence of discrete crumbs or induce a similar result by segmenting the soil with a network of divisions. Excrements of soil fauna may become exceptionally stable individual crumbs; *Oribatei* excrements can even pass intact through the intestinal tract of an earthworm. Excrements may join together to form larger aggregates (fig. 3.1). For example, under a Douglas Fir cover, soil aggregates 100–500 µm diameter are seen to be built up of highly compacted excrements 20–30 µm in size. Earthworm casts are particularly important as a source of stable soil crumbs. Biological processes involved in creating planes of division within a soil are the decay and growth of plant roots (fig. 4.2d) and the burrowing and tunnelling activities of the soil macrofauna. Growth and distribution of plant roots is often restricted to the surfaces of structural units in the soil.

The sizes of peds or crumbs is variable, but micro-aggregation may be recognized when crumbs are smaller than 0.25 mm diameter. Only micro-aggregates, consisting largely of clay and humified organic matter, have some stability. Only slightly disruptive forces are sufficient to break down larger crumbs, and well-developed micro-aggregation can ensure adequate moisture capacity in the soil, and adequate water and air permeability through it. In the plough layer of thick Chernozems a proportion of structural units, larger than 250 µm diameter, of 94–97 per cent has been calculated (fig. 4.1c), with aggregates 1–3 and 0.25–1 mm diameter being predominant. The size of soil crumbs is particularly significant in areas susceptible to erosion. Wind erosion is not marked, for instance, where peds are predominantly larger than 2 mm diameter, but crumbs in the 0.1–0.5 mm size range may be especially vulnerable to destructive wind erosion.

Soil crumbs may have some recognizable shape. For example, platiness may be the result of rapid surface drying of a clay, forming cracks parallel to the surface of drying. Clearly defined double-wedge shaped peds appear to be a characteristic of Vertisols throughout the world, the ped faces apparently developing by displacements along lines of stress within the soil, set up by expansion and contraction forces.

A discrete ped may be some concentric arrangement or layering of its constituents. The cores of some structural units may have a higher clay content than the peripheral zones, suggesting their more intense weathering. Soils in the South Wales coalfield area commonly display an irregular, prismatic structure, with each prism containing a yellowish-red iron pan, approximately 1 mm thick. In fact, clay coatings or clay skins have been described on ped surfaces and on walls of voids in a variety of soils (fig. 4.3). Brewer (1960) proposed the term *cutan* for such coatings. Cutans are identified and classified most readily in thin sections, as they may show birefringence if the colloidal fraction is preferentially orientated. Apart from illuviation, clay films may also form when stagnant water evaporates entirely or when a soil mass shifts, in adjustment to pressure. However, clay skins cannot accumulate in horizons where shrinking and swelling is highly pronounced. In semi-arid soils, turbation due to wetting and drying cycles disrupts illuvial cutans (Smith and Buol 1968). Thick, grey 'flood coatings' are widely developed in floodplain soils in the Brahmaputra, Ganges, Indus and other river alluvial spreads (Brammer 1971). They differ from fine-clay cutans by the rapidity of their development, their penetration into the finest fissures and to depths of 1.5 m or more, and by consisting of all clay-size fractions, together with silt and humus. The development of ped coatings introduces distinctive microvariations into a soil (fig. 4.3b). For instance, in the B horizon of a Wisconsin soil, clay cutans were much higher in organic matter, clay, phosphorus, manganese and iron than the bulk of the soil, and skins with about three times as much clay as the ped interiors have been observed (Buol and Hole 1961). The presence of cutans may, therefore, influence or change the course of soil formation as the coatings can protect the bulk of the soil from many processes, especially leaching. Increasingly, the influence of these microvariations in physical and chemical characteristics of soils is seen to affect engineering and hydrological properties of soils. Clay skins may reduce plant growth and nutrient uptake, especially of P and K, either by barring root growth or by slowing ion diffusion from the soil peds into the inter-ped soil solution (Khalifa and Buol 1969). Thus cutans, although they may comprise less than 1–2 per cent of the total soil volume, may have a substantial influence on plant growth, particularly in horizons such as fragipans where the rooting volume is almost exclusively restricted to polygonal ped interfaces (Miller *et al.* 1971).

4.1.4 STABILITY OF PEDS

The characteristic of soil crumbs of most critical significance in agricultural soils remains their resistance to mechanical disruption, particularly their stability in water (fig. 4.4). This largely determines a soil's erodibility and the infiltration rates of water and air (Allison 1968). One of the first stability indices, termed the 'dispersion ratio' by H. E. Middleton in 1930, was defined as

$$\frac{(\text{clay} + \text{silt in undispersed soil})}{(\text{clay} + \text{silt in mechanical analysis})} \times 100.$$

Another index is the value of the breaking load, calculated per unit of ped weight (0.01 g) (fig. 4.1c). For example, applied to crumbs in the 3–5 mm diameter range, indices of 100 g/0.01 g of ped, show that Sod-Podzolic and Light Grey soils have a low mechanical stability (Kuznetsova 1967).

In agricultural practice, one of the main disruptive forces on soil crumbs is the application of irrigation water. If the water has a low salt content and the soil a high proportion of montmorillonite clays, the swelling processes are particularly destructive and reductions in soil permeability accompanies the slaking of the soil peds. The presence of exchangeable sodium may have strong dispersing effects, but the critical percentage of total exchangeable bases for severe dispersion varies widely between 10 and 30 per cent. For many temperate-zone soils, the critical level is about 15–20 per cent. Solonetzic structures only lose their high stability with exchangeable sodium contents of 20–40 per cent.

4.2 Porosity

4.2.1 MORPHOLOGY

The amount and kind of pore space (fig. 4.2) is highly significant because it determines the nature of the living space within the soil. For instance, the mesofauna inhabiting a soil may be related to the average size of the soil interstices and most bacteria do not enter pores less than 1–3 μm diameter. Plant root hairs do not penetrate into pores less than 10 μm diameter, although deciduous trees, which have more delicate root tips than pines, may penetrate into soil pores more easily. Studies of the root distribution of *Molinia caerulea* and *Erica tetralix* in relation to the size of soil pores showed that the roots of both the species were confined to pores exceeding 150 μm diameter. Porosity also has a fundamental influence on the humidity and gaseous conditions of the soil environment. In general most plants require that 10 per cent of a soil be air-filled voids and as much as 15 per cent for oxygen-sensitive plants such as potatoes.

Ideally, a soil should have an extensive group of pores small enough to resist gravitational drainage and yet large enough to release significant quantities of water to plant roots. Approximate values may be assigned to these two criteria (Cary and Hayden 1973). Some pores must be greater than 200 μm diameter to encourage root elongation. Pore size distribution must also be such that saturated conditions do not occur for more than a few minutes following surface wetting, so that the soil returns within a few hours to an air-filled void state.

Pore sizes may be related to mineral particle size, particularly in sandy soils where porosity is largely texture-determined. In most soils, however, porosity is related to soil structure and two distinctive groupings of sizes are recognizable. These originate as either between-crumb pores or within-crumb pores. This division corresponds approximately with that between pores and micropores, between non-capillary and capillary porosity. In agricultural soils between-crumb porosity is the result of recent soil cultivation. Within-crumb porosity represents the cumulative effects of management, cropping, or soil processes over a longer period (Currie 1966). Bimodality of pore sizes has been observed in thick Chernozems of fine-clay loam texture (Sokolovskaya, 1966) where openings in the macroaggregates are predominantly either greater than 60 µm or less than 5 µm, the latter being within-crumb (intra-aggregate) pores. In this case the volume of pores for which the diameter falls in the 5–60 µm range constitute only 9–12 per cent of the total porosity.

Sizes of pores in clayey soils with intermittently wetting and drying may be small. Soils of the Gezira have most of their pores smaller than 0.2 µm and only 18–23 per cent of their volume may be made up of pores with radii in the 0.2–30 µm size range (Zein el Abedine *et al.* 1969). The smallest pores, in the size order of 100 Å could arise from steps in a crystal surface as well as from separations between neighbouring clay crystals, but may account for a high proportion of the total pore volume in certain natural soil crumbs (Aylmore and Quirk 1967). At the opposite end of the scale the porosity of some soils includes large, irregularly shaped cavities. In soils developed on Recent alluvium on Romney Marsh and at a depth of 120 cm, about 25 holes ranging from 5–30 mm were observed on a horizontal bench 60 cm^2 (Green and Askew 1965).

Soil pores may have some regularity in their shape. Large, lamellar-shaped pores, parallel to the drying and wetting front, have been observed in the top 5 cm of an arable layer (de Leenher 1971). However, the shape of much of the 'porosity' in soils is in the form of tubes not pores. A network of tubular pores is a distinctive feature of loess soils in particular. Apart from weathering and solutional processes, organisms are the basic cause of tubular cavities in soils.

The total volume of pore space in the soil is another critical characteristic. A theoretical minimum value for porosity is 9.3 per cent, the tightest possible packing for spherical sand grains. As soil particles are joined into micro-aggregates which in turn may build up to larger crumbs, porosity usually falls in the 40–60 per cent range. Actual minimum porosities around 30 per cent are found in sandy and puddled soils, with a maximum porosity of about 80 per cent in well-structured loam Chernozems.

4.2.2 INFLUENCE OF PLANT ROOTS

Most roots proliferate in the humic horizon, with approximately 70–90 per cent of roots of herbaceous and woody plants occurring in the top 20 cm of a soil. Tree roots penetrate to greater depths. Roots of vines, for instance, grow in the 20 cm to 1.3 m depth range. The roots of rubber trees in

Malaya are well-distributed to a depth of 2 m, and roots in coffee plantations range down to 5 m. Although roots penetrate unevenly through a soil, the root hairs are sufficiently fine to become an integral part of the soil mass.

Soil porosity can be affected by live roots when trees are exposed to gusting winds. The effect may be sufficient to open up brittle, indurated layers. Such an increase, of approximately 2 per cent in the soils of the Countesswells Association, has been noted in granitic till soils in northeast Scotland (Romans 1959). More widespread is the effect of dead roots on soil porosity. When plants die, some of the bacteria living on the root surface and other micro-organisms decomposing it are likely to produce gums which spread into the mineral wall of the soil channel. Thus the fillings gradually replace dead roots to become channels with stabilized walls. This process accentuates the marked influence of old root channels in modifying the characteristics of the deeper horizons in a soil. Many pores about 1 mm diameter have the appearance of old root channels and may be common at deeper, untilled levels. The adjacent soil is finer and more friable owing to the effects of organic matter, bacterial action and air movement directed along such pores. In Holocene marine clay soils on the Surinam coast, when an *Avicennia* forest dies, a deposit is left behind which is porous. During the dry seasons, oxygen penetrates the soil and oxidizes part of the iron compounds along the root channels. Ferric hydroxides formed in this way are called 'iron pipes' (Augustinus and Slager 1971).

4.2.3 INFLUENCE OF SOIL FAUNA ON POROSITY

For refuge or in pursuit of their prey, the burrowing soil fauna tunnel cavities in the soil. Chambers, voids, and macro-pores created by the macrofauna are termed krotovinas. In the tundra 1.5–4 polar fox burrows/ 1000 ha have been estimated with one burrow occupying an area of 500 m^2. The density of marmot and ground squirrel burrows is generally much greater. Of the latter, the species *Citellus parryi* is active for 3 months each year in mollisols above permafrost layers. In temperate woodlands, burrowing animals include a large assemblage which migrate from tree-tops or pastures to use the soil as a protective cover for hibernation. More permanent inhabitants (fig. 1.7b) include the badger, for which burrows of about 400 m^2 in size have been identified. Moles are capable of excavating tunnels nearing 100 m length within 24 hours. They create a dense network of passages, 5–7 cm diameter. These are mostly in the top 10 cm, created in the pursuit of earthworms. Permanent tunnels lie at 10–20 cm depth, connected to the near-surface runs by oblique shafts. The mole often enlarges part of a tunnel to make an oval nesting cavity about 30 cm long. In all, an average length of passageways of 255 cm/m^2 or 15.3 per cent of the area and 7.2 per cent of the volume in the top 10 cm of soil has been calculated, with maximum figures of a length of 600 cm/m^2, accounting for 36 per cent of the area and 17 per cent of the volume. Beneath deciduous woodland or grassland, earthworms create a dense network of passages. In dry or cold weather many species retire to considerable depths of 1–1.5 or even 2.5 m. In favourable weather they are active in the first 15–20 cm of soil. *Lumbricus terrestris* may burrow as deep at 1.5 m, but other species like *L.rubellus* are essentially non-burrowers. Several species do not come to surface but remain in the 0–15 cm layer of the soil. Earthworm burrows studied on Romney Marsh penetrated to about 1.2 m and averaged 6–8 mm diameter. A maximum number of earthworm channels of over 100/m^2 has been observed. Beneath ants' nests, extensive systems of burrows and galleries radiate downwards. As much as 1.4 m^3 of soil below a nest may be modified by such a network and in all may amount to about 12 per cent of the total soil volume. Depths penetrated by *Lasius flavus* in Lowland Britain approach 1 m and in the North American prairies, the excavations extend to depths of 2 and even 3 m. In addition, ants may excavate chambers up to 1 m in length and with breadths of 30 cm. In warm, drier areas soil burrowing activities of termites may predominate in the top 1.2 m of soil. Laterally, foraging tunnels of over 30 m in length have been observed. Termite channels, 6–7 mm diameter penetrate to depths of 15 m in the Golodnaya Steppe.

5 Physical properties of the soil

5.1 Introduction

A soil incorporates expressions of the local climate as a non-solid, liquid or gaseous phase within the solid framework. In part, the air, water, and temperature regimes are influenced by the arrangement of soil particles. More significantly, they are important, independent soil characteristics, expressing the 'climate' of the soil. In many instances or localities, the nutrient uptake from a soil is limited by insufficient or an excess of water, oxygen deficiency or impeded root proliferation due to high bulk densities. With impeded water movement, moisture deficiency may become the main limiting factor in plant growth. In the case of soil air, restriction of the exchange of oxygen and carbon may reduce the plant roots' ability to translocate nutrients to the leaves. Low oxygen concentrations also may retard the development of nitrifying bacteria. Soil nitrogen, in consequence, remains as unavailable protein instead of breaking down to release ammonium and nitrate ions.

An important characteristic of many physical properties of soils, typical of both the basic framework and of the soil climate parameters, is small-scale variation or soil variegation. Such small-scale changes arise from any initial heterogeneity within the soil framework and also from the dynamic nature of processes taking place in the soil. They include lateral changes over short distances due to microrelief, root spread, animal activity, cracks, with associated changes in the texture and structure of the soil and subsoil. Widely encountered is the passage, within a meter or so, from soil with imperfect hardpan to a fully cemented pan in soil on a better drained site. Desiccation fissures produce a range of lateral changes, depending on climate and weather conditions. Cultivation

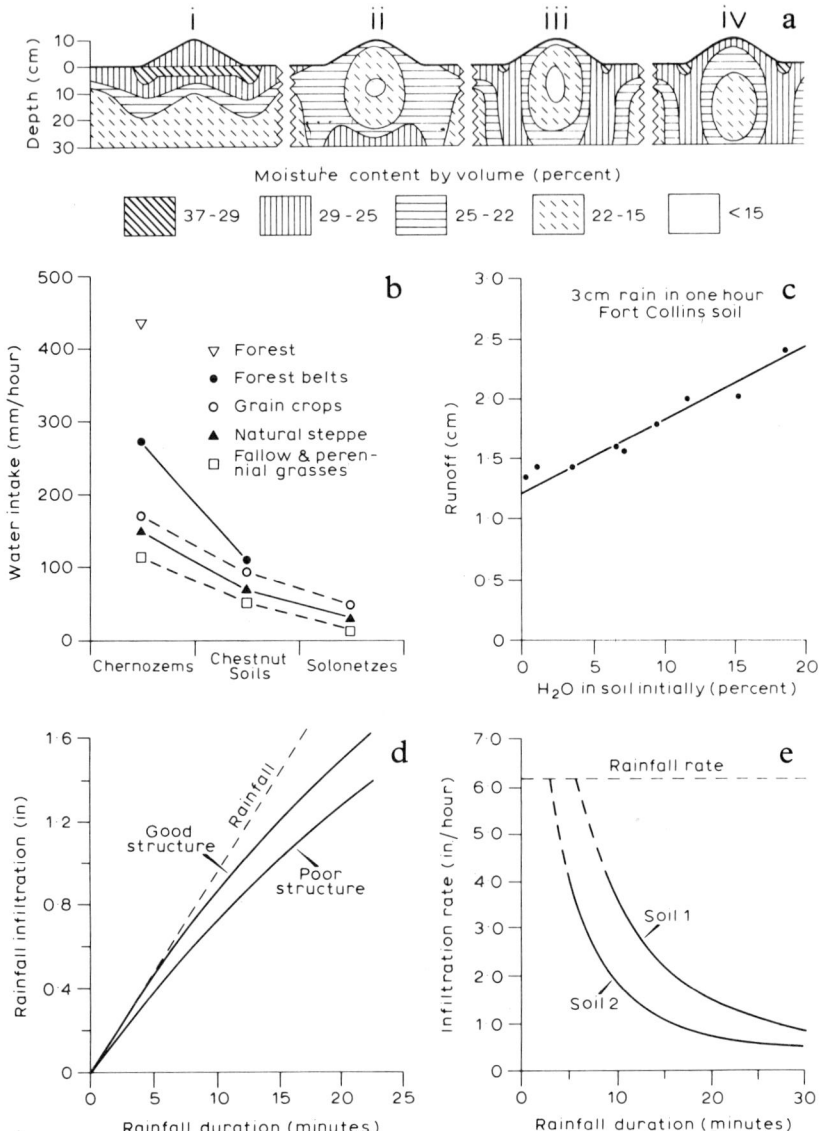

Figure 5.1 Factors affecting infiltration into soils.
a Influence of mole-hill age and shape on distribution of infiltrating water (Abaturov and Karpachevskiy, 1966). Observations followed an August precipitation of 10–13 mm on podzols and the mounds of the European mole *Talpa europaea* L. range in age from fresh ejecta (i), 3–5 years (ii), 5–10 years (iii), and more than 10 years (iv).
b Comparison of water permeability of soils in the European U.S.S.R. and the influence of agricultural usage (Nazarov, 1970).

(Continued on facing page)

techniques introduce local changes. Mouldboard ploughing leaves long continuous strips of loose soil between relatively compact ridges. These strips are often full of relatively persistent post-harvest plant residues. Pore space may vary significantly over distances of only 5–10 cm, and growth and development of plants may be retarded on the compact strips. Other localized effects of compaction or disturbance may be caused by heavy machinery or land drainage operations.

5.2 Soil water

Water is an integral part of most processes in the soil, a major influence on several soil properties, and the critical resource in crop production. Water movement within the soil is the most important process leading to profile differentiation. Water moves suspended and dissolved solids from the ground surface into the profile. It may transport these into the lower horizons or remove them from the profile altogether. In drier environments, profiles are distinguished by the effects of movements of materials upwards in the profile in soil solutions. Apart from the demands from plants, which make up 85–90 per cent of their actively growing shoots from water, micro-biological life in the soil increases with moisture supply, but reaches an optimum when about 40 per cent of the pore space remains air-filled.

In the agricultural utilization of soils, the tasks of improving water-retaining capacity of lighter soils, or alternatively making some of the appreciable quantities of moisture bound up in heavy clays more available to crops, are vital but difficult problems. In irrigation agriculture, the rate of soil-water movement is a crucial factor in the reclamation of salt-affected soils. If excess salts are to be leached or harmful exchangeable sodium replaced by calcium, water of correctly calculated amounts must move into soils and pass through at certain rates. In the engineering use of soils, much practice is based on the concept of soil existing in various states which depend on moisture content (fig. 5.1). The greater the amount of water which a fine-grained soil contains, the more the engineering soil behaves like a liquid.

In many ways the character, volume, and rate of movement of soil water are more vital considerations than the structural frame of the soil itself.

---

c Effect of initial water content on runoff from a clay loam (Kemper and Noonan, 1970).

d and e. Effect of disturbed structure of a lateritic soil on infiltration (Rose, 1962). Comparison of (d) with figure 4.1d shows that disturbance accelerates infiltration rates in the Kikuyu Red Clay Loam. Infiltration rates are therefore greater (e) into a continuously cropped soil 1 compared with soil 2 on an adjacent ridge, recently cleared of grass.

5.2.1 INFILTRATION

The rate at which water enters the soil is not usually considered as a natural phenomenon. The conditions which control infiltration, being those at the soil surface, are those which have been most modified by tillage and deforestation (fig. 5.1a). Also, infiltration rates are particularly of interest to workers investigating irrigated soils where it is logical to study the effect of the sudden and artificial introduction of a large volume of water at the soil surface. In consequence, infiltration is usually described by numerical and analytical techniques. Observations on actual infiltration rates under more natural circumstances are not numerous but are relevant since many soil characteristics persist which reflect an origin closely linked to natural infiltration rates. In particular, under forest conditions, litter characteristics are the major control on infiltration rates of water entering the mineral soil. In the Central Don region, the litter of a 50-year-old steppe pine forest, after drying out during drought, becomes hydrophobic and swells slowly and unevenly during succeeding rainfall. 95 per cent of a shower of rain (3–4 mm) is retained in the litter, and 60–90 per cent was retained when 7–11 mm fell. Under the litter, 40–50 per cent of the top horizons of the mineral soil remained dry after a shower in which 13–15 mm of rain fell. In the tropics, plant litter remains on the surface in moister, woodland areas. Although the quantity is small, not commonly exceeding 3–4 tonnes/ha, it favours a high degree of infiltration into woodland soils, in marked contrast to drier areas. Thus, under natural regrowth of bush fallow in western Nigeria, equilibrium infiltration rates were established at 20–25 cm/h (Wilkinson and Aina 1976).

On arable land, the nature of aggregation and porosity influence the rate at which water infiltrates into the soil (fig. 5.1b). Infiltration rates fall off markedly whilst bare soils are exposed to intense rainfall. In sandy clays and clays from tropical soils, laboratory infiltration rates fell below water-application rates of 10 cm/h within 14 minutes. Infiltration capacity was exceeded in 7 minutes under a 15 cm/h application rate (fig. 5.1d,e). Collapse of soil aggregates seals the surface, fine particles are removed, and increasing surface runoff may lead to soil erosion. Material is sheared off large clods by raindrops and the amount of material removed increases as progressive wetting weakens the clod (fig. 5.2c). The breakdown of soil crumbs also releases fines which clog surface pores, and form a crust on the surface of the soil. Below the thin compacted crust, washed-in fines create a zone of decreased porosity. Simulated rainfall experiments suggest that the permeability of deeper horizons may be five or more times greater than infiltration rates through such crusts. The scale of contrast owing to land use is seen in Hawaii, which remains half-forested. Under forest cover, infiltration rates range from 0.03 cm/h to 39.7 cm/h,

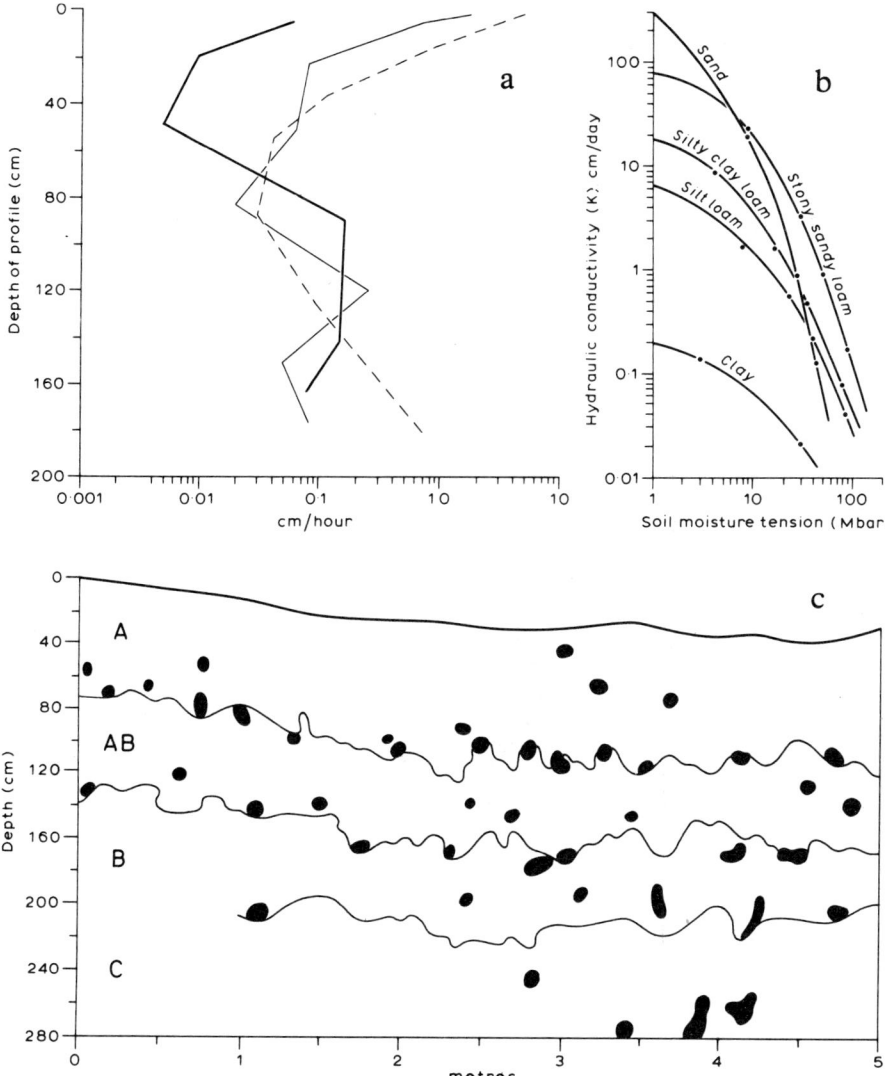

Figure 5.2 Factors influencing hydraulic conductivity in soils.
a The very slow hydraulic conductivity typical of Vertisols (Zein el Abedine *et al.*, 1969), observed in the Central Clay Plain of the Sudan. Vertisol is an FAO (1967) and U.S.D.A. Soil Taxonomy term for tropical clayey soils which change seasonally in volume, formerly termed Grumusols, Tropical Black and Grey Clays, Regur, Black Cotton Soils.
b Influence of soil moisture tension on hydraulic conductivity at several field sites Wisconsin (Bouma *et al.*, 1971).
c Influence of burrowing animals in creating large pores in the soil (from Vashenin *et al.*, 1969), a thick Typical Chernozem on unmowed steppe near Kursk which also had numerous molehills on its surface.

compared with rates of less than 2.5 cm/h for sugar cane plantations and 0.05–4.5 cm/h under pasture (Wood 1977).

Compared with average infiltration rates, which are of the order of 1–2 cm/h, those for compact clays may be much slower (fig. 5.2a). For the grey Compact soils of Cuba, infiltration rates of only 0.008–0.0125 cm/h are reported. Low infiltration rates are typical of soils in many semi-arid areas. Rates of 0.12 cm/h or less are reported from areas in the western USA where furrow irrigation is practised (Williams and Doneen 1960). As the uptake rate by plants is about 0.8 cm/day, these infiltration rates are barely adequate to maintain the quantities of soil moisture required. Furthermore, a distinctive feature of infiltration rates of applied volumes of water is the decline of the rates observed with time, both during a single application and also through repeated irrigations. Thus, on Southern Chernozems, an infiltration rate of 27.6 cm/h at the end of the first 10 minutes of an initial irrigation, fell to 3.3 cm/h at the end of the second hour, and to 1.3 cm/h at the end of the fifth hour of an experiment. With repeated irrigations, the infiltration rate was 0.84 cm/h (Gusenkov et al. 1966). Some artificially induced infiltration rates fall off more dramatically during irrigation.

Soil characteristics which influence infiltration are essentially those which create porosity. In addition, the mechanical stability of soil crumbs is significant (fig. 5.2d,e) and variations according to clay type have been observed. Under rainfall kaolin crumbs break down rapidly and the clay is dispersed mechanically by raindrop impact. In contrast, practically no breakdown of soil crumbs and no dispersion was observed under comparable rainfall conditions for calcium illite, montmorillonite, or mixtures of kaolin and calcium illite. Layering in a soil may arrest infiltration (fig. 5.1b). In the Euphrates River valley, for instance, infiltration into fine clay loam occurs at 0.83 cm/h, but where clays exist with interlayers and lenses of sand, infiltration is only 0.04 cm/h.

5.2.2 PERMEABILITY

The permeability of the soil largely controls most aspects of soil wetting processes and the water balance as a whole, including surface runoff. Rates of water movement are commonly in the 1.5–5.0 cm/h range. In comparative terms, rates less than 0.5 cm/h are very slow, whilst those above 15 cm/h are very rapid. Typical of the last category are Thick Chernozems in the Kursk Oblast, where the soil's permeability is 13.0–18.0 cm/h. Clayey soils in semi-arid areas tend to be at the opposite extreme. For alluvial soils of central Iraq laboratory tests on horizons of minimum permeability in gilgai are less than 0.60 cm/h, and rates for bad-structured horizons are about 1.45 cm/h.

Associations of permeability and particle size characteristics may be noted. For instance, in the steppe zone of the European U.S.S.R. rates of 2.4 cm/h are typical of loamy Dark Chestnut soils, whereas the rates for coarse-textured Chestnut soils is 17.6 cm/h (Grin and Nazarov 1967, p. 1671). Where soil structure is poorly developed, correlations with particle size may be more obvious. Rates of over 15 cm/h are observed where the sand content in alluvial soils in Iraq is 80 per cent, falling to less than 5 cm/h where the sand content is less than 5 per cent. Permeability rates in silt soils may also be high.

After water infiltrates into the soil, the rate of advance of the wetting front depends on changes in porosity in particular. An exponential increase in permeability with increased pore size has been suggested. According to Poiseuille's equation, volume of flow increases with the 4th power of the pore diameter. Theoretically, therefore, a pore 1 mm diameter conducts downward 10,000 times as much percolating water as a 0.1 mm pore. Thus, soil water permeability is not determined by all pores but only by the largest ones. These are mainly non-capillary pores and are essentially cracks, earthworm and insect tunnels, and the tubes left by decaying roots (fig. 5.2c).

Much of the water infiltrating into wet clays will move through structural boundaries or through soil-filled shrinkage cracks much more quickly than it moves through the larger volume of smaller pores. Even a podzol on morainic loam has been observed to have a permeability within structural units of 0.07–0.1 cm/h compared with a rate of 0.8–1.2 cm/h between the structural units (Lebedeva 1969). The permeability of the calcium-montmorillonite surface layers of the Gezira Clay soil depends to a large extent on cracks. The higher the clay content, the more extensive are the cracks, which increases permeability. Such a greater penetration of water reduces losses from surface runoff.

Conversely, during rainfall, certain clays approach zero permeability. Continued percolation depends on the continuity, stability and persistence of the macropores with time, and it is therefore a reduction in their size which controls the degree of reduced permeability. Changes in permeability are controlled directly by the swelling of clays until their dispersion and movement begins. In addition to involving the stability of dry soil to rapid wetting, permeability equally is influenced by the stability of wet soil to dispersion by moving water. However, the possibility of dispersion of clay plates and their movement to block pores and channels in the soil, although it exists, is much less critical than periodic constriction of the macro-pores by swelling.

Because of the existence of natural, preferred paths of percolation, a wave of increased soil moisture permeates the soil as a series of wetting 'fingers' rather than as a 'front'. Free surface water is thus distributed

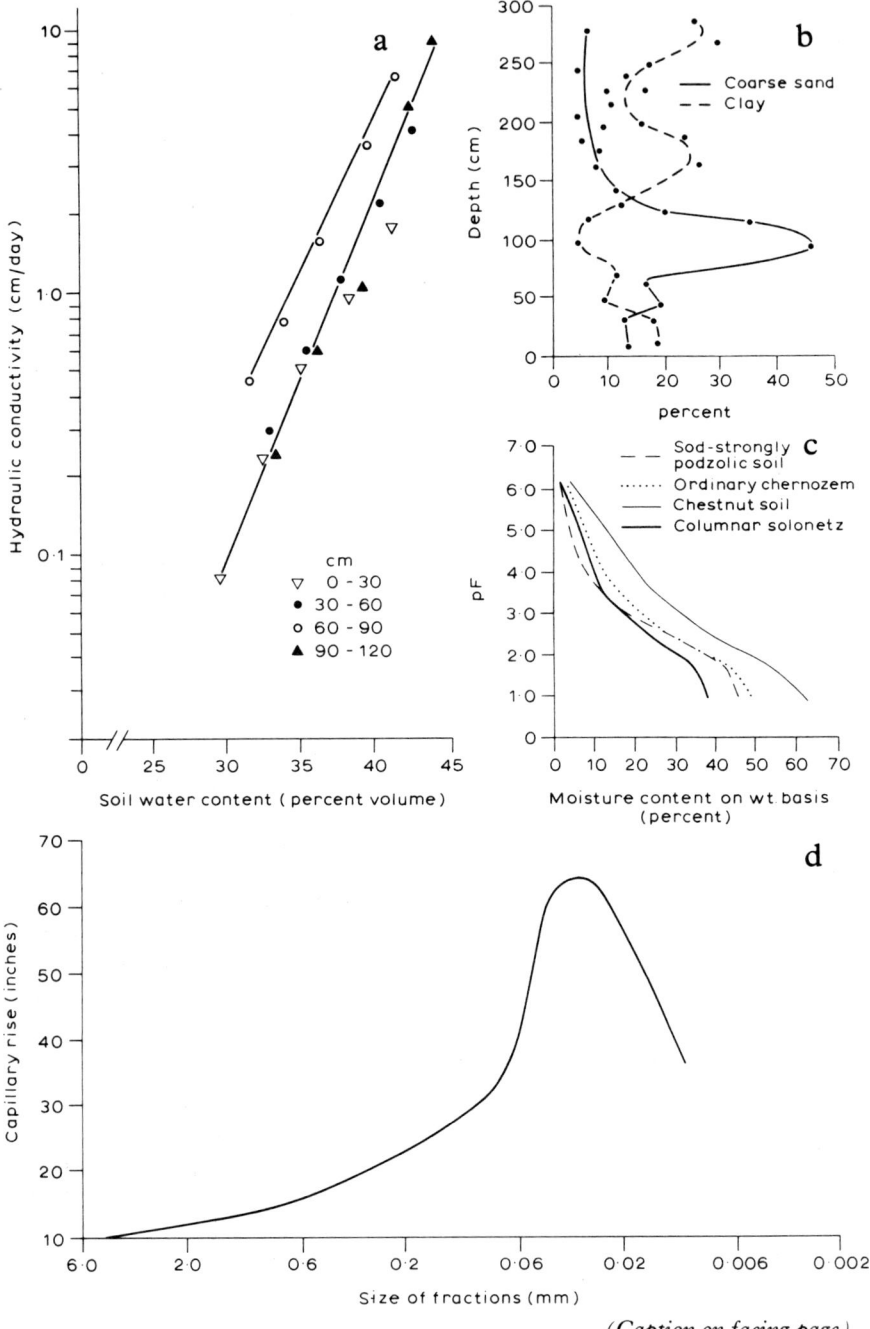

(Caption on facing page)

Physical properties of the soil 123

rapidly along this arterial network of macropores and the same channel system serves for exhausting air displaced by the wetting fingers' advance. Water is observed to spread rapidly in all directions from the main arteries. The point can be illustrated with reference to data for an Ordinary Chernozem, for which ploughed soils have a moderately rapid permeability of 10 cm/h. In an adjacent shelterbelt, permeabilities of 78 cm/h were attributed to the presence of unsevered zones of macropores (Grin and Nazarov 1967).

The dominance of the macro-pores on soil permeability has several important implications. Since soil solutions are moving essentially through macrocavities between structural units or along biopores, soil formation processes are consequently most marked in the peripheral edges of the structural units and along the walls of biopores. Increasing knowledge about the role of wetting fingers facilitates predictions about the amount of rain or irrigation required for groundwater recharge. An increased appreciation of the depth reached helps to explain the deep penetration of mobile pollutants like nitrates. Also farming practice has been modified to simulate the natural process more closely. Thus, a 'vertical mulch' enables water to percolate readily to greater depths more effectively than wetting downwards from the soil surface. The vertical mulch approach also promises to limit, to some extent, high water losses by evapotranspiration in sub-humid areas (Fairbourn and Gardner 1972).

Variations with depth in a soil profile are as significant a characteristic of water permeability as the lateral changes in macroporosity. A soil mass which has no distinct stratification, but where the porosity declines with depth, will show anisotropic permeability. Uniform profiles, in fact, rarely occur, and it is more common for soils to have well-defined layers, each differing from the other in permeability characteristics (fig. 5.3a,b). One of the most-studied layered soils are the fine-textured soils which cover some 5 million ha in the Southern High Plains. Here the Pullman silty clay loam has a moderately permeable surface horizon, but overlies a dense B_{22} horizon. This horizon, found at a depth of 20–60 cm, has a high content of montmorillonite and illite clays which reduce permeability to less than 0.25 cm/h after several hours' flooding (Unger

---

Figure 5.3 Influence of soil layering, type, and texture on soil moisture characteristics.
a and b. Dependence of soil hydraulic conductivity on water content in an alluvial sandy loam (Hillel *et al.*, 1972), yet with a significant difference at 90 cm depth (a) due to a predominantly coarse-textured layer at 75–105 cm depth (b).
c Influence of moisture content on moisture retention characteristics in A_1 horizons of four soil types (Voronin, 1974).
d Well-defined maximum of capillary rise at particle sizes of 20–60 mm (Whiteside *et al.*, 1967).

1970). The soil becomes more permeable again beneath the dense horizon. One of the most commonly encountered hindrances to permeability is the presence of a hardpan. These are widely distributed in soils in Pennsylvania and a mean permeability of 2.3 cm/h has been calculated. Soils without hardpans have a mean rate of 22.9 cm/h (Deer *et al.* 1969). In many parts of upland Britain, the thick organic surface layer as well as the iron pan impede free water movement in the peaty gleyed podzols. In fact, very few of the podzols in Wales, apart from those on very light-textured parent materials, can be classed as essentially freely drained. In podzols generally, the nature of the B horizons is particularly influential in controlling their permeability.

Air in soils, trapped in voids or more porous subsoils or layers, may diminish permeability. For example, when water levels were within 0.5 m of the ground surface in Worcester soil profiles, and upper horizons were waterlogged, air void contents of more than 4 per cent were common in the coarse, angular blocky structures of the B horizon (Thomasson and Robson 1967). However, most of the studies on water movement in soils is based on the assumption that air in the soil offers negligible resistance to flow. This simplication, and some others, greatly facilitates the solution of Darcy-based flow equations. Nonetheless, in the artificial case of border irrigation, experiments confirm that air pressure develops and impedes percolation.

A final physical influence on permeability of soil water is water itself as an initial component of the soil system prior to rainfall or irrigation, irrespective of whether short or longer periods of wetting are involved. The rate of advance of the wetting fingers increases with larger initial water content, and exponential rates of increase have been observed.

Striking changes in permeability rates over short periods of time are also critical. After a downpour or application of irrigation water, resistance to continued percolation through the soil gradually develops, due to swelling of clays, illuviation of silt particles, the mechanical breakdown of sandy soil particles, or the build-up of pressure in entrapped air pockets. For example, in a compact Vertisol in the former Ural River delta, water intake was observed at a rate of 6–17.4 cm/h during the first 10 minutes of a controlled experiment. After 20 minutes permeability had fallen to 3 cm/h. Finally, with swelling of the clay, permeability was closed down to a constant rate of 1.2 cm/h by the fourth hour of observation.

5.2.3 CAPILLARITY

An important mode of water movement occurs in tubes of hair-like diameter, due to capillary action (Latin, *capilla* = hair). Capillary phenomena appear in tubes of minute dimensions where solid, liquid, and

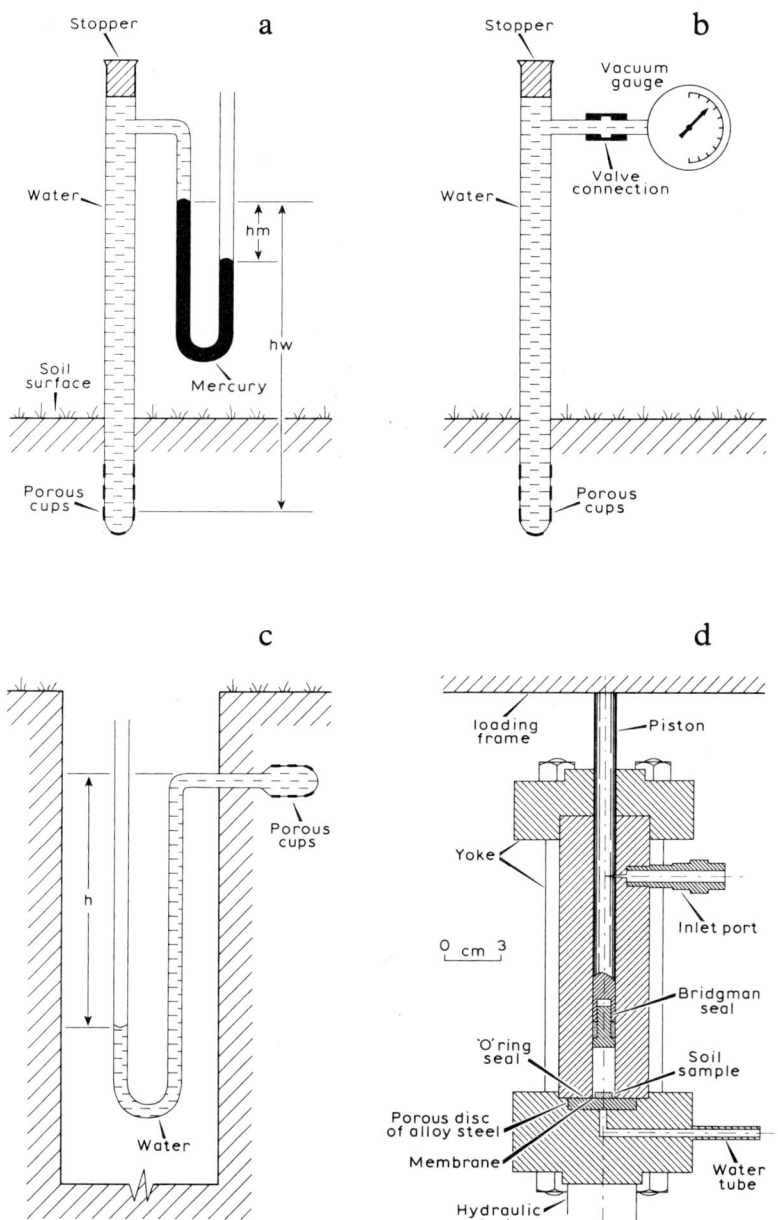

Figure 5.4 Methods of measuring soil moisture suction.
a–c Tensiometers in which suction is measured by mercury manometer (a), vacuum gauge (b), and by the height of a water column (c).
d Membrane apparatus for producing high pressures corresponding to suctions in the soil sample equivalent to pF 6.2.

gaseous interfaces meet. In fine tubes the inter-molecular adhesion of water to the soil exceeds the cohesion of water molecules to each other. A tapering edge of water molecules is thus attracted to the capillary walls and the curved interface or miniscus develops between water and air. Surface tension acts to flatten the curve of the meniscus and, as a result, the column of water creeps upwards until the force of surface tension is balanced by the weight of the liquid in the capillary tubes. Root hairs and microscopic biopores create and maintain a capillary network whilst particle size remains an important control. The maximum height of capillary rise is directly proportional to the radius of pores, in the ideal case. In soils, capillary rise is sufficiently pronounced to be a major factor of soil fertility, especially well-developed in fine silts. Thus in columns of Chernozem soil, the height of capillary rise observed was 1.3, 3, and 7–8 m for particles in the diameter ranges of 50–100, 10–50, and 5–10 μm, respectively. In sandy soils suspended moisture cannot, in general, move to the evaporating surface (fig. 5.4a). In pores with openings in the 15–300 μm diameter range, the water moves to the consumption zone steadily regardless of the distance if a pressure differential exists between the soil water and the moisture adsorber. At smaller particle sizes, however, only a portion of suspended moisture rises by capillarity. Much of the soil moisture is adsorbed on to clay particle surfaces. In clay soils, once capillary moisture has been utilized or evaporated, movement ceases. In Chernozems, as the soil moisture content falls below 35 per cent of the dry soil weight, the water moves towards the consumption zone in pores less than 15 μm only. Gradually the capillary bond is severed, firstly in the larger pores. The Soviet soil scientist M. M. Abramova has termed this state the capillary-ruptured moisture content. At this state, conductivity is sharply reduced and consequently so is the availability of water to plants.

5.2.4 ADSORPTION, RETENTION, AND SUCTION

Capillary and sorption influences are active in all soils. In clay and loamy soils, the influence of adsorption begins to predominate over capillarity. The boundary appears to lie around porosities of about 5–10 μm diameter. Thus moisture in a clay soil is retained partly by capillarity, but as the percentage clay content increases, moisture is mainly retained as thick films adsorbed between the negatively charged surfaces of the clay particles. The phenomenon of retention of water by soil, as a result of the capillarity and adsorption characteristics of the soil matrix, against the external forces, like gravity or atmospheric pressure gradients, is recognized as one of the distinctive, primary contributions of the soil to the sustaining of plant growth under the naturally intermittent supplies of precipitation.

Observations have been made on freely draining laboratory columns of Thick Chernozem clay loams. Water occupied pores with an aperture of less than 60 μm or less than 30 μm to some degree but, in the main, filled pores with a diameter of less than 5 μm (Sokolovskaya 1967). The maximum thickness of the water film influenced essentially by adsorption is 9–10 molecular layers (25 Å). Beyond this critical thickness, the effect of adsorption is less than that of gravity, and excess water drains off. In the absence of drainage, particularly in the capillary fringe, the thickness of the water film is inversely proportional to the specific surface area of the soil particles. In sandy soils it is equal to 260 molecular layers (1000 Å) in the lower part of the capillary fringe and 9–10 molecular layers in the upper part (Michurin and Lytayev 1967).

The retention of water in soil by the combined effect of capillarity and adsorption is termed soil suction (fig. 5.4) by workers following R. F. Schofield's lead. The availability of soil water for plant growth decreases progressively as soil suction increases. The first signs of decreased growth are recognized at quite low suction values. For instance, in Lower Greensand soils near Cambridge, hydraulic conductivity decreased dramatically with increased soil suction. There was a fivefold reduction in hydraulic conductivity when a suction of 5 cm only was applied to the soil at zero suction (fig. 5.5a). The moisture content of the soil fell by 1 per cent in the topsoil and 3 per cent in the subsoil. The abruptness of the reduction of hydraulic conductivity was attributed to the immediate desaturation of macropores (fig. 5.5c), attributed to the tunnelling of earthworms.

The intensity with which retention of water increases as a soil dries out, prompted the use of a logarithmic scale for its measurement. Schofield's index is, therefore, the logarithm of the height of a water column, measured in centimeters, that would exert the same influence as the suction which accounts for a given amount of water retention by the soil. In designating the index 'pF', p is the power to which 10 is raised and F represents freedom. A pF of 3, therefore, indicates that soil suction is an influence comparable with that of a column of water 1000 cm high, the height held up at a pressure of 1 atm. Various significant boundary zones in soil water behaviour can be specified in terms of pF. For instance, the limit at which percolating water slows down appreciably when all the noncapillary pores are largely emptied, corresponds to the tension that would be exerted by a column of water 50 cm high or a tension of 1/20th atm. This corresponds to a pF of 1.7. Field capacity as the equivalent of $\frac{1}{3}$-atm tension is expressed as pF 2.53. This value can be compared with field sampling on a range of Polish soils which gave values ranging from 2.2 to 2.55, or with experiments on a selection of Belgian soils for which the pF-values measured, ranging from 2.32 to 2.50, were in close

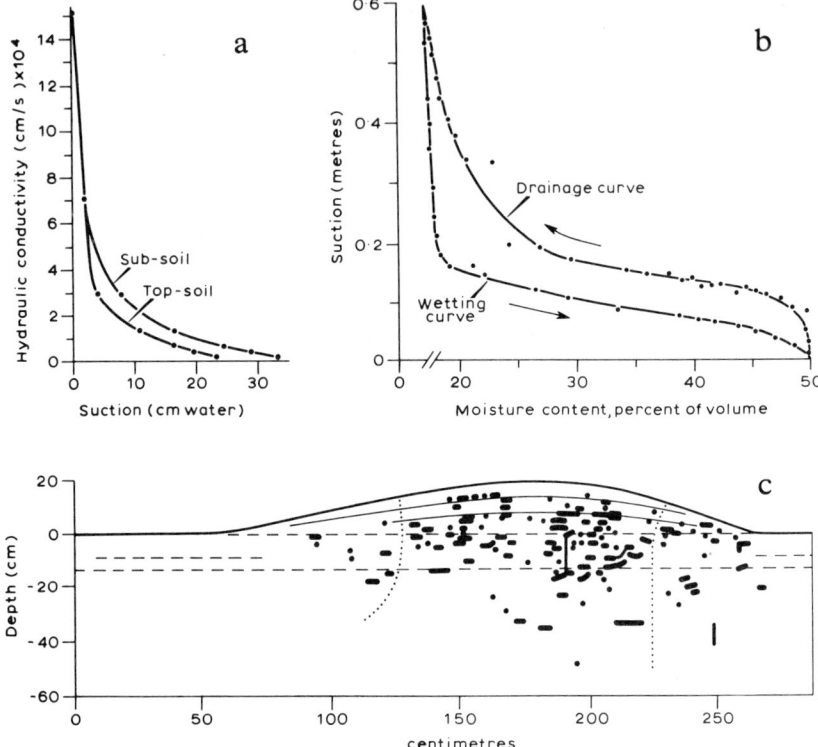

Figure 5.5 Coarser porosities and their influence on soil water properties.
a Rapid decrease of hydraulic conductivity with small increases in suction due to earthworm activity (Poulovassilis, 1973), observed in a 2 m deep undisturbed monolith of Cambridgeshire Lower Greensand soils, with early desaturation following intact earthworm channels.
b The hysteresis loop in a sandy soil (Maslov, 1967).
c Cross-section of an ants' nest, showing that 17 per cent of the A horizon is tunnelled with biopores (Greenslade, 1974), in lateritic podzol in the Mount Lofty Ranges near Adelaide. In addition, a 1.9 mm diameter gravel covers the nest, a near-optimum size to act as a mulch, also the work of the ecologically important meat ant *Iridomyrmex purpurens* which is the dominant ant in South Australia.

agreement with figures at field capacity (Verstraeten *et al.* 1971). Wilting point, for which the 15-bar percentage is now commonly used, corresponds to 14.8 atm or a pF of 4.18. If intermediate values are plotted, there is a hysteresis effect when wetting and drying pF curves are compared for the same soil. A soil brought to a given pF by the drying out of a wet soil will have a higher moisture content than if brought to the same pF by the wetting of a dry soil (fig. 5.5b).

As a compact, numerical summary of tension exerted by capillarity combined with the attractive forces that exist between water and solid soil particles, the pF scale has been widely adopted in some areas, such as Britain and Australia. In other regions, its utility has been questioned and the concept rejected (Richards and Weaver 1944). The confusion seems to have arisen because the concept was applied, without appropriate modification, to the irrigated soils of the south-west U.S.A. Here dissolved salts in the pore water control vapour pressure by their osmotic effect. In consequence, a more general concept, that of 'soil moisture potential' has been extensively developed and applied in the U.S.A., with the movement of soil water described mathematically by Darcy's equation. Soil moisture potential is based on the energy of retention of water by the soil and in this way, it is argued, is more directly related to the energy which a plant must exert in absorbing water from the soil. Soil moisture potential is the total effect of all types of energy acting on water in the soil, including gravity, capillarity, surface adsorption, and osmosis. The moisture and osmotic potentials are often treated as two independent components of free energy, but this usage has also become a source of some confusion as it is doubtful whether they are, in reality, independent. According to the theory of soil moisture potential, soil water moves to points of lowest potential.

5.2.5 SOIL WATER AND PLANT MOISTURE REQUIREMENTS

Many plants are exceedingly sensitive to a deficiency of water. The water regime and associated physical characteristics of the soil are therefore of particular concern in crop productivity studies where soil conditions are being studied (fig. 1.11e). Even the bulk of trees is more than half water, which is an essential constituent of protoplasm, and indispensable for maintaining turgidity and photosynthesis. Thus in cases like the sub-humid northern Great Plains, the addition of 2.5 cm of water to supplement a precipitation of 20–25 cm can increase spring wheat yields by 2.5 hectolitres or more. It has been suggested that the addition of as little as 12 mm to soil water may represent a margin between good and poor crop yield.

After infiltrating into the soil, water is subjected to three main influences of adsorption, gravity, and evapotranspiration. First, a portion of the soil is adsorbed by soil particles or retained by surface tension in small pores. Secondly, some portion of soil water may be free to percolate downwards through the macro-pores and beyond the soil profile. Thirdly, there will be a demand for water from the plant cover, varying according to season.

There are certain ranges of soil moisture content and borderlines between them which have specific significance for plant growth. 'Field capacity' is defined as the amount of water, following heavy rain, snow-melt, or irrigation, which is retained in the soil after free gravity water has drained from the soil. Field capacity is conventionally regarded as the upper limit to water available to plants. The second concept is that of the 'permanent wilting point'. Normally it is regarded as the lower limit of available water in soils. Thirdly, 'available water capacity' is usually assumed to be that quantity of water held in the soil between the limits set by field capacity and permanent wilting point. These concepts are expressed as a volume percentage or as centimeters of moisture per centimeter of soil. Soviet soil scientists use a fourth concept, that of full moisture capacity. This is the equivalent of porosity in non-swelling soils.

There are perhaps four noteworthy reasons why the characterization of apparently simple soil-water properties eludes precise summarization. First, since their formulation reflects plant requirements as the over-riding concern, it is not the purpose of many studies on plant physiology to recognize and characterize pedological phenomena. Indeed, the pioneer work of F. J. Veihmeyer and A. H. Hendrickson appeared in the early volumes of the journal *Plant Physiology* in 1927 and 1931. Secondly, since plants are adapted to their own particular environment, their degree of intolerance or resistance to drought or deluge varies from plant to plant. Thirdly, it has become standardized practice to characterize a plant's physiological limits in moisture requirements mechanically in the laboratory. Both permanent wilting percentage and field capacity are conditions simulated by certain pressures in a pressure membrane apparatus. This practice was introduced after a close correlation between water content at permanent wilting point, as established with sunflower seedlings, and equilibrium moisture contents at 15 bar was established. Several workers have shown that the 15 bar percentage moisture content, provided the soil samples are undisturbed, approximates to the permanent wilting point. A fourth consideration is the soil profile. Different depths in a real soil may provide differing values. The surface soil only attains the maximum moisture content when it has been wetted to some depth, about 0.5 m for many agricultural soils. The time taken for percolation to such depths, however, is sufficient for clay in surface soils to swell. Conversely, there are rather shallow depths at which water ceases to be within reach of plant roots anyway. If root penetration is shallow, as in pine plantations where water 10 cm below the roots is not used by the plant, soil moisture information from depths of more than 0.5 m would be irrelevant.

Field capacity was defined in 1931 by Hendricksen and Veihmeyer, although the Russian equivalent, minimum moisture capacity, was defined in the 1870s. The concept is one of the most important in the soil hydrology of irrigated agricultural soils. It is assumed that field capacity is that moisture content and that tension at which water conductivity in the soil becomes insignificant. It is generally recognized that the term is not precise since drainage by gravity does not cease altogether below some particular moisture content and most workers standardize sampling to a time 2–3 days after the soil profile has been thoroughly wetted. A tension of $\frac{1}{3}$ bar is generally used for the indirect estimation of field capacity, although tensions of 0.2, 0.1 and 0.05 bar have been effective in certain tests. For example, in Zambia, where the difference between 0.3 and 15 bars underestimated the available water by 35 per cent compared with field conditions, soil moisture tension between 0.1 and 0.2 bar approximate actual field capacity better than the 0.3 bar (Maclean and Yager 1972).

An important aspect of field capacity is the proportion of total pore space occupied by moisture at field capacity. The amount of air in the pores, to be adequate for normal plant growth, should be about a third of the total. For example, in Chernozem-like soils in the Crimean mountains, water at field capacity occupies some 60–70 per cent of total porosity, reaching 80–88 per cent in the underlying clays. If evapotranspiration is high, compaction of the soil leads to vapour loss of soil water, if the moisture content is greater than 80 per cent. Such losses are ten times greater when soil moisture contents are 95 per cent of field capacity than when it is 50 per cent. Another agricultural factor, as field capacity approaches 100 per cent, is the risk of soil tubular pores and capillaries being destroyed.

Permanent wilting point was originally determined in 1929 by the sunflower seedling method of Henrickson and Veihmeyer. Subsequently, two opposing tendencies have developed in the interpretation of wilting point. Some investigators consider that the difference in the wilting moisture level for most species of crops does not differ, and that wilting depends chiefly on soil properties, especially texture. Other workers consider that plants do wilt at significantly different moisture contents. Certainly, the soil moisture content at which vines wilt is much lower than the wilting moisture for young sunflower plants. The difference amounts, in Chernozem soils, to 2.5 per cent in clay loams and 4 per cent on clays. The capacity of plants to take up 'non-readily available' water in such cases is not a peculiarity, but simply an expression of their physiological adaptation to resist drought. This is not a soil phenomenon, as it relates to the activity of the leaf apparatus.

Wilting percentages have been established clearly in the distinctive location of Culbin sands, where coarse texture prevents excessive evaporation and the absence of silt and clay reduces capillary pore spaces. The maximum capillary rise in these sands which include no organic matter was 5 cm, field capacity was 5–6 per cent, and the wilting point was 0.5–

0.9 per cent. This compares with a wilting point of 1.8–2.1 per cent observed in Middle Don sands, in the U.S.S.R., which were also free of organic matter. In contrast, in Thick Chernozems in the Kursk Oblast, wilting points of less than 13 per cent are regarded as comparatively low. Here the field capacity in the A_p(0–20) horizon is 32–35 per cent, declining only slightly in the 20–40 cm layer to 30–32 per cent. The importance of local influences, particularly textural differences, is illustrated by data from south Arabia where uniform fine silts have a wilting point of 8.8 whereas alluvial sands have a wilting point of 1.6 per cent (Western 1972, p.276). The corresponding figures for field capacity were 35.4 and 10.3 per cent.

Many observations show that useful estimates of available water, in a form readily taken up by plants, can be made, and that these estimates can be correlated with soil properties such as texture, including percentage coarse fragments, organic matter content, soil structure, and bulk density. Specific examples of available water calculations can be compared with generalization that crops may suffer from inadequate water supply in soils with less than 20 per cent available moisture capacity. In a range of Warwickshire soils, for instance, available water capacity ranged from 0.39 mm/cm in a sand to over 2.5 mm/cm in a silt loam. The importance of organic matter has been demonstrated in the same soils. The available water capacity in the top 45 cm of a sandy loam soil was increased by a third over a 6 year period by annual applications of farmyard manure. Contrasts in pastureland are also related to organic matter. In a 2-year ley soil, an available water capacity of 1.26 mm/cm compared with a figure of 2.64 mm/cm in the top 15 cm of an old pasture on the same soil. The wilting points were 7.3 and 13.2 per cent, respectively (Low 1954). In Chernozems, when the wilting point is not very high, available water of 18–23 per cent has been observed in 0–40 cm layers. At depth, values may change as in Warwickshire soils with a mean available water capacity of 1.8 mm/cm in the 0–15 cm horizon which declines to 1.4 mm/cm in the 45–60 cm horizon (Salter *et al.* 1966).

5.2.6 TEXTURE AND SOIL MOISTURE CHARACTERISTICS

The correlation between soil texture and soil moisture characteristics may be sufficiently good to provide critical information about moisture characteristics of soils which influence plant growth. In approximate terms, sandy soils can hold about 8 cm/m whilst loams and silty loams may hold twice this amount for the same soil depth. Clays have intermediate values. Correlations have been established in a wide range of soils and environments. For example, in the Kulunda steppe there is a very significant correlation between silt and clay and estimated field

Physical properties of the soil 133

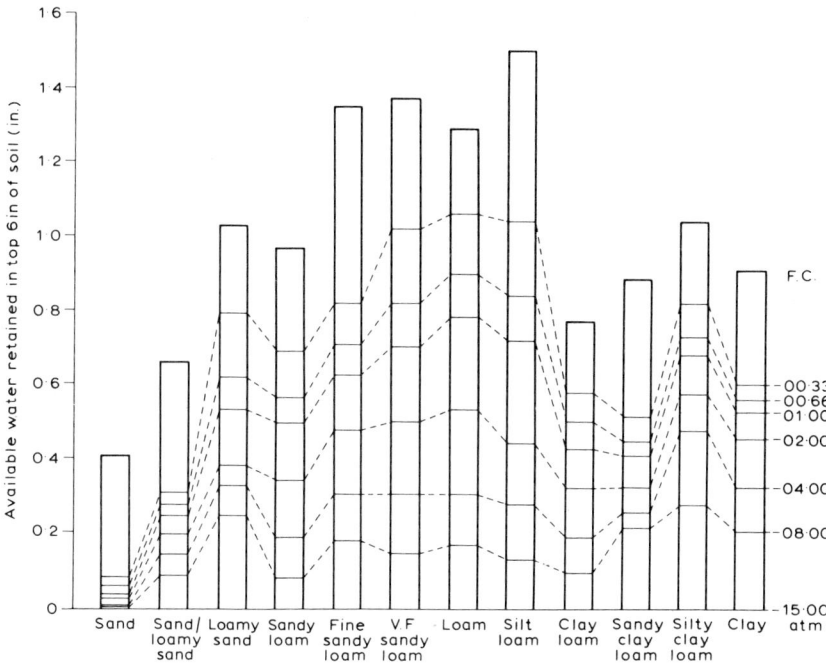

Figure 5.6 Influence of texture on the moisture characteristics of a range of soils in Lowland Britain (Salter and Williams, 1965).

capacity, $r = 0.91$, $N = 52$ (Lebedeva 1969). A range of soils with different textures in Warwickshire (fig. 5.6) has been shown to yield available-water capacity data which was negatively correlated with the percentage coarse sand and positively correlated with USDA silt (Salter *et al.* 1966). Calculated regression equations estimate the available-water capacity of this range of arable soils with a mean error of approximately 16 per cent. For example,

$$AWC = 1.86 - 0.011a + 0.018b,$$

where AWC = available-water capacity in inches per foot, a = per cent coarse sand (2.0 − 0.2 mm), and b = per cent silt (0.05 − 0.002 mm) (Salter and Williams 1967). Similarly, for selected soils in North Dakota, the relationships between silt and silt + very fine sand fractions and available-water capacity were statistically significant, with the highest coefficients being $r = 0.68$ (Rivers and Shipp 1972). More specifically, from correlation analysis on lab tests on silt loams from Pennsylvania, the 5–20 μm silt was the most important particle size for controlling available-water capacity in a very wide range of soils (Peterson *et al.* 1968).

134   Geography and soil properties

Compared with the negative influence of the sand fraction, and the controlling influence of fine silts, clay percentages do not usually suggest linear relationships (fig. 5.7a). The surface area of clays exerts only a certain effect, and available-water capacity is frequently described by a parabolic curve. For a range of Hungarian soils the maximum lies in loamy soils, perhaps due to the favourable air/water ratio in loamy soils. Typically, an increase in clay content up to 25 per cent correlates with an

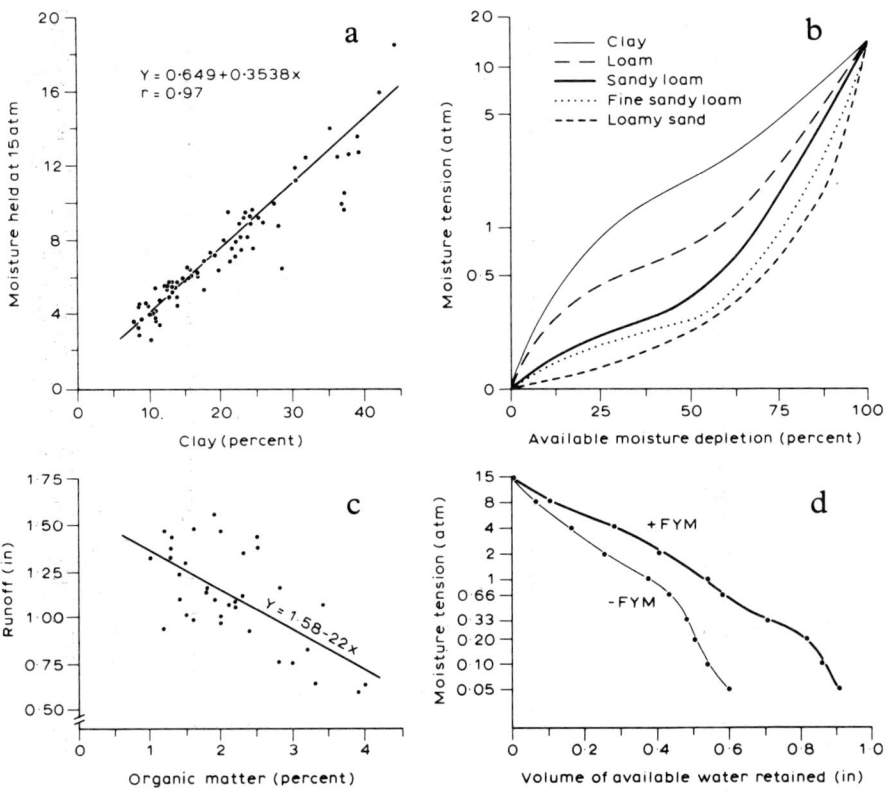

Figure 5.7   Influence of clay and organic matter on soil water properties.
a Relationship between percentage clay and soil water retention in the semi-arid Hissar district of India (Abrol *et al.*, 1968–9), as reported by several workers for moisture held at high tensions.
b Moisture retention curves for soils with increasing clay contents (Richards and Marsh, 1961).
c Relationship between organic matter and runoff (Wischmeier, 1966), following 140 mm of rainfall on soils from Agricultural Experimental Stations throughout the U.S.A.
d Effect of farmyard manure in increasing the volume of available water (Salter and Williams, 1963) in alluvial sandy loam of the Newport Series, Warwickshire, receiving 8 tonnes/ha/year FYM for 6 years.

increase in available water capacity, beyond which there is a tendency for the available water capacity to decrease (Bartelli and Peters 1959). In silty soils in semi-arid regions of the Hissar district in India, clay contents greater than 25 per cent result in decreased available water capacity, and a similar decline occurs in soils which have clay + silt contents of over 50 per cent. Clay type influences water retention, with montmorillonite retaining more water than illite or kaolin. The degree to which water-retention capability changes for mixed-ion montmorillonite depends on the relative proportions of the cations occupying the exchange sites on the clay surfaces. For example, it has been found that samples of this mineral retain approximately 10 per cent more water than the calcium-saturated clay.

Correlations between texture and soil moisture characteristics are particularly clearly established for tropical soils. Equations obtained by multiple linear regression analysis of Zambian soil data gave closer predictions for available water than found in Canada or in England. The most significant fraction was USDA silt, $r = 0.71$, $N = 143$ (Maclean and Yager 1972). Similar studies on Ugandan soils also gave better fitting equations than the equivalent equations for Warwickshire soils by Salter and Williams (Pidgeon 1972). In studies of 84 profiles in Puerto Rico clay content alone accounted for 77–81 per cent of the variation in estimated wilting point (15 atm tension) in four of the Major Soil groups (Philipson and Drosdoff 1972).

5.2.7 ORGANIC MATTER AND SOIL MOISTURE CHARACTERISTICS

Organic matter influences the capacity of soils to retain available water (Salter et al. 1966). In woodlands leaf litter is an excellent medium for water storage. On the floor of hardwood forest in eastern Tennessee, for example, the average litter volume was 9.4 metric tons/ha in midsummer. This surface layer on the soil had a field capacity of 135 per cent, the equivalent to a water-holding capacity of 12.5 mm. Within the soil profile, the organic content may correlate with hydrophysical properties (Hollis et al. 1977) either due to its direct influence or to its indirect influence on soil structure and porosity (fig. 5.7c). For example, data from a wide range of Zambian soils showed a significant correlation between organic carbon and available water, $r = 0.57$, $N = 143$, with the amount of carbon in the 20–30 cm layer being particularly significant (Maclean and Yager 1972). In cultivated soils, moisture characteristics of manured and unmanured soils are often significantly different (fig. 5.7d). In sandy loams in Warwickshire, the addition of manure modifies the structure of the soil and alters the pore-size distribution. More water is released at low tensions and increases in available water capacity compared with

136 Geography and soil properties

unmanured soils of the same series of up to 75 per cent were observed.

If soils within the same texture class vary in available water capacity, soil structure is an important source of variation. The effect is more pronounced at low tensions because the larger pores, holding water under low tensions, are more affected by structural changes than the smaller capillary pores. In field soils in the European U.S.S.R. the importance of soil structure is reflected in the trends in water permeability and content of water-stable aggregates. Both these decrease to the north and south of the Typical Chernozem zone.

5.3 Soil temperatures

One of the main elements of the soil climate is, of course, its temperature (fig. 1.13a). Soil temperature at 20 cm depth is an average index of the thermal state of the root-inhabited layer, and for many purposes is an adequate expression of the thermal state of a soil. Conditions obviously range with latitude, from arctic soils which are frozen for about 9 months of the year, to tropical soils where the water which percolates through soils is always warm, at about $25°$ C, with intermediary areas, like central and northern Europe experiencing average soil temperatures of $10°$ C or less.

5.3.1 DIFFERENCES BETWEEN ATMOSPHERIC AND SOIL TEMPERATURES

Mean annual soil temperatures are always higher than the corresponding air temperatures. In fact, soil temperature for the 50 cm depth can be approximated by adding $1.1°$ C to the mean annual air temperature. Actual observation shows that in the red clay soils of Cuba, for instance, the mean annual temperature in the 3 m soil layer is $1.2°$ C higher than the prevailing mean annual air temperature. There are areas where observations show that the soil temperature is partially a separate element of the climate, significantly distinct from the climate of the ground layer of the atmosphere. Across the U.S.S.R., the difference between the mean annual soil and air temperatures ranges from $2°$ C to $7°$ C, increasing in a northerly direction (Dimo 1967).

One reason for soil temperatures being, on average, higher than air temperatures is their exposure to the full glare of the sun whereas meteorological instruments enjoy the continual shade of the Stevenson screen. Vegetation cover has an insulating effect, recognizable even in Antarctic soils, against extremes of cold. In the arctic there are substantial differences in the freezing pattern of mineral compared with peat soils. The peat has the same insulating effect observed beneath wet moss patches in the

Antarctic. For example, on Vaygach Island in the arctic tundra, soils in exposed places thawed out to a depth of 93–97 cm in midsummer, whereas the depth of permafrost in bogs was 25–30 cm. In the taiga zone in general, therefore, the minimum depth of seasonal thaw is found in peat beds while the maximum depth occurs in sandy deposits. In podzols in such areas, freeze–thaw cycles penetrate more rapidly and deeply in cut-over areas than where a natural cover remains, and in sand loams of an oak stand in Minnesota, freezing temperatures penetrated 20 cm deeper in plots without litter than in litter-covered controls plots. The removal of vegetation and the exposure of bare, arable soil to very high temperatures is most marked in tropical areas where soil temperatures may average $4°C$ more in the open than beneath forest, where they remain close to air temperatures.

In cooler environments a major factor in differentiating soil temperatures from air temperatures is a cover of snow which protects the soil, whether it is bare or vegetation covered, from sudden changes in air temperatures. The earlier the soil is covered by snow and the deeper the cover, the less is the extent of soil freezing. In fact, if the accumulation of snow is sufficiently early, deep, and persistent, the soil will not freeze. In the northern states of the U.S.A. field observations suggest that snow depths of 40 cm or more often prevent freezing of the underlying soil (Hart 1961). In the oak stand in Minnesota, the frozen soil was deeper and lasted longer in snowless plots. For several weeks, frozen soil was 38 cm deeper on average in snowless plots (Thornd and Duncan 1972) and in spring frozen soil persisted in the snowless plots about 10 days longer.

In patches or places where there is no vegetation, moss, litter, or snow cover, drying may lower soil temperature owing to its evaporative cooling effect. The flow of heat from a soil during evaporation is such a complicated process that sufficient data are not normally available to explain field observations. However, laboratory experiments on Edina soils showed a vertical difference of $30°C$ for radiative drying and as much as $5°C$ for wind drying (Selim and Kirkham 1970).

Soil temperature is the critical factor determining germination and growth of seeds. For example, maize germination is very low at $6°C$ and seedling growth is delayed until temperatures of $13°C$ are exceeded. For root growth, a range of soil temperatures has been reported. Optimum growth of roots of fruit trees occurs at soil temperatures of $15–20°C$ and their growth stops at $35°C$. The effect of soil temperatures on the assimilation of water and minerals is critical. Plants gradually lose their abilities to absorb nutrients at low temperatures. In the growth of rice it is the association with low temperatures of high carbon dioxide concentrations, volatile organic acids, and ferrous iron in soil solutions which is unfavourable (Cho and Ponnaperuma 1971). In cooler environments, the penetration of freezing temperatures to the root zone may have several effects on plant growth. For instance, freezing and thawing of bare, saturated soils of fine-texture may cause frost heaving of such crops as alfalfa.

5.3.2 TEMPERATURE GRADIENTS AND SEASONAL WAVES IN THE SOIL PROFILE

Reduced amplitude of seasonal changes characterize progressively deeper horizons within the soil. Initially, at the soil surface, amplitude of temperatures tend to be greater at higher latitudes since in tropical zones, diurnal temperature ranges may be the dominant contrast. For instance, at Hallett Station in Antarctica, a soil surface seasonal range of temperature from $-45.6°$ C to $32°$ C has been observed. The temperature drops from the soil surface to the upper boundary of the frozen zone may be about $20-30°$ C in the summer and sometimes as much as $50°$ C during the day. These figures represent mean gradients in soil temperatures of about $2-3°$ C/cm. For example, during an austral summer in Wheeler Valley in Southern Victoria Land, maximum temperatures were $1°$ C air temperature 1 m above the ground, $15°$ C at the soil surface, $7°$ C at a 5 cm depth in the soil and $1.5°$ C at 15 cm. The icy, hard permafrost layer lies at about 15 cm depth (Cameron *et al.* 1970). In tropical soils, in more continental locations, surface layer temperatures may have a range of over $50°$ C which decreases rapidly with depth. Observations by Braak at Jakarta, where the annual difference between absolute maxima and minima in the atmosphere was $17.5°$ C, narrowed to differences of 14.8, 12.0, 9.4, 8.1, 6.8, 3.2, and $3.6°$ C at depths of 5, 10, 15, 30, 60, 90, 110 cm, respectively.

The significance of changes of soil temperatures with depth and the narrowing of their seasonal range is particularly felt in the 0–20 cm root-inhabited layer in cooler environments. Most of the soil micro-organisms live in the top 5 cm layer, and the propagating waves of diurnal temperature changes are significant in their activities. In cold soils the presence of permafrost will damp down a propagating wave at shallow depth. In a wet sphagnum bog in Manitoba where the mean annual temperature is $-0.5°$ C, temperature changes advance at about 1 cm/h and the temperature course is reversed at a depth of less than 15 cm (Lettau 1971, p.174). Towards the opposite extreme, diurnal fluctuations of temperature in Poona soils are insignificant below a depth of 20 cm. For tropical soils the retardation in the advance of daily maxima is about 1–2 h for every 5 cm depth of soil, and reversal takes place at depths of 20–30 cm. Where seasonal changes are recorded at greater depths, the lag may be greater. In coarse loamy sands in the Adirondacks, maximum and minimum temperatures were reached 2–3 months later at 3 m than the near-surface soil layers (Leonard et al. 1971). Just as diurnal waves are significant for smaller, shallow rooting plants, so too are these seasonal lags significant for the growth of deeper rooting, higher plants.

Heat transfer in a moist porous medium such as soil may take place by conduction, convection, radiation and distillation. The relative importance of these depends on the relative depth and velocity of the temperature wave in question. In daily fluctuations, vapour movement can be assumed to be proportional to the temperature gradient and to be the probable mechanism which creates diurnal fluctuations in soil temperatures. Heat conduction by vapour transfer is explained by the assumption that vapour flux due to temperature differences is approximately proportional to the temperature gradient which exists across an ideal gas-filled pore. Thermal conductivity down a soil profile is, over the season, more influenced by water flow. The amount of water contained in the soil is also an important regulator. In North Dakota, high soil water content in autumn may lower the soil temperature at 30–120 cm depth by 1–2° in the next year's growing season (Willis et al. 1977). If soil temperatures fall to the degree that soil water freezes, the whole complex of phenomena of the soil thermal and moisture regimes changes.

The effect of temperature in processes of soil formation is inseparable from water in the soil, seen particularly at the distinctive stage when soil water freezes. This commences in autumn, from the surface downwards. Ice crystals begin to form in the pore spaces as the soil water begins to freeze. Commonly, in field soils, a liquid-like film about 10–40 Å thick remains at the ice–air interface and comparable liquid films of variable thickness remain at the other interfaces between ice, soil particles, and air (Cary and Mayland 1972). If freezing continues, ice crystals completely fill the pore spaces and the initiation of physically disruptive expansion is imminent. As the soil particles are pushed apart by the continued ice crystal growth, the continuous liquid film along the surfaces of the soil particles separates. With the movement of additional water from the liquid to the solid phase, as described by Hoeckstra (1966), there is a recognizable hardening of the frost with time. If negative temperatures prevail long enough in cold environments, ice lenses begin to separate out within the soil. Water may move from warmer areas in the soil to such lenses, but since little can move past them the lenses continue to grow.

The effect of water freezing on fissile or porous gravels is splitting and sand grains may be comminuted to particles as small as 10 μm. In the much shorter term, freezing of the soil water initiates changes in the properties of soil clays, which tend to flocculate under the influence

140   Geography and soil properties

of persistent negative temperatures. This effect is experienced by other colloids in the soil, particularly the organic humus fraction.

The low temperature of a wet soil is due to its high specific heat and to evaporation. Vaporization of soil moisture is achieved by the absorption of heat from the surrounding environment, which cools the soil, especially at its surface where most evaporation occurs (Fluker 1958).

5.3.3 INFLUENCES OF PARTICLE SIZE ON SOIL TEMPERATURES

As heat transfer is conditioned by the porosity of the media, soil temperatures are influenced by the control of particle size on porosity. Outwash gravel terraces, for instance, present favourable conditions for maximum penetration of seasonal temperature fluctuations, having a low moisture content and a high coefficient of thermal conductivity. In addition, the infertility of most gravels supports a sparse vegetation which allows a wide spread of temperature range at the ground surface (McDole and Fosberg 1974). For example, observations in a forest near Volgograd showed that coarse-textured soils heat up to 40–45° C on hot summer days to a depth of 10 cm. The contrast is seen even in loamy sands, which warm up to only 23–30° C (Vashchenko and Gayel 1967). In West Africa, owing to the low thermal diffusivity of the preponderantly sandy soils with gravel in the topsoil, soils can reach 42° C at 5 cm depth, inhibiting growth (Lal 1974).

The influence of sands on soil temperatures is most closely studied where the frost-line moves up more quickly when sandy soils thaw compared with those of finer texture. Observations in south-west Wisconsin revealed that, during the January thaw, the frost line moved up by 60 cm in a sandy soil but by only 15 cm in a silt. During the spring thaw, the sand started to thaw at the surface a week before the silt, and was completely thawed a fortnight later. At that time the silt was still frozen between depths of 25 and 90 cm (Sartz 1970). The same two soils freeze at the same time, but frost was observed to penetrate more quickly and to greater depth in the sandy soil. Sand froze to a depth of 105 cm during the first freeze–thaw cycle, whereas the comparable depth reached in the silt was only 75 cm.

The depths to which temperature effects may be felt are marked in coarse-textured soils. In the case of coarse loamy sands under pines in the south-east Adirondacks, the range of temperature extremes was still approximately 7° C at a depth of 3 m. This figure compares with a range of 20° C at the 5 cm depth (Leonard et al. 1971). In sandy areas of the south-east European U.S.S.R. seasonal temperature fluctuations involve the layer of soil and underlying soil materials to depths of 13–19.5 m. Daily temperatures affect a zone 0.5–0.6 m deep (Kulik 1967).

5.3.4 EFFECT OF TEMPERATURE ON SOIL ORGANIC MATTER

Within certain limits, increased temperature favours both the rate of production and decomposition of soil organic matter. The relationship is not simple if the modifying influences, such as rainfall distribution, are taken into account. Thus, in East Africa, with its marked wet and dry seasons, the effect of high temperatures on the activity of soil micro-organisms is nullified during part of the year by the adverse effect of dry conditions. This slows down decomposition of organic matter in the surface soil without necessarily slowing down vegetation growth and organic matter production, provided moisture is available in the subsoil. Also, the sharp increase in organic matter levels at altitudes over 1500 m in Malawi, although they may be accentuated by the change from cultivated to untilled soils, probably reflect some change in the balance of decomposition of organic matter where the mean annual temperature falls below 17–18° C (Young and Stephen 1965).

The freezing of different solutions of humic acids is generally accompanied by their fractionation. The fractions which separate depend on the molecular size and degree of complexity in the original colloids, but such changes in the nature of humic acids following freezing are difficult to specify. However, it has been observed that the freezing of a very stony soil in the Far East of the U.S.S.R. resulted in an increase in the water-soluble organic matter. This increase was more than two-fold in the upper horizon and by a factor of 1.4–1.8 in the 20–30 cm layer (Nechayeva 1967). It seems that freezing induces the hydrolytic conversion of complex humic substances into more mobile, less condensed substances.

One of the most significant effects of temperature on organic matter is on rates of nitrification. The process of nitrate production virtually ceases below 4.5° C. Above 10° C, however, nitrogen in the ammonium form is oxidized quite rapidly to nitrate, and continues to increase until around the 32° C level. The temperature effect on the rate of nitrification creates one of the essential differences between temperate and tropical soil processes. In temperate regions the slowness of nitrification is often a limiting factor on plant growth, whereas in the tropics soluble nitrates are formed but leached from the profile before they can be absorbed by plants.

5.3.5 EFFECTS OF FREEZING ON SOIL STRUCTURE

Even where permafrost is absent in cool environments, such as Iceland, frost-heaving plays an important part in the formation of crumb structure. One third of the soils in the U.S.A. are affected by the freezing of soil crumbs in the plough layer. Soils with a portion of organic matter freeze readily to shallow depth and a granular structure is created in the soil. Continued freezing, particularly in soils again high in organic matter, changes the granular structure into a loose and porous structure. This is typical of forest soils and is referred to as honeycomb structure. If the surface soil of such a structure thaws then re-freezes, ice-needle structures may form. Ice needles have been observed to push upward and lift the surface crust or sod by as much as 15 cm. Ice-needles as long as 2 cm were noted in the investigations on Malham Moor (Matthews 1967). Where freezing temperatures can penetrate easily into mineral soil, as in bare or heavily grazed pasture soils low in organic matter, concrete-like structure may form. Generally, concretely frozen soils are impermeable.

In permanently frozen ground, the dynamic element in the soil structures of the layer above the permafrost is in such constant motion during the summer season that it is described as the active layer. The depth of this constantly churned active layer depends on texture and wetness and the degree of surface insulation by snow or vegetation. In the Rocky Mountains, the active layer above patches of permafrost is within 30 cm of the surface under peat. In Spitzbergen, the active layer is as little as 15 cm thick where a thick plant litter persists as beneath mossy grooves or on banks between mud polygons. One of the particularly significant features of the active layer, the absence of pores as continuous tubes, is also developed to some extent in temperate soils. Freezing may rapidly destroy non-capillary porosity, particularly in unstable crumbs in coarse textured soils, an effect which becomes more pronounced after each thaw.

5.3.6 EFFECT OF TEMPERATURE ON HYDRAULIC CONDUCTIVITY

Water conductivity increases with increasing temperature, but the rate depends on the proportion of water moving as vapour. The effect of temperature has been found to be small at low suctions but to increase as the soil becomes drier. This increase in hydraulic conductivity at a given water content, due to temperature rise, is generally attributed to the decrease in viscosity of water with a rise in temperature.

Under natural conditions, temperature gradients always exist, and liquid water tends to move from warmer to cooler areas (Ferguson et al. 1964). Temperature variations with depth give rise to thermal gradients which tend to move water vertically, both in the liquid and vapour phases. Philip and de Vries (1957) account for soil water movement in response to temperature gradients by a theory based on refinements of Fick's Law. It is assumed that water moves, in both vapour and liquid phases, in response to thermal gradients and to soil water pressure as well. In the field, experiments over winter months indicated that more than 6 cm of water moved upward into soil profiles in response to changing thermal gradients. In particular, freezing of the soil surface induces upward movement of water which saturates the surface soil just below the line of freezing temperatures. For instance, in the 1963–4 winter, stony soils in forest areas of Manchuria were observed to increase by 20–50 per cent in their moisture contents in the 5–15 cm depth horizon. The increase was by 5–15 per cent at the 20–30 cm depth, and rate of moisture gain increased as the freezing became more intense (Nechayeva 1967). The thawing of such a layer induces waterlogging in the soil which may persist for a long time and a direct relationship between depth and rate of freezing, and the type and degree of soil drainage is often recognized. Water may move gravimetrically through a frozen soil, as freezing does not always reduce infiltration completely.

Hydraulic conductivity is much affected if freezing and thawing break down larger soil aggregates. In general, the magnitude of the structural breakdown increases with greater water contents, increased aggregate size, and lower freezing temperatures. The partial clogging of water-conducting pores with ice becomes more pronounced after each thaw. Some infiltration may continue if some of the larger pores remain undisrupted. In untilled soils, ice crystal growth not sufficient to clog larger macropores is observable and many earthworm tunnels have been noted to persist. In general, however, freezing and thawing tends to increase drainage problems on wet soils. In addition the processes of infiltration of irrigation water into frozen soil, particularly where urban waste water is used, remains an unresolved problem.

5.4 Soil air

5.4.1 MOVEMENT OF SOIL AIR

(a) *Diffusion in dry soil*
Soils may be regarded as a system of gas-filled pores and water-saturated pockets. The soil air contains water vapour and the water contains air in solution. Gases consumed and produced within the soil enter and leave primarily by diffusion. H. L. Penman recommends that the diffusion coefficient of a gas and of vapour through a medium with a free porosity of up to 60 per cent be determined by

$$D = \alpha p D_o$$

where p is the porosity of the medium, D_o is the diffusion coefficient of the corresponding gas or vapour in the air, and $\alpha = 0.66$, a coefficient allowing for the mean free path of gas and vapour molecules in the porous medium. Many investigators believe that the coefficient should be increased for vapour. In a dry soil, however, the average effective diffusion coefficient is probably close to the coefficient of molecular vapour diffusion in air. Gaseous diffusion in soil is ordinarily sufficiently rapid to renew the soil atmosphere several times a day. Theoretically, 10 per cent by volume is the lowest value at which air can be exchanged in the soil, with cross-sectional area, tortuosity factors and the probability of continuity between adjacent pores being important determinants. Otherwise, as a result of random movements of gaseous molecules, gas diffuses from areas of high to areas of low partial pressure. The partial pressure of a gas in a mixture is the pressure which this gas would exert in a gaseous mixture having a total pressure of 1.0 atm, a function of the number of molecules in a given volume or concentration. For instance, if O_2 makes up 21 per cent of air having a total pressure of 1.0 atm, its partial pressure is 0.21 atm. For water vapour, the gradient is determined by the difference between the vapour pressure of the moisture-saturated soil air and the vapour pressure of atmosphere humidity. At suctions of 31 atm and less, soil water is sufficiently free to maintain the relative humidity of the soil at or close to 100 per cent. Vapour flow begins between pF 4.2 and 5.1 and is the dominant form of moisture transfer between pF 5.1 and 5.8.

As in the liquid phase, water vapour flows through soil from warmer to cooler areas, due to the exponential dependence of vapour pressure on temperature. Thus, during summer drying of a dark clay loam, 20–60 per cent of vapour diffusion below 20 cm depth was observed to move downward into the soil, in the direction of decreasing soil temperature (Abramova 1968). Generally, and at most times of the year, there is a temperature-induced vapour tension gradient in soil air, the magnitude and direction of which changes rhythmically through the year and through the day. A diurnal fluctuation of $10°C$ could induce sufficient O_2 interchange in the upper 15 cm to supply 10–150 per cent of the daily O_2 requirements of most roots and micro-organisms. At approximately constant temperatures, partial vapour pressure in soil pores rises with increasing soil water content. Vapour then moves in the direction of the lower soil water content in any unevenly wet soil, and towards a dry soil surface (Onchukov *et al.* 1972).

Vapour transfer cannot carry ions with it. Much of the downward vapour flux could, after re-condensing, be returned to the surface by capillary flow and consequently would act as a salt pump serving to move plant nutrients from the root zone toward the surface. In semi-arid areas, this entails the possible movement of undesirable salts, a risk accentuated in fine-textured soils, from which twice as much water evaporates compared with sandy soils. The suspended moisture cannot move to the evaporating surface in most sandy soils.

(b) Gas diffusion through soil-water films

Since soil particles and live tissues are hydrated, transfer from soil air to active sites must be by diffusion through water films. For example, the link between atmospheric oxygen and that taken in by plant roots is the soil solution, its oxygen being replenished from the soil air. Theoretically, diffusion of O_2 or CO_2 will be about 10,000 times greater in air than in water. The concentration gradient across a water film 1 mm thick could be ten times greater than that across a metre of air-filled pore space. Water films thus greatly complicate the nature and rates of diffusion in soils.

(c) Mass flow

Vapour transfer is not purely diffusion. Air mixing due to wind movement may contribute to gaseous exchange, ten-fold increases in evaporation being observed with increases of wind speed from 0 to 40 km/hour over bare soil. Such turbulence is particularly effective in litter, mulches, and at very shallow depths in coarse-textured soils. Also, air may be displaced when a soil is wetted which is then followed by the re-entry of fresh air as the soil pores drain. When a dry soil is wetted, evaporation at the liquid front continually supplies a gas phase which, by displacement and diffusion, moves as a front immediately ahead of the wetting front (Perrier and Prakash 1977). Another aspect of mass flow is the transfer of dissolved gases in soil water movements. Water at $25°C$ and in equilibrium with air at 1 atm pressure, contains about 6 ml/l of oxygen, nearly that amount used by 1 gm/hour (dry weight) of very active roots (Grable 1966). Similarly, percolating water acts as a CO_2 pump, drawing this gas from upper soil layers. Although only a small amount of CO_2 enters the soil dissolved in water, a large amount may leave, due to the high solubility of CO_2, being 87.8 cm^3 in 100 ml of water at $20°C$.

(d) Soil crumb structure and gas movement

Gas-filled pore space is not uniformly distributed along the length of the soil column, nor is an 'effective diffusion path' readily predicted because of anisotropic pore orientation and decrease in air permeability with increasing depth. In naturally aggregated soils, all the micro-organisms and most of the active roots occur and respire within, rather than between, the soil crumbs (Currie 1961). Diffusion in crumbs is much retarded due to their consolidation and partial saturation. Diffusion of gases within the crumb becomes confined to the liquid phase where the diffusion coefficient is four orders of magnitude lower. In the centres of larger saturated crumbs, critical values of O_2 are reached as they become essentially anaerobic foci. From such centres, scattered irregularly through the profile, characteristic products of anaerobic microbial activity can diffuse outwards. Even where air diffuses readily through well-developed intra-aggregate pores, the diffusion rate within the aggregates is not affected immediately. The size of aggregates thus emerges as the main control on the degree of anaerobiosis in soil (Smith 1977).

5.4.2 GASEOUS CONSTITUENTS OF THE SOIL

(a) *Oxygen in the soil*

Oxygen is important as the final electron acceptor in respiration. If gas exchange is impeded, biological processes deplete soil oxygen to the detriment of plant growth and microbiological activity. The benefits of soil aeration include stimulation of root growth, expressed in greater root branching, length, area, more and thicker roots, all of which increase the feeding power of the plants. The oxygen requirement during a growing season may amount to some tonnes per hectare. Even brief soil oxygen deficiency can reduce electron transfer and root respiration. Fermentation occurs, producing only a fraction of the usable energy produced by respiration. Glucose utilization is 200 per cent greater but energy liberation is only 24 per cent that of respiration (Grable 1966). Persistent soil oxygen deficiencies influence the entire metabolic activity, cell permeability increases, and roots rot and die. Plant physiological responses to different levels of O_2 vary greatly. Root nodules and mycorrhizae are sensitive to oxygen levels, whereas some plant roots can oxidize an anaerobic medium. Requirements range from the 21 per cent needed for maximum growth of tomatoes, to those of rice which grows in flooded soils devoid of oxygen. Most agricultural crops have very stringent oxygen requirements. Grass growth ceases when air content of the soil is less than 5 per cent. Normally, air content of more than 20 per cent is adequate for the growth of most crop plants and an air content of more than 30 per cent favours vegetable growth. The critical index of oxygen uptake is the oxygen diffusion rate (O.D.R.). For many plants an O.D.R. of 0.20 $\mu g/cm^2/min$ will arrest root growth, and roots rot at values lower than this. With O.D.R. in the 0.20–30 $\mu g/cm^2/min$ range, several plants will suffer some oxygen deficiency. However, in all cases where the gaseous composition of a soil may appear satisfactory for plant growth, the liquid film and the anaerobic core of saturated crumbs may control the true aeration status at the respiring root surface. With complete flooding, oxygen declines to negligible quantities. Free oxygen is absent from the soil water in all types of bogs and even in temporarily flooded woodland soils. The water surface eliminates soil-atmosphere exchanges, and oxygen diminishes to near-zero levels.

(b) *Carbon dioxide*

Even in well-ventilated soils, carbon dioxide diffusing in the aqueous phase is commonly 10–100 times its concentration in the atmosphere.

In the form of carbonic acid it therefore has a significant role in the development of distinctive soil-profile characteristics owing to leaching of more soluble mineral elements. Its effect is not as immediate, vital and direct as oxygen on plant growth. However, by increasing the solubility of calcium, magnesium, and phosphates it renders such minerals available to plants in the longer term. In the anaerobic conditions of flooded paddy soils, carbon dioxide has an over-riding influence on ionic concentrations in the soil.

The source of soil carbon dioxide is the oxidation of all the organic substances created by photosynthesis which ultimately produces carbonic acid with some weak organic acids as intermediate products. About $\frac{2}{3}$ of the carbon dioxide released into the colloidal water film is released by micro-organisms whilst $\frac{1}{3}$ is the direct result of plant-root respiration. However, since micro-organisms cluster around plant roots, the points of carbon dioxide release are clearly defined.

Four different units of measurement are required to specify the quantities of carbon dioxide in the soil. Its initial release from plant root or micro-organisms into the soil water film which surrounds plant roots can be expressed as $\mu l/g/h$. From a wide range of soils, the lowest rate of 1.7 $\mu l/g/h$ was observed from an almost pure beach sand of basaltic origin. In the same study rates of evolution from bare glacial tills were 3.6–5.5 $\mu l/g/h$, rising to 6.5 and 20 under moss and plant cover, respectively. The evolution of carbon dioxide from peat soils studied was highest, at 120–130 $\mu l/g/h$. In Arizona desert soils, a ten-fold increase in carbon dioxide evolution to a rate of 50 $\mu l/g/h$ may occur immediately after rainfall (Macgregor 1972).

Secondly, the actual concentration of carbon dioxide in the soil water film is conveniently expressed in mg/litre. Concentrations are particularly high in the soil water of temporarily and permanently wet forest soil types, owing to low diffusion. Thus, in the northern parts of the European U.S.S.R., accumulations of as much as 130 or even 150 mg/litre have been observed (Veretennikov 1968). Normally, with even the most vigorous biochemical activity, concentrations in soil water in freely draining soils rarely exceed 100 mg/litre.

Thirdly, carbon dioxide released in air exchange at the soil surface is expressed in terms of ground surface area and commonly referred to as soil 'respiration'. Maximum rates of carbon dioxide release at the soil surface of 23, 27, and 47 kg/ha/h have been recorded in meadow sod, perennial grassland, and birch meadow soils, respectively (Vugakov and Popova 1968). As an average figure for the month with highest respiration, a figure of 6.5 kg/ha/h is comparable with several other observations. During short period observations in mountain rain forest in Java and in lowland rain forest in Sarawak, release of carbon dioxide from the soil surface averaged 1.7 and 2.2 kg/ha/h (Wanner 1970).

Fourthly, there is the percentage of carbon dioxide in the soil air. Land use is a major control on the percentage of carbon dioxide in the soil air. Arable soils have percentages in the order of 0.1–0.5 and may at times be no higher than the atmospheric air concentration of 0.03 per cent. Grassland soils have a soil air with 0.3–3.3 per cent carbon dioxide contents, simply reflecting their higher organic matter content and some impedence to diffusion in an unploughed topsoil.

The influence of water on the character of soil carbon dioxide is expressed in several contrasted ways. Ample moisture supply is necessary for optimum respiration rates in the plant root zone. The most intense rates of carbon dioxide evolution from the ground surface of Red Earths, ranging from 21 to 53 kg/ha/day were observed with moisture contents at 80 per cent of field capacity. Conversely, the amounts of carbon dioxide dissolved in soil water may decrease after rain owing to dilution. A wetting front, however, will tend to push down air with an increasing content of trapped carbon dioxide. In such circumstances, the carbon dioxide concentration in soils beneath Douglas fir, to the south-east of Seattle, was observed as 1.3 per cent. With such movement to lower soil horizons, carbon dioxide may accumulate in large quantities, commonly more than 2 per cent of the total soil air. In cool environments, air exchange is impaired by the layer of sphagnum moss and entrapped dust. Carbon dioxide contents fluctuating between 1.3 and 6.4 per cent have been noted in such circumstances. Similarly, under flooding conditions, large amounts of carbon dioxide in the trapped soil air are to be expected.

Given adequate moisture supply, plant root zone respiration is highly correlated with temperature. Oxygen uptake increases ten-fold with a temperature rise from 5° C to 30° C and carbon dioxide respiration mirrors such increases. Diffusion, however, is proportional to the square of the absolute temperature so concentrations of carbon dioxide within the soil may not represent the full scale of increased respiration at higher temperatures. Since such activity is stimulated by photosynthesis in the leaves, solar radiation is possibly a more fundamental control than merely temperature alone. This may explain why especially large amounts of carbon dioxide release from the soil surface is observed in midsummer in high latitudes. For instance, rates of 2.6–3.8 kg/ha/h have been observed in soils in pine stands in the Kola peninsula during July and August (Repnevskaya 1967).

(c)   *Gaseous forms of nitrogen*

In addition to the 78.08 per cent of N_2 in the atmosphere, inorganic soil nitrogen is often oxidized or reduced to gaseous forms within the soil. The three gaseous forms of nitrogen are nitrous oxide (N_2O), nitric oxide (NO), and nitrogen dioxide (NO_2). The principal reduced form of

Physical properties of the soil 149

N include molecular NH_3 which is normally a gas. In freely drained soils the proteins, amino acids and ammonium salts in the soil organic matter, or added to the soil as fertilizers and manure, are oxidized slowly by air. Gaseous nitrogen is lost largely by the volatilization of ammonium, mostly as nitrous oxide, one of the most abundant constituents of soil air. Elemental nitrogen may also be lost. Gaseous losses can be high, sometimes exceeding 50 per cent, and commonly 30–40 per cent of fertilizer nitrogen escapes from the soil as gas. If anaerobic conditions develop, nitrate is reduced by micro-organisms utilizing the combined oxygen in nitrates to nitrous oxide and nitrogen gas, the nitrate acting as a hydrogen acceptor which can diffuse out of the soil. The reduction changes are

$$NO_3^- \rightarrow NO_2^- \rightarrow N_2O \rightarrow N_2$$

This biochemical reduction of nitrate nitrogen to gaseous compounds is termed denitrification, probably the most common type of volatilization. If nitrogen is readily available in a soil which becomes extremely wet and individual aggregates approach saturation, nitrous oxide is evolved in quantities sufficient to deplete completely the soil of available nitrogen within 10 days (Arnold 1954). For example, nitrous oxide concentrations in an Oxford clay loam under waterlogged conditions were 800–2500 ppm at 90 cm depth, whereas 10 ppm was common at other times (Dowdell and Smith 1974).

(d) Other gases

The concentration of gases such as methane and hydrogen sulphide, which are formed by organic decomposition in damp soils, is somewhat higher in soil air than in the atmosphere. As well as occurring as gaseous H_2S, there are significant combinations containing S in an oxidized state, notably sulphur dioxide (SO_2). Small quantities of rare gases dissolved in the soil water may also have to be considered. Finally, several organic pesticides exhibit fairly high vapour pressures.

5.4.3 RELATIVE PROPORTIONS OF SOIL AIR GASES

The composition of the soil gas phase tends continually towards the 78 per cent N_2, 21 per cent O_2 and the 1 per cent of other gases, with varying quantities of water vapour. The diffusion coefficients of CO_2 and O_2 are similar in both the gas and in the liquid phase, and the amount of CO_2 respired under aerobic conditions is roughly equivalent to the amount of O_2 used. Thus any decrease in O_2 concentrations may be accompanied by an equivalent increase in CO_2. Although the individual concentrations of O_2 and CO_2 in soil air change continuously, the combined concentration is often close to 21 per cent. The volumetric ratio of CO_2 evolved to O_2 consumed is termed the respiratory quotient (R.Q.) and is higher than 1.0 under aerobic conditions. However, because the solubilities of O_2 and CO_2 in water are 0.039 and 1.45 g/l at 25°C and 760 mm pressure, respectively, the concentration gradient for O_2 and CO_2 will reflect this 37-fold difference at the air-water surface. Only a small fraction of oxygen is present in the soil water, whereas the bulk of CO_2 is contained in the liquid phase. Because diffusion is much faster through gas than through aqueous phases, oxygen gradients will be less in the gas than in the aqueous phases whilst CO_2 remains largely dissolved in water and may move downwards into the subsoil with percolating water. Concentrations can reach 95 per cent of their equilibrium value within 3 minutes for oxygen, compared with one day for carbon dioxide (Radford and Greenwood 1970).

When CO_2 dissolves and is removed in percolating water, the percentage of O_2 and N_2 in the soil atmosphere increases. The per cent O_2 will still be less than the atmosphere percentage, but the per cent N_2 will be greater. Low O_2 and low $(CO_2 + O_2)$ are inversely related to N_2 percentage. N_2 is not constant in soil air but may vary by as much as several per cent from the 78.08 per cent of the atmosphere. A large increase in the ratio of N_2 to N_2O occurs as the air-filled pore space, and hence the ease of N_2O escape decreases. However, nitrous oxide may be found in the presence of oxygen, even at concentrations similar to those of the atmosphere. Such conditions are attributable to centres of larger soil aggregates being anaerobic and therefore favouring denitrification.

6 Chemical properties of the soil

6.1 Oxidation-reduction processes

Soil formation is determined in many ways by the character and intensity of oxidation-reduction processes. Oxidation processes occur at the expense of either free oxygen, or soil constituents able to release oxygen or to take in hydrogen. Micro-organisms and active plant roots are usually involved in these processes. As long as oxygen persists, other oxidized soil components are not susceptible to biological and chemical reduction. If oxygen is exhausted, which can occur within a day following waterlogging, the requirement for electron acceptors by anaerobic organisms leads to the reduction of several oxidized components. The oxidation-reduction potential, or electron availability, influences the oxidation states of hydrogen, carbon, nitrogen, oxygen, iron, and sulphur, and of trace metals like manganese, cobalt and copper. In consequence, the varying degrees of free exchange of oxygen between the soil and the atmosphere is fundamental in creating distinctive soil properties and profile characteristics. Soil fertility is also affected, since the state of some nutrients and their movement depends on oxidation-reduction, or redox, potential (Bohn 1971) under waterlogged conditions (fig. 6.1a). In paddy soils, the availability of phosphate is affected and toxic effects of sulphide may be experienced. Other toxic compounds appear under extreme reducing conditions and induce diseases like the 'Akiochi' of Japanese paddy soils (Aomine 1961). Striking changes occur in most irrigated soils as well as in rice soils in particular (Yamane and Sato 1968). The redox potential is measured by placing a bright platinum electrode in the soil, connected to another electrode of known potential. Measurements are balanced against the known EMF on a potentiometer and are

152 Geography and soil properties

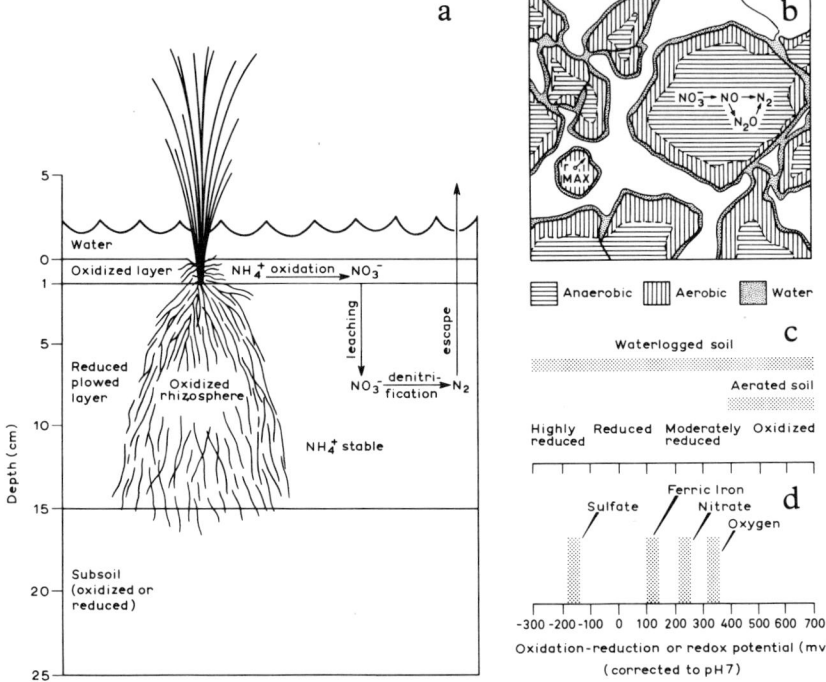

Figure 6.1 Biochemical significance of soil aeration.
a Profile changes in oxidation-reduction conditions of a flooded soil (Sanchez, 1972). Nitrogen dynamics adjacent to rice roots are shown. A thin superficial horizon remains oxidized, being in equilibrium with the oxygen dissolved in the supernatant water. The rhizosphere is also oxidized because active roots exude oxidized compounds and yellowish-red ferric compounds are precipitated on root surfaces.
b Substantial micro-heterogeneity as the fundamental characteristic of soil aeration (Flühler *et al.*, 1976). Compared with the maximum radius at which a soil crumb might remain fully oxidized, the centre of larger crumbs may be anaerobic, and denitrification may occur.
c and d. Range of oxidation-reduction (redox) potentials usually encountered in waterlogged soils (c), and critical or threshold redox potentials at which oxidized components may become unstable (d) (Patrick and Mahapatra, 1968).

regarded as oxidation-reduction potentials because their magnitude varies in accordance with the degree of waterlogging. The electrical work done in oxidation or in the reverse process of reduction is expressed in Eh values, which are proportional to the log of the ratio of reduced to oxidized products (Pearsall 1938) the result of the opposing rates of electron donation and acceptance. Electron donation results mainly from decomposition of organic matter and electron acceptance is primarily the reduction of oxygen. Where oxygen diffusion is restricted, iron, manganese, nitrogen and sulphur compounds, together with water, are the electron acceptors.

The relation between ferrous and ferric iron exemplifies redox phenomena. The oxidation of ferrous to ferric iron involves the loss of one electron

$$Fe^{3+} \rightarrow Fe^{++} + 1 \text{ electron}$$

Consequently, electrical potentials differ between soil solutions containing ferrous and ferric ions. As oxidation is not just literally the addition of oxygen but may simply involve the loss of an electron, the hydrogen ion itself is regarded as an oxidation product where

$$H_2 = 2H^+ + 2 \text{ electrons}$$

The condition of equilibrium

$$H_2 \rightleftharpoons 2H^+ + 2 \text{ electrons}$$

is arbitrarily fixed at a zero mV potential. Iron may contribute to the Eh value over a wide range, with manganese and oxygen as other important contributors. The redox value for a soil may be regarded as the algebraic sum of the Eh values of such individual redox couples. Surface horizons which are rich in fresh organic residues favour vigorous microbial development which tend continually to depress Eh values. In waterlogged sandy soils, frequently low in organic matter, ferric oxide may act as the main oxidizing agent. If Fe_2O_3 is also deficient, soil behaviour is strikingly affected and low fertility results (Jeffrey 1961). In many instances the unique air-water properties of sandy soils maintain a high oxidizing state even under wet conditions. Inevitably, Eh is greatly affected by changes in moisture and temperature. As there is a 1 mV decrease for each degree centigrade rise, most negative redox potentials are observed in warm, water-saturated soils.

Whilst Eh values do not represent the aeration status of well-drained soils adequately, conversely, O.D.R. measurements are difficult at redox potentials of less than 200 mV (Armstrong 1967). Five ranges of redox conditions have been suggested (fig. 6.1d). Aerated soils have redox potentials over 400 mV, moderately reduced conditions lie between 100–400 mV, reduced soils are − 100 to 100 mV, and highly reduced lie between − 100 and − 300 mV. These ranges compare with redox potentials representative of Russian soils, where podzols are 600–750 mV and Chernozems have Eh values of 450–600 mV. If there is turf on the podzol or grass cover on the Chernozem, redox potentials lower to 100–200 mV. (Kaurichev and Shishova 1967). On a wide range of soil types in Northern England Eh values normally lie between 0 and 600 mV, with waterlogged soils normally less than 250 mV and dry-site soils exceeding 380 mV (Pearsall 1938). Nitrate and manganese are reduced at fairly high redox potentials. At Eh values of 400–300 mV nitrates become unstable and are denitrified. When all nitrates have been consumed and at about $+200$ mV, anaerobes reduce Mn^{4+} to Mn^{++} compounds. The boundary between oxidizing and reducing conditions is perhaps most significant about $+120$ mV because iron compounds are usually more abundant than the other electron acceptors and they are reduced at about this Eh value. Sulphate ions are reduced to SO_3^{--} and S^{--} ions at Eh values of about -150 mV and at approximately -180 mV, several organic acids are reduced to alcohols. Sulphides may reach toxic concentrations at potentials of less than -300 mV. Microscale contrasts are a key feature of redox phenomena (fig. 6.1b). The Eh at the centre of a 6–7 mm diameter wet soil crumb may be 100–200 mV lower than at its surface (Kaurichev and Tararina 1972).

6.2 Ion Exchange

6.2.1 CATION EXCHANGE CAPACITY

The best single index of potential soil fertility is its capacity to exchange cations (fig. 6.2). Exchangeable cations can be artificially changed to improve soils for crop production and for engineering purposes, and cation exchange capacity is an index of the soil's ability to adsorb biocides, such as herbicides of the triazine class and radioisotopes from thermonuclear fallout.

The typical, reversible cation-exchange reaction may be written

$$A^+ + B^+\text{micelle}^- \rightleftharpoons B^+ + A^+\text{micelle}^-$$

and

$$C^{2+} + 2B^+\text{micelle}^- \rightleftharpoons 2B^+ + C^{2+}(\text{micelle}^-)_2$$

where A^+ and B^+ are univalent cations, C^{2+} is a divalent cation, and a micelle is a negatively charged colloidal soil particle, usually of clay or humus, or more commonly both. The unit of measurement is the milliequivalent, the amount of matter which will replace or combine with 1 mg of hydrogen, per 100 g of dry soil. Cation exchange capacity is thus expressed by the abbreviations me/100 g CEC.

Virtually all exchangeable cations are adsorbed on the negatively charged surfaces of soil micelles. In the case of clays, where the 2:1 aluminosilicate and silicate platelets are held together predominantly by covalent bonds, there are still excess charges at well-defined sites which are offset by cations (fig. 6.2d). The negative charges on the planar surfaces of expanding lattice clays are the result of isomorphic substitutions, such

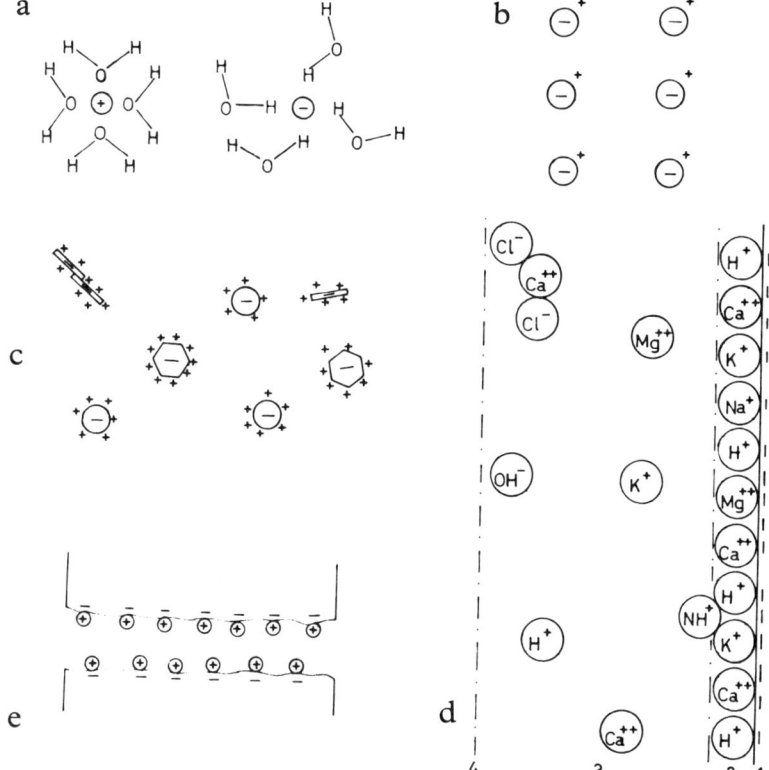

Figure 6.2 Models of ion behaviour at different types of charged surface.
a Orientation of water molecules in the primary hydration shells of a cation and an anion, to which they are weakly attracted.
b Attraction between positively and negatively charged ions in solution. Such electrolytes are commonly used in laboratory experiments due to the impossibility of isolating the adsorbed water layer, shown as zone (2) in diagram (d).
c Suspension—as occurs after dispersion of colloids in saline soils or in streams draining agricultural lands, with cations including lost nutrients and potentially toxic heavy metals adsorbed on to the negatively charged clay and humus particles.
d and e. Distribution of ions in the presence of a negatively charged clay particle (1). Water molecules in the adsorbed water layer (2) tend to be orientated. In the diffuse ion layer (3) of hydrated cations and anions, the anions are repelled from the clay surface, including certain nutrients like the NO_3^- and SO_4^{--} anions. Water molecules are loosely bound but still orientated, unlike free water. Since the negative charge of the particle and the positive charge of the counter ions are spatially separated, this zone is regarded as an electric 'double layer' in theoretical studies.

At a distance some molecules' width away from the clay surface (4), there may be an air-water interface between the soil water film and the soil air, or the surface of an inert particle such as a quartz grain, or a zone in the soil solution where cations begin to be attracted to an adjacent clay surface, as in fine pores or inter-lamellar spaces (e).

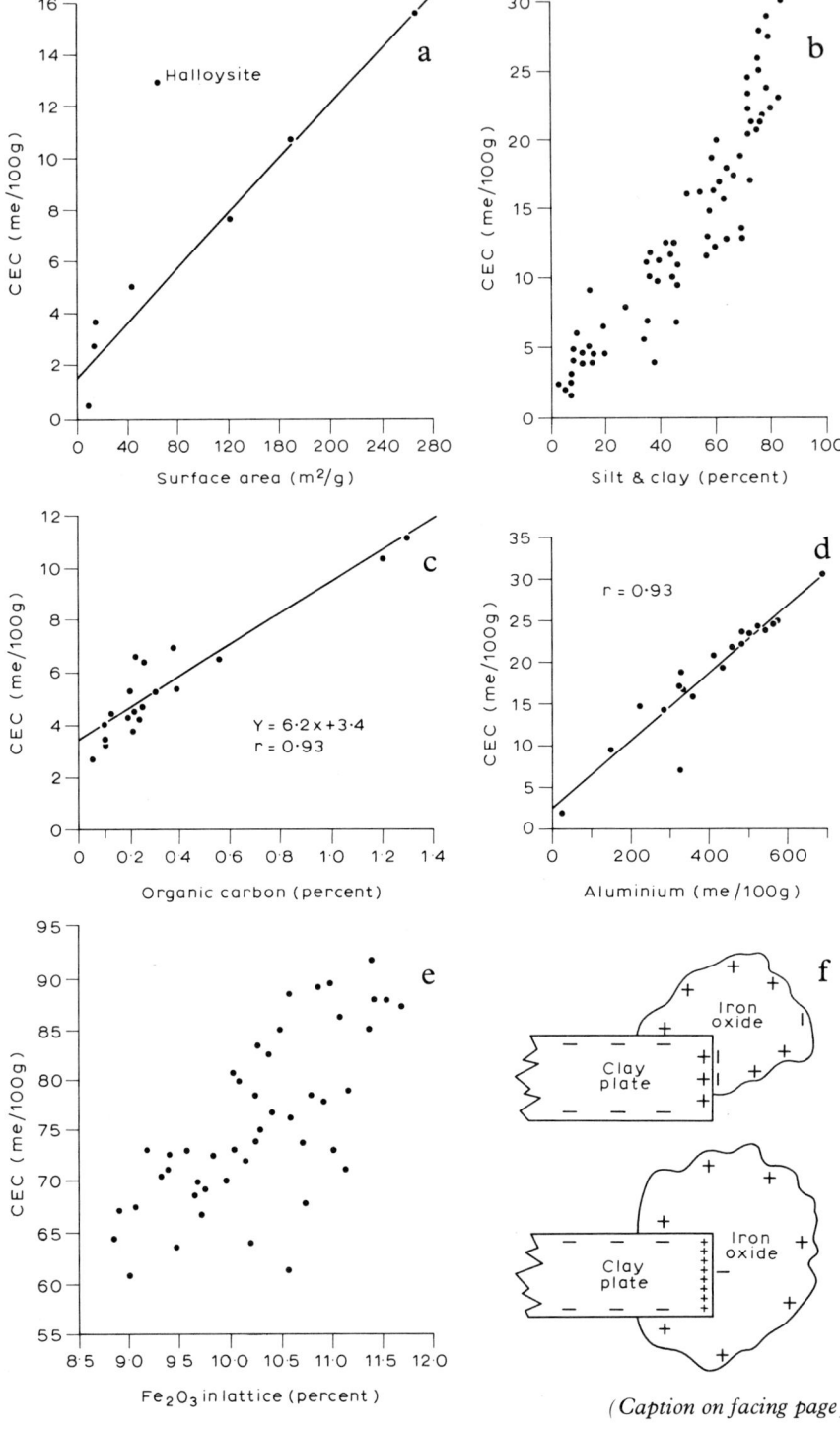

(Caption on facing page)

Chemical properties of the soil 157

as that of Si^{4+} by Al^{3+} in the silicate sheet. The cations which satisfy such charges may be replaced by an equivalent number of other cations from an external pool without destroying the structure of the interlamellar spaces. In addition to planar surface charges, 'broken-edge' sites also occur. Most broken-edge sites are negative in charge, but there are also some positively charged sites which will attract anions. Particle size is a predominant factor (fig. 6.3 a,b). Increasingly fine clay fractions tend to have a higher CEC than their coarser fraction because of the much increased surface area.

Of the specific clay minerals, exchange sites in the case of the comparatively large kaolinite particles are restricted to broken bonds on the particle edges. In consequence, CEC is uniformly low, at 3–15 me/100 g. In the case of the kaolinite in the A_p horizon of Maryland soils, which averaged 41 me/100 g as a whole, the CEC was 8 me/100 g. In the humid tropics of southern Thailand, the preponderance of kaolinite accounted for CEC being under 6 me/100 g in all soils. In the case of illite, where the origin of the negative sites is due to isomorphous substitution as well as to broken bonds on particle edges, CEC is commonly in the 15–40 me/100 g range, owing to the limit placed by the non-expanding lattice. Illites in Maryland soils have a CEC of about 30 me/100 g. In the English Midlands, the Blisworth illite has a CEC of 37 me/100 g of oven-dry clay. The expanding lattice clays have higher values, montmorillonite in the order of 70–100 me/100 g, and vermiculite usually well over 100 me/100 g. In the case of the amorphous allophane, broken bond sites contribute greatly to equally high CEC values.

Mineral particles larger than clay are usually negligible contributors to the CEC of a soil. In fact, the proportion of sand, with a CEC of 1–5

Figure 6.3 Factors influencing cation exchange capacity in soils.
a Linear relationship between surface area of minerals in tropical and subtropical soils (Sawhney and Norrish, 1971), an effect attributable to the surface hydroxyl groups.
b Influence of silt + clay percentage on CEC (Western, 1972), observed in arid-zone soils in southern Arabia.
c Influence of organic matter on CEC (Perkins et al., 1971), observed in the Southern Piedmont of the U.S.A.
d Influence of aluminium interlayers on CEC of clays (Mutwewingabo et al., 1975), observed in glacial till soils near Quebec.
e Influence of iron interlayers on CEC of fine-textured soils (Kornblyum, 1967), observed in soils of the Volga-Akhtuba floodplain where 45–55 per cent of the soils are clays finer than 1 μm.
f Models to explain some charge characteristics of lateritic soils (Davidtz and Sumner, 1965). From observations of Natal soils, in the case (f), the charges at the clay-iron oxide interface are neutralized, and in the case (g) there is limited neutralization but with the physical coating of a large number of charges.

me/100 g, tends to 'dilute' the CEC in a soil derived from the colloidal fraction. However, feldspars do possess charged tetrahedra in which Al^{3+} plus a cation substitutes for Si^{4+}, and silt and even fine sands derived from basic rocks may contribute locally to the CEC, as in basaltic soil profiles in Northern Ireland (McAleese and McConaghy 1957). In soils developed on tills derived from volcanic rocks in Central Scotland, the mean contribution to the total CEC is 34 per cent for clay, 22 per cent for silt, and 35 per cent for fine sand and 9 per cent for coarse sand (Wilson and Logan 1976). Primary micaceous particles may be important seats of CEC, since they are only slightly different from hydromica clays. Thus, in the case of soils in Panama, silts account for 5–30 per cent of the total CEC of the soil.

A significant part of the sorptive capacity of many soils for cations and for proton-accepting molecules resides in the humus fraction of the soil. This is due to the presence of weakly ionizable acidic functional groups. This acidic character of the humic acids has been found to be due to ionization of H^+ from carboxyl groups and from the phenolic hydroxyl groups. In the case of the soil polysaccharides, there are three main classes. These are the neutral sugars which largely contain alcoholic –OH functional groups, the amino sugars, which also have the $-NH_2$ amino groups, and the acid sugars with the carboxylic acid –COOH group. A monolayer coverage of cations will not be found at the surface of the non-rigid humus particles in the same conditions as on some clay platelet surfaces (Burford *et al.* 1964). The CEC of soil organic matter varies from about 150–400 me/100 g dry organic matter, is due mainly to its humic acid content, and increases with degree of humification.

Frequently the organic content variation, owing to its much higher CEC than clays, accounts for most of the variability in the CEC of soils as a whole (fig. 6.3c). Thus, on loam-textured, calcareous till soils in Ohio, the correlation between organic matter content and CEC was $r = 0.74$, $N = 28$ for the A horizon of the Miami soil series (Wilding and Rutledge 1966). In New South Wales, in the Inverell soil practically the whole of the CEC appears to be associated with the organic matter. In the eastern United States, where soils on the Coastal Plain, Piedmont and Appalachian zones have organic fractions with 200, 316, and 201 me/100 g CEC, each additional gram of organic matter in the soil contributes an extra 2.95 me/100 g to the CEC. In West Africa CEC is over 200 me/100 g and largely dependent on humus content. In cultivated fields under well irrigation in Rajasthan, the CEC of humic acids varied between 252 and 570 me/100 g of humic acid (Singh and Bhandari 1965).

In several situations, the correlation between organic matter content and CEC is sufficiently close for CEC to be predicted simply on the basis or organic matter content. In the eastern United States,

$$y = 4.83 + 3.87x$$

where y is the CEC and x is the percentage organic matter, $r = 0.73$, $N = 57$. Within this area the corresponding calculation for Coastal Plain localities only gave

$$y = 1.91 + 4.63x,$$

with $r = 0.76$, $N = 27$. In the Piedmont area too, more than 50 per cent of the variability in CEC could be attributed to differences in organic matter content (Wright and Foss 1972). Similar relationships are particularly clear in higher latitudes. In Alaska, where CEC in A_2 horizons of soils was 15–26 me/100 g of soil

$$y = 4.90 + 3.45x,$$

where y is the CEC and x the percentage of organic carbon in the soil sample. In this case, $r = 0.92$, $N = 44$ (Stephens 1969).

Thus, CEC can be accounted for largely in terms of either clay or organic content of the soil. Their relative importance varies in a given region, even though an essentially organoclay complex will be involved in many mid-latitude soils.

In general, about 25–90 per cent of the CEC of the plough layer of soils is the result of the presence of organic matter. Local conditions, particularly of soil drainage, may be as significant as any broader latitudinal differences. In the case of Maryland soils, organic matter is 1.25 times as important as clay content in predicting CEC, and for the Miami series soil in Ohio, clay can improve the degree to which CEC is accountable in terms of other soil properties by

$$y = 7.2 + 2.2a + b,$$

where y is the CEC, a the per cent organic matter and b the per cent fine clay (Wilding and Rutledge 1966). From tropical environments, Panama surface soils may have 10–85 per cent of their CEC linked with organic matter but equally 10–83 per cent may be associated with their clay contents (Martini 1970). Similarly, in laterite-like Oxisols in Georgia, with mean values of 1.28 per cent organic carbon and CEC of 11.2 me/100 g, there was a significant correlation with carbon but not with clay, $r = 0.93$, $N = 18$ (fig. 6.3c). Israel soils are typical of arid zones where the organic matter in the soil is usually low, with most of the variability, up to 75 per cent of the CEC, being accounted for by variations in clay content alone. This overall pattern is established despite the relatively high CEC of the organic matter itself at 300 me/100 g.

In general, it seems that in cool, damp environments like Alaska or Iceland, the small amount of clay produced by slow weathering processes, and the presence of sesquioxide interlayers (fig. 6.3d,e), means that the amounts and degree of humification of the organic matter are the overriding control. In the humid tropics, by contrast, clays are so weathered that the insignificant CEC of kaolinite and the sesquioxide coatings (fig. 6.3f,g) leaves CEC heavily dependent on whatever fraction of organic matter persists in such conditions. In arid areas, where not only is there little organic matter, owing to low biological productivity and high oxidation rates, but there is the likelihood of large but variable amounts of relatively unweathered clays contributing to the CEC. In mid-latitudes where Chernozem-type soils have a uniformly high content of humified organic matter, possibly the differentiating factor becomes the mineral fraction, depending on the degree to which the soil is sandy or clayey.

Local conditions include the degree to which the topsoil has been eroded away. As B horizons of soils differ profoundly in their CEC characteristics, surface patterns will change considerably if these are exposed at the surface. In the B horizon, CEC usually declines with depth in the profile due to decreasing humus content, and the CEC measured is often correlated with clay content. Thus, in the Miami series soil in Ohio, were the covariance of total clay and CEC is, in fact, $r = -0.22$. the relationship is converted to $r = 0.82$ in the B horizon, with the correlation being $r = 0.85$ for fine clay (less than $0.2\,\mu m$) and only $r = 0.41$ for coarse clay ($0.2-2\,\mu m$).

The percentage of cation exchange sites occupied by bases in non-saline soils is referred to as the per cent base saturation, where base saturation = (me/100g of basic cations/total CEC) × 100. Exchangeable hydrogen, negligible in many agricultural soils, dominates the CEC of acid, moorland or pine woodland soils. In podzols, exchangeable hydrogen may account for over 80 per cent of the base saturation. In agricultural soils, calcium commonly accounts for 80–85 per cent of the base saturation. If calcium is less than this percentage of the exchangeable bases, liming quickly improves the soil condition. Calcium is naturally predominant in per cent base exchange because it is adsorbed more strongly than any of the other basic cations. Equally significant is that calcium is released more rapidly by weathering than any other cation in non-saline soils because its parent mineral or calcareous rock weathers readily. The per cent base saturation of other cations is in fairly consistent proportions, with magnesium accounting for 12–18 per cent, potassium for 1–5, and sodium for about 1 per cent.

As expressed in per cent base saturation, it can be seen that cation exchange capacity is the principal buffer mechanism in the soil. A buffer is a solution which minimizes changes which would otherwise occur towards either increasing acidity or alkalinity in the soil, and the presence of calcium and magnesium too is important in maintaining near-neutral conditions in the soil. In the tropics, however, minimal calcium in the soil solution means that soils in the humid tropics are poorly buffered against changes. The practical significance relates to fertilizer applications which, particularly in the poorly buffered soils, must be carefully balanced in both major and minor nutrients if critical excesses or deficiencies are to be avoided.

6.2.2 ANION EXCHANGE

If the sesquioxides are present in large amounts in acid soils, a small but net electro-positive charge may characterize the oxide coatings. The positive charge arises from the addition of hydrogen ions (protonation) to hydroxyl groups on the surfaces of hydrous oxides, and anion exchange capacity (AEC) may equal or exceed CEC. Even poorly crystallized kaolinitic clays may have many exposed OH ions available for anion exchange. The positive charge attracts anions such as the important nutrients $H_2PO_4^-$, SO_4^{--}, and NO_3^-, and also Cl^-. These anions can exchange with each other at the sites of the positive charges, just as cations are exchanged by negative charges. Such anions may interact with the oxide surface 'specifically' or 'non-specifically'. In non-specific adsorption they are held in the diffuse layer opposite the positive charge on the oxide surface. Alternatively, anions such as SO_4^{--} or PO_4^{---} can be 'specifically' adsorbed, with the anions actually displacing an OH^- group from the oxide surface and becoming co-ordinated with the oxide metal ion. This chemisorption provides a very strong bond and such anions are then difficult to extract. In either case, soils with a net positive charge are unable to retain cations, which accounts for poor responses to liming in tropical soils. Humid tropical soils are dominated by sesquioxides which, in varying stages of hydration, may account for a quarter of the total soil. In mid-latitude soils, anion exchange is usually masked by the much larger CEC which characterizes clay-humus crumb structure. Some British acid soils, however, show increases in surface area in the B horizon which are not attributable to either clay or to humus. Since the increases coincide with the depth at which extractable aluminium increases substantially, amorphous aluminium oxides may also be significant in the exchange properties of these soils (Pritchard 1971). In higher latitudes, colloidal iron compounds, together with organic compounds, may be sufficiently active to mask the exchange effects of the limited amounts of weathered clays in podzols.

For the tropical case, it is now clear that the unique electrochemical properties of weathered soils suggest that management techniques different from those used in temperate regions will have to be employed (Gillman and Bell 1976). In particular organic matter must be conserved or even increased, because its contribution to the negative charge is vital to the retention of nutrient cations. In many highly weathered tropical soils, maintenance of organic matter is almost synonymous with maintenance of CEC.

6.3 Soil acidity

6.3.1 SIGNIFICANCE OF SOIL ACIDITY

Ion exchange and other adsorptive reactions largely control soil acidity and soils are predominantly acidic at smaller and larger scales on the earth's surface. For example, of cultivated lands in Estonia, about 40 per cent is acid in reaction and, at higher latitudes, about 80 per cent of Finnish soils are equally acid. Acidity occurs wherever percolating water transports the more soluble bases from the soil profile. Acid soils limit fertility both by their depleted nutrient status and by the presence of toxic substances, with the toxicity of aluminium and manganese and deficiencies of calcium, magnesium, and molybdenum being the most commonly encountered. Many micro-elements may reach toxic levels as, with the exception of Mo, they are all more soluble in acid soils. Soil microbiological activity drops off markedly with increasingly acid soil conditions which are also one of the most important controls on enzymatic activity in the soil. Soil acidity also influences engineering decisions connected with the emplacement of metals and concrete in the soil.

6.3.2 HYDROGEN ION CONCENTRATION

H^+ is a major source of soil acidity, influencing the presence or precipitation of many mobile elements in the soil solution. It is also an important weathering agent, disrupting crystal structures. A characteristic expression of acid soil processes is the replacement of basic cations at exchange sites by cation exchange with H^+.

Although water is primarily a molecular substance, slight ionization tends to occur even in pure water. Two water molecules interact to create acidic and weak base tendencies

$$H_2O + H_2O \rightleftharpoons H_3O^+ + OH^-$$

The concentrations of the hydronium ion (H_3O^+) and the hydroxyl ion (OH^-) are equal in pure water. Increasing levels of acidity are reflected in greater hydrogen ion concentrations, designated simply as H^+, the proton combined with the water molecule. The H^+ concentration increases exponentially, measured conveniently by the logarithm of the reciprocal of active H^+ ions. This unit of measure is called the pH scale,

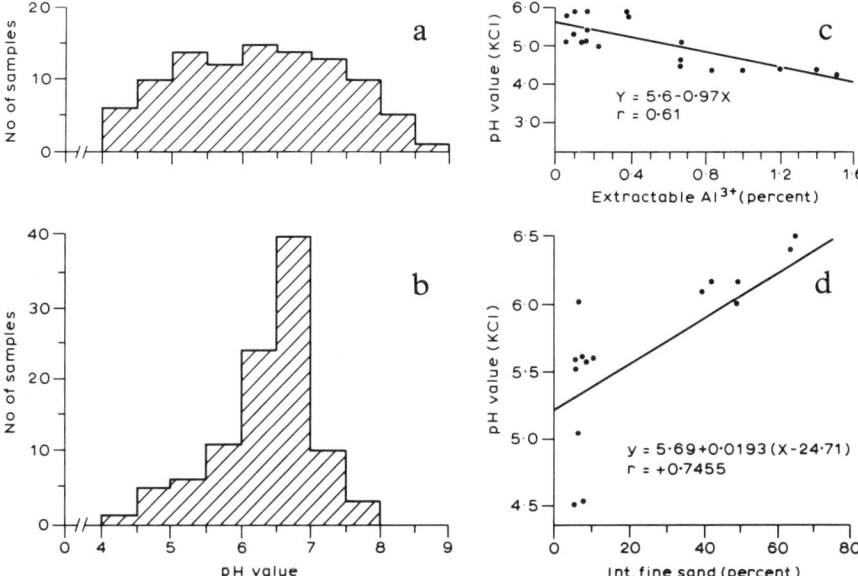

Figure 6.4 Some influences on soil pH.
a and b. Frequency distribution of pH means in August (a) and May (b) (Groenewoud, 1961), observed in 1958–59 in forest podzolic soils in the Candle Lake area of Saskatchewan.
c Relationship between fine-sand content and pH in a Nigerian soil (Unamba-Oparah, 1972). The fine sand fraction is probably the source of unweathered minerals in such tropical soils.
d Influence of extractable aluminium on the pH of a highly weathered soil from the Southern Piedmont (Perkins et al., 1971).

$$\mathrm{pH} = \log\frac{1}{(H^+)}$$

A change of 1.0 in pH corresponds to a 10-fold change in the H^+ concentration. Compared with neutrality, where pH equals 7, there are 10 times as many H^+ ions at pH 6 and 100 times as many at pH 5. Conversely, there are 10 times the number of OH^- ions at pH 8 compared with neutrality, with a 1000-fold increase by pH 9. In the field, pH varies substantially over short intervals of space and time (fig. 6.4a,b).

One of the main sources of H^+ is that supplied from CO_2 dissociating in water to the hydrogen ion and to the bicarbonate ion (HCO_3^-)

$$CO_2 + H_2O \rightarrow HCO_3^- + H^+$$

Life processes and the decomposition of organic matter also add hydrogen ions. Most organic compounds are acidic in nature. The hydrogen covalently bonded to carboxyl and phenol groups dissociates, leaving negatively charged sites. The hydrogen ions produced by organic matter decomposition are, however, unstable in mineral soils because they react with layer silicate clays, releasing exchangeable aluminium and siliceous acid (Coleman and Thomas 1967). Acidity in the soil is also related to the presence of carbonic, nitric, and sulphuric acids and depends on their solubility in water and the degree and readiness with which they release H^+ by ionization. In acid soils an equally significant characteristic is the capacity of clay minerals for adsorbing cations. This contributes indirectly to acidity by adsorbing basic elements in a non-exchangeable form. At the negatively charged exchange sites, the overall acidity or alkalinity of a soil is defined by the relative proportions of positively charged cations on the one hand or by H^+ or Al^{3+} on the other.

pH is clearly not an independent variable, but rather a function of several inter-related factors. H^+ concentration is determined by soil respiration, organic matter decomposition, the CEC of clays and the type of cations present, as influenced by parent material decomposition (fig. 6.4c) and rate of loss by leaching. Finally, per cent aluminium saturation of the effective CEC is noted as an integral factor in measures of soil acidity.

6.3.3 ALUMINIUM AS A SOURCE OF SOIL ACIDITY

Adsorbed Al^{3+} increases the acidity of the soil solution (Jackson 1963), providing a major supply of H^+ ions in acid soils because of its tendency to hydrolyse

$$Al^{3+} + H_2O \rightarrow Al(OH)^{++} + H^+$$

In very acid soils Al^{3+} is in equilibrium with the soil solution, but as pH increases, it is converted to aluminium hydroxy ions

$$Al^{3+} + OH^- \rightarrow Al(OH)^{++}$$

which can produce further hydrogen ions by hydrolysis

$$Al(OH)^{++} + H_2O \rightarrow Al(OH)_2^+ + H^+$$

Thus, aluminium contributes to soil acidity not merely by a large supply of H^+ ions at very low pH values but also by reducing the OH^- concentration in the soil solution of less acid soils which may continue until gibbsite, a non-ionized molecule, is precipitated (fig. 6.4d)

$$Al(OH)_2^+ + OH^- \rightarrow Al(OH)_3^\circ$$

6.3.4 INFLUENCE OF pH ON ION EXCHANGE

In ion exchange there is a second, variable charge in addition to the 'permanent' negative charge on silicate clays. This is termed the 'pH-dependent' charge and is closely related to the varying average charge on soluble Al species. At pH 5 and above, the Al^{3+} ion combines with the hydroxyl OH^+ ion to form first $Al(OH)^{++}$, then $Al(OH)_2^+$ and finally $Al(OH)_3$. These complexes exist as polymers of indefinite size and in differing degrees of hydroxylation. The electrostatic charge has a pH component because hydroxylation involves the progressive release of H^+ ions. In addition, the sesquioxides block negative charge sites in very acid soils, especially the inter-layer spaces of vermiculite, which may be held rigid. As pH rises, these coatings are removed as insoluble $Al(OH)_3$ and $Fe(OH)_3$, thus releasing exchange sites.

Aluminium hydroxide is amphoteric, in that it reacts with both acids and bases. The change from a cationic species, such as $Al(OH)_2^+$, to an anionic species like $Al(OH)_4$ on going from an acid to an alkaline environment is a distinctive property of soluble Al. At high pH values the hydrogen ion tends to dissociate from the oxygen in the hydroxyls, leaving negative charges on the surface of the hydroxy coatings. The variable charge minerals are the intergrades, those 2:1 clays with admixtures of iron and aluminium hydroxides, several species of aluminium and iron oxides and hydroxides, and allophane. In oxide systems the charge is entirely pH-dependent, depending on whether the additions at the oxide surface are preponderantly OH^- or H^+ ions, a fact largely controlled by local environmental conditions. 'Specific' adsorption is also strongly pH-dependent.

The pH-dependent properties of organic matter (Pratt 1961) are due to the reactions which occur at the edges of organic matter-aluminium complexes if the soil pH increases. A net negative charge on the carboxyl radicals is left as the complexed aluminium ions are precipitated.

6.3.5 pH AND REDOX POTENTIAL

The hydrogen ion is also an oxidation product, since oxidation simply involves the loss of an electron. The pH and Eh concepts are not independent, and estimates of the oxidation-reduction potential of wet soils are usually corrected and expressed at a constant pH level. Eh and pH are not invariably linked since independent variations in pH may cause side reactions which change the relative amounts of oxidized and reduced forms of organic and mineral compounds. Notably, while the pH increases after flooding of paddy soils by 1.5 pH, it decreases after the soil has been drained with drops of nearly 2 pH units. These changes are a function of the Fe^{++} ion concentration. The pH of acid soils increases because of the release of the OH^- ions when $Fe(OH)_3$ and similar compounds are reduced to $Fe(OH)_2$ (Ponnamperuma 1972). They displace cations like ammonium from exchange sites. Similarly, peaty soils are more acid when well-aerated. The pH values of alkaline soils decrease because the increase in partial pressure of CO_2 causes a net release of H^+ ions. After 15 weeks of submergence, pH values have been noted to correlate with CO_2 partial pressure ($r = -0.65$, $N = 56$, Ponnamperuma et al. 1966).

6.4 Interaction of organic matter with metals, oxides, and clays

The formation of stable combinations between organic matter and metal cations strongly influences the availability of micro-nutrients to micro-organisms and to higher plants. Complexing of metal ions by humic substances is significant in leaching and podzolization processes, since organomineral derivatives of humic acid play a major role in the migration and accumulation of substances in the soil profile. The stability of soil aggregates is also affected. Humic substances complex trace elements, inactivate pesticides and toxic heavy metals and store ammonia and nitrite.

Chelation is regarded as an equilibrium reaction between a metallic ion and an organic molecule during which more than one bond links the compounds. The word chelate (Greek, *chele* meaning 'claw') refers to the pincer-like action of the two bonds on the metal ion. These may be an ionic and a covalent bond or an ionic and co-ordinate bond between the metal and the organic molecule. The metal is termed the complexing agent and the chelating organic molecules are called ligands. Chelation produces a cyclic structure, closed by co-ordinate bonds of the end atoms. Complex formation may also involve ion exchange, surface adsorption, coagulation and peptization processes.

Metal cations tend to be strongly bonded to carboxylic acids, which are strong, negatively charged ligands. The lowest molecular weight fractions, the fulvic acids, appear to be the most efficient component in complexing metals since they contain more functional groups with the highest total acidity. Copper is fixed in the soil in a form of low availability to plants, since it combines with humic acid as a organometal complex. Zinc is also present in the soil solution largely as a organometal complex (Hodgson et al. 1966), especially in calcareous soils. Chelation of inorganic phosphate cations by organic acids in the soil is probably important.

Iron and aluminium ions are bound to humic and fulvic acids, as many carboxylic and phenolic groups are strong ligands for these ions. Iron and aluminium displace hydrogen from some of these functional groups and, in combining with the anion part of the organic molecules, are no longer manifest as cations. In natural solutions in acidic soils, water-soluble aluminium and iron are more bound to the fulvic rather than to the humic type of water soluble organic matter as the fulvic acids contain more functional groups. Aluminium ions are much less strongly bound than iron ions. In acid soil profiles, however, the amount of aluminium bound to organic fractions increases with depth in the profile, reflecting the degree to which aluminium migrates as a organo-metal complex. Organo-iron and organo-aluminium complexes are characteristic of freely drained acid soils, particularly podzols. They are not a feature of soils like Chernozems, with a high content of exchangeable calcium and magnesium.

A major characteristic of soil properties is perhaps the existence of 'bridges' which link the two electronegative substances of clay and humus. Hydrogen bonding with adsorbed water molecules is one of the adsorption forces linking organic anions and clay surfaces. Such bonds have been termed 'water bridges' (Mortland 1970). This hydrogen bonding is possibly due to the inability of the single valency electron of the hydrogen atom to 'screen' completely the small postively charged hydrogen nucleus from the attraction of the electronegative substances of clay and humus. Many studies have shown that larger amounts of humic acids are adsorbed by clays when iron and aluminium are present on the exchange sites and on the surfaces of clays (Baker 1973). Thus, where hydrogen bonding has no positive effect, iron and aluminium 'bridges' may be important in the formation of organo-clay complexes. Calcium and ammonium also can act as 'bridges', but it is hydrated iron oxide which is considered to be the major factor in the binding action of humus on to clay particles and is important in the formation of stable structure. In acid soils organic colloids are also tied up with clay through aluminium oxides which, if present in large amounts, also favour stable granulation. Where exchangeable calcium is present in sufficient quantities to act as a bridge at lattice edges, the organo-clay complexes formed are more stable and less easily dispersed.

In addition to anion exchange reactions in the linking of clay and humus particles, a 'specific adsorption' or 'ligand exchange' of anions by iron and aluminium oxides and by kaolinite has been demonstrated (Hingston et al. 1967). Unlike anion exchange reactions, this type of adsorption involves more than displacement. The anion is incorporated within the surface hydroxyl layer after penetrating the co-ordination shell of an iron or aluminium atom in the surface of the hydroxide.

6.5 Iron, aluminium, and silica

6.5.1 MOBILIZATION OF IRON

If soil water fills most or all the soil pores, micro-organisms requiring oxygen may be forced to obtain it from iron compounds. Thus iron may be reduced from the immobile ferric or trivalent form to the ferrous or divalent form. The ions of iron are in various degrees of hydration owing to the ready solubility of iron salts in water. In the case of ferrous iron (Fe^{2+}), $Fe(OH)^+$ (ferrous hydroxides) may be present as well as the Fe^{2+}. In the profile, iron may move in the solution phase or it may be transported as a colloid. Clay coatings on pores, the void argillans, may be uniformly red throughout their thickness, suggesting that iron oxides moved in suspension with the clay minerals. Secondly, iron may be mobilized by the action of organic compounds, the fermenting products of plant decomposition or the secretions from living roots. Mobilization of iron is unusually intense during a gleying period. Iron can be reduced during gleying as a result of the action of dehydrogenases in the course of anaerobic respiration of bacterial cells. In peat soils, fulvate complexes of iron are more stable than humic and are readily transported in drainage water far beyond the borders of a peat bed.

6.5.2 IMMOBILIZATION OF IRON

Ferrous iron may move in the soil solution until it is immobilized by oxidation back to the ferric form, including oxidation by microorganisms. Thus, less soluble iron may form in upper soil horizons where warm, dry summers ensure a seasonal increase in oxidation processes. In lower horizons too, as soon as ferrous iron comes in contact with an oxygen-containing rhizosphere, iron precipitates as hydroxides. Microbiological activity separates the iron from the organic matter (Aristovskaya 1975) which is chiefly responsible for its movement in the upper horizons. The formation of insoluble hydroxides is particularly conspicuous where oxidation processes and microbiological activity accompany the drying of a peat bed. Iron can also be deposited as a gel by flocculation of organic complexes by bonding to metals, commonly calcium and magnesium. Crystallization depends on seasonal changes in moisture regimes and on fluctuations in biological activity. For example, tubular limonites from gleyed paddy soils can become well-crystallized goethite. Good crystal development is favoured by absence of interfering organic matter or Al ions.

6.5.3 AVAILABILITY OF IRON

Chelating agents maintain Fe in solution at concentrations much higher than those normally found in soils. The higher concentration in solution increases the supply from diffusion by increasing the concentration gradient of Fe to plant roots. Since the iron content of soils usually ranges between 0.5 and 5 per cent, deficiency does not develop. Deficiencies may

Chemical properties of the soil 169

occur in irrigated arid areas where pH values of soils are high. Also, there is antagonism in availability to plants between iron and manganese and iron deficiency is a problem in such soils, as in Hawaii, which are high in manganese. Wet soils at low pH values may contain so much ferrous iron that the concentrations are toxic.

6.5.4 MOBILIZATION OF ALUMINIUM AND PODZOLIZATION

Aluminium is a major component of most inorganic soils and when released in weathering processes undergoes hydrolysis with a resultant increase in soil acidity through the release of protons (Miller 1968). A preponderance of hydrogen ions on the negative exchange sites of a clay results in dissolution of aluminium from the clay. In a reducing environment, the surface of clay minerals is freed from iron hydroxide films, thus facilitating the separation of aluminium and its dissolution. Clay minerals are also destroyed by exposure to the strongly reduced products of anaerobic fermentation and reducing enzymes of the anaerobic micro-organisms themselves. Organic acids percolating downward in the soil form complexes with Al as well as with the Fe released by weathering.

Podzolization involves the reactions concerned in the movement and accumulation in the soil profile of iron and aluminium sesquioxides and of organic matter too, in freely drained conditions and under a raw humus horizon (Mackney 1961). Initially, breakdown of clay minerals in the $Å_o$ horizon leads subsequently to differentiation of silica and sesquioxides. The less mobile silica remains behind in the eluvial horizon. Weatherable ferromagnesian minerals are also depleted. In addition, 'lessivage' occurs, the downward movement of inorganic colloids alone, which are translocated without the participation of acidic organic protective colloids. Finally, ferric and aluminium oxides are reprecipitated and organic matter deposited in B horizons.

6.5.5 INFLUENCE OF TEXTURE

Podzols do not form in textures heavier than sandy loam since capillary action ensures that most water is evaporated at the surface in heavier textures and plant nutrients remain within the soil–vegetation–litter–soil cycle. Streaked gleying is typical of cohesive soils rich in colloids in which the deeper layers crack as a result of temporary drying. Organic matter providing the energy source for micro-organisms is deposited on such crack surfaces from which gleying gradually extends inwards.

6.5.6 SIGNIFICANCE OF ORGANIC MATTER

The humic substances in the soil apparently promote high mobility and equally the accumulation of elements of the iron family in the form of organomineral compounds in the soil profile. For example, in dark grey soils of the Central Russia forest steppe, 60–75 per cent of the mobile Fe_2O_3 was found to shift in the form of organometallic complexes (Akhtyrtsev 1968). Rain bearing leaf washings may be an important source of polyphenol for the shift of iron in soil profiles (Coulson et al. 1960). Water which has passed through peat is also very rich in humus colloids. It also has a very low oxygen content and under these anaerobic conditions, iron occurs as ferrous iron, is soluble up to pH 9 and over, and is leached beyond the B horizon.

As with iron, the behaviour of aluminium, firmly bound to organic matter in the soil solution, depends on the presence in the soil solution, and the effect of, such components of the water-soluble organic matter as the low-molecular organic and fulvic acids. The presence of organic matter is a prerequisite for the development of the gley process which does not develop in soils or in soil horizons where organic matter is minimal. In Bog soils gleying may involve the entire profile.

6.5.7 AIR AND WATER

A major control on iron and one which differentiates its behaviour from that of aluminium is the degree to which excess water is present, permanently or temporarily, in the soil. Excess water limits air exchange between the soil and the atmosphere. Oxygen depletion is greatest where organic matter has accumulated, particularly the most active fractions like fulvic acids, low molecular organic acids and compounds like polyphenols. H (atomic hydrogen) and H_2S (hydrogen sulphide) are the most powerful reducing agents and CH_4 (methane) is commonly evolved by anaerobic processes. Such gases reduce iron hydroxides and destroy aluminosilicates, releasing mobile forms of aluminium and other elements.

Reduced iron moves to zones with greater oxidation–reduction potential where it is oxidized and forms a shell of rusty, ferruginous formations. The gleyed zone becomes bleached and the interior boundary of the shell is sharp but the exterior diffuse. When seasonal excessive wetness is relatively short-term, gleying is rust-coloured blotches of $Fe(OH)_3$, but with more prolonged wetness, secondary alumino-silicates form in waterlogged horizons. Such mottles, created by very local migrations of iron, are characteristic of poorly drained soils subject to periodic reducing conditions. With freer drainage iron may be translocated further, to concentrate in a B horizon within the same soil or in other soil profiles downslope. Complete gleying is seldom observed in cultivated soils.

In contrast to iron, reduction is not essential for aluminium translocation as its solubility is not affected by redox conditions.

6.5.8 THE INFLUENCE OF pH

The oxidation–reduction potential that appears necessary to convert sesquioxides from the divalent to the higher state of oxidation depends largely on the pH of the environment (Collins and Buol 1970). For example, in soils in the Himachal Pradesh in India a highly significant negative correlation ($r = -0.88$, $N = 39$) between pH and exchangeable iron has been reported (Takkar 1969).

When the concentrations of aluminium in soil solutions are as low as 5 ppm, toxic effects can be readily noted. Aluminium is apparently released as the Al^{3+} ion from the crystal lattice of clay minerals and moves to exchange position sites when the pH drops below 5 (fig. 6.5). Solubilities rise from 0.2 ppm at pH 5.5 to 15 ppm at pH 4.5, the latter

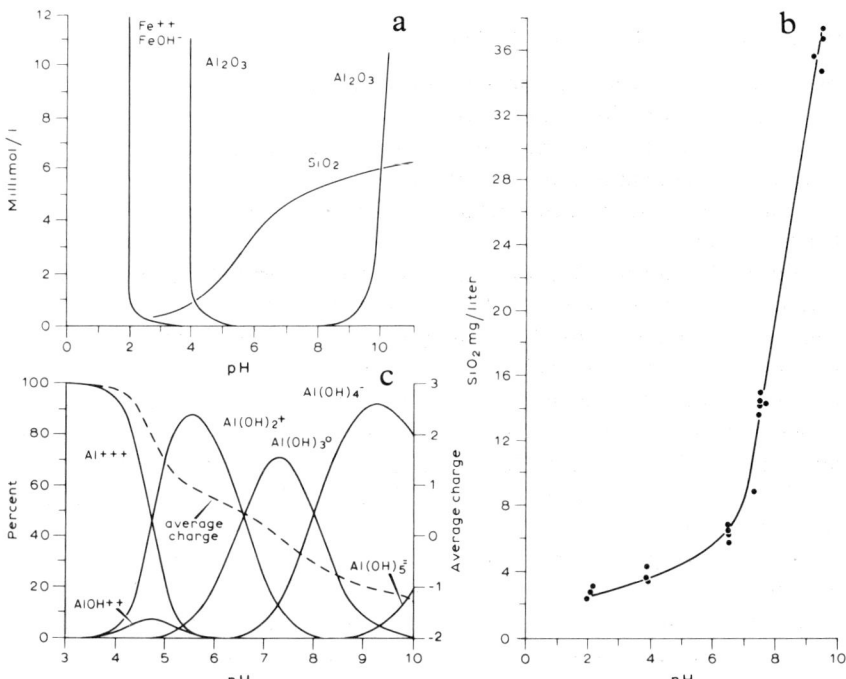

Figure 6.5 Influence of pH on the solubility of aluminium, iron, and silica.
a Solubility of amorphous silica and sesquioxides (Larsen and G. V. Chilinger, 1967).
b Solubility of amorphous silica of phytoliths (Aristovskaya and Kutuzova, 1968). The pH is that of the growth medium for a 20-day culture of micro-organisms, the SiO_2 being extracted under the influence of micro-biologically produced alkalis.
c Solubility of alumina and various hydroxyaluminium compounds as a percentage of the total soluble Al at different pH values (Marion et al., 1976).

being a level which most plants cannot tolerate. Thus, in Puerto Rico, where essentially no exchangeable aluminium was found above pH 5, the aluminium saturation of base exchange sites rose from less than 5 per cent at pH 5 to about 40 per cent at pH 4 (Abruña-Rodrigues et al. 1970, p. 631). The relationship established was $r = -0.87$, $N = 98$ for the covariance of exchangeable aluminium and pH. Similarly, in eastern Nigeria the role of exchangeable aluminium is closely tied up with pH, when the latter falls below 5 (Ekpete 1972). In the case of several Wisconsin soils, where the covariance of pH with aluminium was much closer for exchangeable than for non-exchangeable forms, with correlation coefficients of $r = -0.77$ and $r = -0.49$, $N = 127$, aluminium complexed in organic matter has been proposed as the non-exchangeable form (Pionke and Corey 1970).

6.5.9 INTERACTION WITH OTHER IONS AND WITH PLANTS

Soil clays can never become saturated with hydrogen ions because the soil solution contains other cations, mainly calcium and magnesium, as well as hydrogen. An important consideration is the ratio of hydrogen to calcium plus magnesium in a soil solution above which the concentration of exchangeable hydrogen becomes high enough to dissolve aluminium from the clays.

Basic aluminium chloride salts are readily formed since the substitution of anions for OH is facilitated if the anion and OH ion are similar in size and form. There may be an inverse relationship between the rate of $Al(OH)_3$ crystallization and the tendency of anions to remain in the solids. This tendency followed the order Cl^-, NO_3^-, ClO_4^-, which corresponds with the order in which these anions increase in size and structural complexity (Ross and Turner 1971). Thus, a decrease in the Cl^- concentration is followed by a fairly rapid crystallization of aluminium hydroxide.

Plant roots are important influences on iron behaviour in soils. Iron-enriched material is usually located around root channels, suggesting that the variation in iron content down the profile is mainly a function of the root distribution. Lower parts of humic horizons may be aerated by plant roots. If, therefore, roots are unable to penetrate into the mineral soil once true peat has formed, gley mottles do not develop and iron is lost from the profile.

6.5.10 CRYSTALLIZATION OF SILICA

Secondary quartz exists in soils in a range of phases from freshly precipitated amorphous Si to crystalline quartz. Amorphous silica is an unstable intermediate and soils are known to adsorb silica from monosilicic acid

Chemical properties of the soil 173

solutions under certain conditions. Freshly precipitated aluminium hydroxide and ferruginous clays are particularly effective in removing Si from solution. Trees and other vegetation may concentrate silicate ions from the external solution which are deposited in leaves as amorphous siliceous phytoliths. On the death and decomposition of the host organisms, the phytoliths are released. Silica in the phytoliths of several grasses is $SiO_2.nH_2O$ (opal) (Smithson 1958). In alkaline soils, various species of diatom, by using silica to construct their shells, accumulate a pinky white deposit of fine silica in the surface horizons.

6.5.11 MOBILIZATION OF SILICA

Numerous estimations of silica in natural waters show up to 35 ppm silica in river waters and up to 50–60 ppm in groundwaters (Siever 1957). However, in soil samples in Eastern Australia with soil solutions maintained at field capacity, silica present is entirely $Si(OH)_4$ (monosilicic acid) and is commonly present in the range 14–19 ppm. In field soils, biogenous acids and alkalis formed by microbial metabolism is an important extractant of SiO_2 from poorly soluble compounds. Silicon is liberated from alumino-silicates most intensively by acid metabolites and from quartz by biogenous alkalis. Phytoliths are dissolved most intensively by alkali-forming micro-organisms (fig. 6.5b) and may account for the substantial percentage of SiO_2 draining from podzols.

6.5.12 INFLUENCE OF ORGANIC MATTER AND pH ON SILICA

The release of silica is enhanced in the presence of organic matter and its decomposition products. Where silica is adsorbed on soil colloids solubility is a function of pH and with a change from acid to alkaline conditions in soils, there is a decrease in solubility with a minimum between pH 8 and pH 9. The influence of pH varies, however, according to the type of silica present. In the case of phytoliths, solution is greater at higher pH. It seems that the activity of alkali-forming micro-organisms helps plant silica to dissolve, whereas the activity of acid-forming micro-organisms inhibits this process.

6.5.13 INTER-RELATIONSHIPS BETWEEN SILICA AND OTHER IONS AND WITH PLANTS

In addition to oxides of iron and aluminium, the presence of magnesium oxide and dolomitic lime may also influence the amount of silicon found in soil solution. Silica from silicates and also from fine quartz has been observed to dissolve readily in water containing Ca and Na ions (Hallsworth and Waring 1964). A number of anions particularly phosphate, displace silica from the solid phase, although sodium salts of chloride, nitrate and sulphate do not affect a significant release of silica from soils or kaolinite.

Silica is not necessarily accumulated by plants. Uptake may be 'passive', with the concentration of monosilicic acid remaining practically constant as water is lost by transpiration. It is usually accepted that Si (silicon) does not satisfy the criteria for essentiality. Under some conditions, Si may increase significantly the growth of graminaceous species and sugar cane (Fox *et al.* 1967). Any stimulatory effect, however, is secondary as Si may alleviate toxicities of micronutrients, particularly manganese, and improves resistance to fungal and insect attack.

6.6 Potassium, calcium and magnesium

In most soils, K, Ca, and Mg are distinct from other nutrients in being originally derived from the weathering of primary minerals and occurring in significant quantities in exchangeable form. Calcium occurs in highly variable amounts, ranging from traces of less than 0.05 per cent to quantities amounting to over a quarter of the bulk of some soils in arid areas. Although concentrations of exchangeable K are much less, K is required and utilized by plants in much greater amounts.

6.6.1 IMMOBILIZATION

Unlike the other dominant soil cations, potassium is fixed by most soil clay minerals in a form non-available to plants because the potassium ion fits precisely and is held in the hexagonal holes in the oxygen sheet of the silicate layers of 2:1 type clay minerals. Adsorbed potassium is present in wedge-shaped zones between collapsed mica layers and the expanded 14Å layers of weathered micas (fig. 6.6 a,b). These wedge-shaped zones are too small to accommodate most ions and the K^+ ions can only be displaced by small ions such as NH_4^+ (Rich and Black 1964). Thus, some illites may be formed from vermiculite and montmorillonite by their fixation of potassium. Further, since illite is very similar to mica in structure, it can change to a mica by adsorption of K^+ ions just as micas weather to clays with the loss of these ions.

As a result of repeated dissolution and partial reprecipitation of carbonates during the course of soil evolution, between 10 and 50 per cent of carbonates may be newly formed (Salmons and Mook 1977). Reprecipitation is due either to the movement of the soil solution towards an environment with a lower carbon dioxide partial pressure or to the evaporation of the soil solution. The reprecipitation of secondary calcium compounds is a process common to many soils in drier areas (fig. 1.8) (St. Arnaud and Herbillion 1973). There are many similarities between

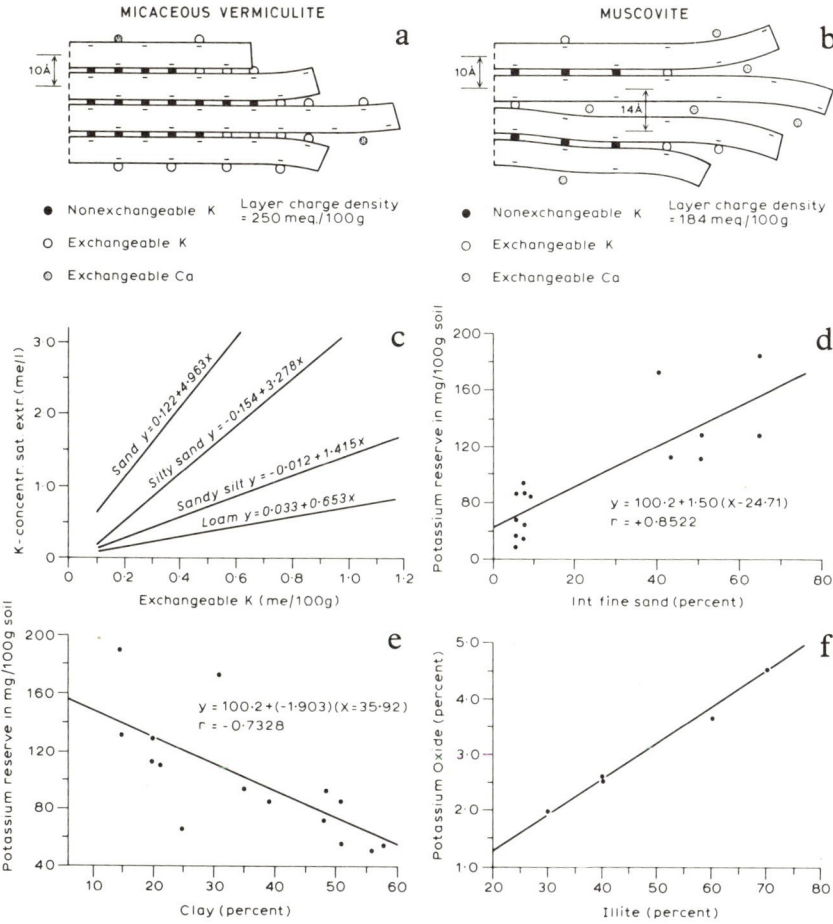

Figure 6.6 Influence of clay type and clay percentage on the adsorption of potassium in the soil.

a and b. High potassium selectivity of micaceous clays (Dolcaster *et al.*, 1968). The geometry of mica layers influence the selectivity. The holes in the tetrahedral sheet are approximately the same size as, and provide a close fit for, the K cation (as they do also for NH_4 and microelements such as Rb, Cs, and Ba).

c Influence of texture on the relationship between exchangeable K and K concentration in the soil solution (Nemeth *et al.*, 1970), for a range of soils in northern West Germany. For a given amount of exchangeable K, the K in the soil solution will be high when the silt-clay is low.

d and e. Relationship between soil texture and potassium reserves in tropical soils in Nigeria (Unamba-Oparah, 1972).

f Relationship between potassium oxide content and illite percentage in a range of major soil groups in Australia (Karim, 1954).

Ca and Mg, but Mg released from primary minerals often becomes part of secondary silicate compounds in soils. As these are relatively stable, Mg is less readily leached from soils than Ca, and amounts may exceed those of Ca in humid regions.

6.6.2 AVAILABILITY

Three conditions of potassium availability are frequently mentioned (Welte and Niedderbudde 1965). First, potassium contained as a constituent of primary minerals is clearly unavailable. Secondly if K is added to the soil, fixation may occur within minutes as it penetrates the interior crystal structure of 2:1-type clays, and this may become available to plants only slowly. Thirdly, and by contrast, potassium adsorbed on the surfaces of expanded clays and at edge sites is exchangeable.

In general, it is the proportion of potassium ions on the exchange sites and in the soil solution rather than the absolute amount of exchangeable potassium in the soil which controls potassium nutrition. Although a critical lower limit for plant growth is about 0.2 me/100 g potassium in a soil's exchange complex, the optimum supply (fig. 1.11) occurs when exchangeable potassium constitutes 1.8–3.0 per cent of the total adsorbed bases (Avakyan 1969). The greater the amount of exchangeable potassium in relation to the total exchange capacity, the greater is the K fixation.

Most cultivated soils have 80 per cent or more of their exchange capacity satisfied by Ca^{2+} ions. Calcium may be adsorbed on to clay particles and by humus. The Ca ion causes flocculation and clays in this condition consist of stable packets containing 5 to 20 platelets each. As water containing dissolved carbon dioxide percolates through the soil, the carbonic acid so formed displaces calcium in the exchange complex. Calcium is the most mobile element, being rapidly displaced by downward flow of water in the soil profile (fig. 1.21b). If considerable percolation of such water through the soil profile takes place, soils gradually become acid. This condition is accentuated because trivalent Al is more energetically adsorbed than bivalent Ca. For calcium there is no form comparable to fixed or slowly available potassium. Like any other cation, however, the exchangeable and solution forms are in dynamic equilibrium.

Magnesium is an essential component of the chlorophyll molecule. Silicate minerals serve as the main reservoir for soil Mg (fig. 2.3a). The same general principles apply to its behaviour as apply to calcium and potassium. It occurs in water-soluble, exchangeable lattice forms as well as a primary mineral (Prince *et al.* 1947).

6.6.3 INFLUENCE OF TEXTURE

Clay mineral composition directly influences K supply to plants (Karamanos and Turner 1977). Because potassium is fixed by most clays, the concentration of potassium in the soil solution and its supply to plant roots is controlled by soil texture (fig. 6.6). Kaolinite fixes very little potassium and the amounts fixed by montmorillonite are very variable. However, fixation is increased in soils rich in illite (fig. 6.6f). In consequence, such clay soils may show little response to potassium fertilization.

The type of clay influences the degree of calcium availability. 2:1 clays require a much higher degree of saturation than 1:1 clays for a given level of plant utilization. Thus, montmorillonite clays require a calcium saturation of 70 per cent or more before this element is released sufficiently rapidly to the growing plant. In contrast, kaolinitic clays are able to satisfy the Ca^{2+} requirements of most plants at saturation values of only 40–50 per cent. In tropical soils, per cent clay and calcium reserve may be inversely correlated, as in Nigeria. The association with magnesium is similar and arises because reserves of nutrients in tropical soils decline as soils weather and increased clay content expresses this advance in age. In temperate zones, total and available Mg status of soils increases with increasing percentage of clay, as in Swedish soils (Salmon 1963).

6.6.4 INFLUENCE OF ORGANIC MATTER AND WATER

If humus enters the expanded clay crystal lattice, it reduces the amount of contraction on drying. Potassium fixation is less as fewer ions are entrapped by contraction on drying. Calcium and magnesium reserves are often correlated with amounts of organic matter, as their maintenance depends on nutrient cycling and litterfall. For example, in cacao plantations in western Nigeria, the surface concentration of calcium correlates with the quantity of organic matter ($r = 0.94$, $N = 15$) as does magnesium ($r = 0.88$) (fig. 6.6d,e).

Potassium ions may move inside the clay crystal lattice when wet and expanded and are then trapped in the interior lattice structures on drying. For example, in Oxisols in Trinidad, fixation was not affected by moisture contents between 0.5–1.6 field capacity, but lower moisture levels resulted in increased fixation. Sufficient soil moisture is also vital to maintain adequate mass flow conditions for available potassium to move to the plant roots.

6.6.5 INTER-RELATIONS WITH OTHER IONS

In acid soils considerable amounts of exchangeable Al may affect their ability to fix and release K. In general, readily exchangeable K can be exchanged by Al ions while slowly exchangeable K cannot, but Al^{3+} is strongly preferred to K^+ on montmorillonite. Selectivity by sites on the exchange complex for K ions, as opposed to Ca + Mg, changes with soil type. Mg^{2+} is held more weakly than Al^{3+} to the surrounding oxygens of the crystal structure, owing to its divalency compared with the trivalency of Al^{3+}. Mg^{2+} is, therefore, more easily dislodged from its position than Al^{3+} by the electrical potential difference which is created in the crystal when H^+ enters into its interior. The possibility of displacement of magnesium by iron is governed by the similarity of their ionic radii in sixfold co-ordination (0.74 and 0.73 Å, respectively). Applications of potassium and ammonium fertilizer salts aid in the release of more magnesium from the soil's exchange complex. This explains the tendency of heavily fertilized soils eventually to become deficient in magnesium.

6.7 Nitrogen, phosphorus and sulphur

All living cells contain nitrogen and each molecule of chlorophyll includes four atoms of nitrogen. Although N_2 constitutes 78 per cent of the atmosphere, it is useless to plants in this elemental form. Conversion of inert atmospheric nitrogen to various organic and inorganic forms in the soil is mainly the work of soil micro-organisms, termed 'fixation'. The natural supply from the soil as microbial, proteinaceous tissue decomposes is often too slow for optimum plant growth. Conversely, cultivated crops may recover less than half the nitrogen added as fertilizer which is easily lost due to the mobility of nitrogen in the soil.

Phosphorus is vital early in plant growth, being a constituent of nucleic acid in which genetic patterns are encoded. There is, however, no efficient mechanism for holding exchangeable and available anions like $H_2PO_4^-$, nor does the phosphate anion participate in isomorphous substitutions. Further, phosphorus is the one major element in soil organic matter which must be supplied almost entirely by parent materials in unfertilized soils because of low atmospheric returns (Walker and Syers 1976).

Sulphur is an active element which combines directly with most other elements as well as being one of the six major nutrient elements needed by plants. Plant requirements for S are similar to those for P, with SO_4^{--} anions being the principal plant-available form. It is present in organic structures in the sulphhydryl ($-SH$) group in compounds with C-S linkages and, like nitrogen, is an important constituent of protein with deficiencies causing serious malnutrition.

6.7.1 IMMOBILIZATION AND ADSORPTION OF NITROGEN

Several processes decrease the availability of soil nitrogen. NH_4^+ (ammonium) ions may be incorporated within the crystal lattice of 2:1 silicate clays since they are almost identical to the K^+ ion in size and valence. More than 10 per cent of the total nitrogen in surface soils and as much as 60 per cent in subsoils may be in the form of clay-fixed ammonium (Rodrigues 1954). The term 'fixation' is again loosely used since, as with 'fixation' of atmospheric nitrogen in microbial tissues, nitrogen in such forms is not available to plants. Defined as the NH_4^+ in clay minerals which does not exchange with K^+, 'fixed' ammonium may be theoretically available to plants, but in practice growing organisms have difficulty in taking it up. This is because fixed NH_4^+ is incorporated within the clay crystals rather than being merely adsorbed on their surfaces. Fertilizer ammonia is quite readily replaced with cations which expand the clay lattice, like Ca, Mg, Na, or H.

Nitrogen may be present in the soil as the negatively charged anion, NO_3^- (nitrate ion) as well as the cationic form, NH_4^+. Adsorption sites for NO_3^- are the positively charged surfaces of amorphous aluminosilicates and hydrous oxides (fig. 6.7a). Young volcanic ash soils and much weathered tropical and subtropical soils are high in such amorphous materials, and in such countries as Mexico and Hawaii and in South America, positive adsorption of NO_3^- occurs (Singh and Kanehiro 1969).

Organic matter can fix and immobilize ammonium and its availibility to plants is low, but the common mode of conversion of nitrogen to an organic form is by assimilation by micro-organisms. These are predominantly heterotrophic, depending on organic compounds already synthesized by plants for their carbon and their energy. Heterotrophic bacteria obtain their organic carbon either as symbiotic parasites or as free-ranging or non-symbiotic organisms. Parasitic bacteria which invade plants to feed on carbohydrates and proteins in the cell tissues do so by producing appropriate enzymes. The injury induces root cells to grow round the bacteria and a symbiotic relationship is established. Certain strains of the genus *Rhizobium* establish a nitrogen-fixing, symbiotic relationship with legumes. Several symbiotic, legume bacteria are named after the host plant, such as *Rhizobium trifolii* which nodulates red and white clover. For example, nitrogen fixation by white clover *(Trifolium repens)* is an important factor in grassland production, and over 90 kg N/ha/year have been reported from New Zealand. Similarly, nitrogen fixation by clover in grasslands in Ireland is sufficient to maintain highly productive pastures without the use of fertilizer nitrogen. Bacteria in symbiosis with leguminous crops can contribute 150–300 kg N/ha/year. There are also leguminuous tree species like acacia which are also involved in fixing appreciable amounts of nitrogen. Nitrogen-fixing bacteria can also live in symbiosis with non-leguminous plants, a fact particularly significant for tropical agriculture. Atmospheric nitrogen is fixed in the root zone of rice plants *(Oryza sativa L.)* by bacteria, a process which accounts for continued fertility in flooded paddy fields (Yoshida and Ancajas 1973).

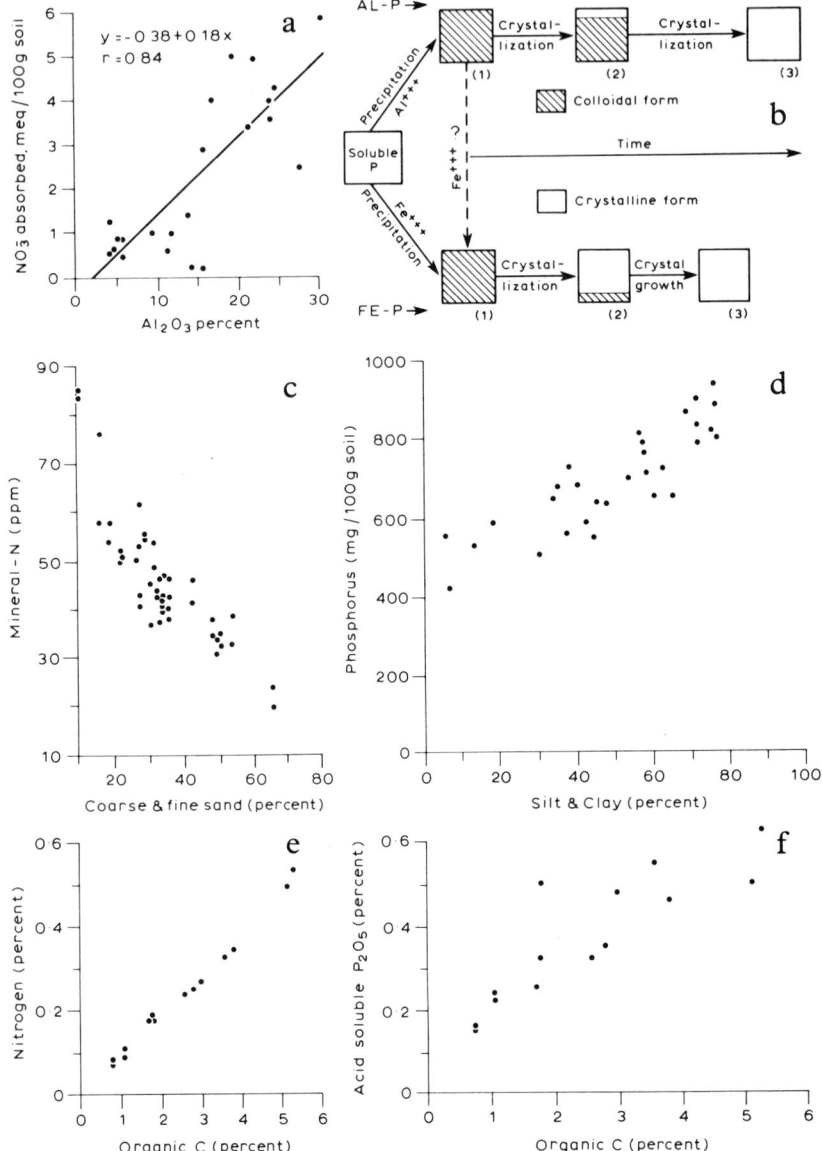

Figure 6.7 Factors affecting nitrate and phosphate behaviour in soils.
a Positive adsorption of nitrate by amorphous inorganic material (Kinjo and Pratt, 1971), the aluminium being extracted from soils of volcanic ash origin (Andepts of the U.S.D.A. Soil Taxonomy) from Mexico and South America.
b Transformation of soluble P into less available forms (Juo and Ellis, 1968).
c Silt-sized particles as the primary centres of N mineralization in Kenyan tropical red earths (Robinson, 1967).

(Continued on facing page)

Of the free-living, or non-symbiotic heterotrophs, some obtain the organic carbon required for growth from living tissue. However, the majority are saprophytes, living on decaying organic matter. These type of bacteria include the nitrogen-fixers, *Azotobacter* and *Beijerinckia*. Azotobacter is an aerobic saprophyte, depending for respiration on an air supply with an oxygen component. Anaerobic bacteria obtain oxygen by reducing such compounds as sugar or certain ions, and include the nitrogen-fixing saprophyte, *Clostridium*, which is responsible for the immobilization of much nitrogen in poorly drained soils.

In the majority of cultivated soils in the temperate zone, nitrogen-fixing algae are important agents of nitrogen turnover, since they utilize atmospheric nitrogen. Blue-green algae help to maintain the nitrogen level in paddy soils, the amount accumulated being 15–80 kg N/ha/year. Nitrogen-fixing blue-green algae are, therefore, used as a fertilizer for rice fields, and more nitrogen may be fixed than is removed by the growing crop. In desert soils, algae are the dominant micro-organism and nitrogen fixation by the algal crust may be the major source of nitrogen for the desert ecosystem as a whole.

6.7.2 FIXATION OF PHOSPHORUS

Fixation of phosphate by soil minerals involves mechanisms quite different from those by which nitrogen is immobilized in the soil. Fixation rate of P applied to cultivated soils is high and the ensuing low recovery in crops is an important problem. Fixation proceeds in two stages, an initial rapid reaction with a subsequent, slower if not indefinite reaction. There are two types of reaction, related to the positive surface charges which occur on corners, edges, and lattice disturbances of clay minerals and on the surfaces of iron and aluminium compounds (Syers *et al.* 1971). There is, for example, a high correlation ($r = 0.84$, $N = 60$) between exchangeable aluminium and phosphate retained by North Carolina soils.

First, adsorption of phosphate refers to the process of concentration at surfaces of a solid phase of phosphate ions from solution in forms that are exchangeable or replaceable. If amorphous sesquioxides are crystallized, the adsorption of phosphate is sharply reduced. Secondly, chemical precipitation refers to the removal of phosphate ions from solution and their chemical bonding to the solid phase. Initially, in weathering and after the application of phosphate fertilizer, calcium and aluminium

---

d Influence of soil texture on P content of arid-climate soils in South Arabia (Western, 1972).

e and f. Association of organic carbon with N and P content of rendzina soils in the Seychelles (Lionnet, 1952).

phosphates are more likely to be formed. Gradually, these change into iron phosphate, which is the least soluble form (fig. 6.7b). Inositol phosphates are the main organic phosphate compounds in soil, but are not readily available to plants. Over a long period of time organic forms are an important, potential source or sink of plant-available P (Islam and Mandel 1977). For example, organic P constitutes between 20 and 78 per cent of phosphorus present in soils of southern Nigeria (Omotosu 1971). In West Africa as a whole, non-availability of P is often the most severe constraint on growth (Nye and Bertheux 1957).

6.7.3 IMMOBILIZATION OF SULPHUR

Sulphur occurs in three principal forms in soils. First, sulphur in the plough horizon is mainly in the stable form of organic matter, protected from rapid release and hence non-available to higher plants. Thus stabilized, organic sulphur shares a unique similarity with organic nitrogen. Sulphur can also exist in both positive and negative oxidation states. Like nitrogen, it assumes a number of valence states, varying from S^{--} to S^{6+}. Most of the range of secondary sulphate compounds formed are reasonably soluble. Immobilization of inorganic forms of sulphur occurs when energy-rich, low-sulphur compounds are added to soils, inorganic sulphur being synthesized into tissues during the stimulated microbiological growth. The mechanism appears to be the same as for nitrogen. Most of the small sulphate retention by soil minerals is associated with kaolinitic 1:1 clays and with hydrated iron and aluminium oxides in lower horizons. The mechanism involves hydroxyl groups in the hydrous oxides and silicate clays which are replaced by sulphate ions. The adsorption mechanism is much as for $H_2PO_4^-$ except that strength of retention of the SO_4^{--} anion is weaker than that of phosphate. Phosphate will displace or reduce the sorption of sulphate, so S remains available to plants.

6.7.4 MINERALIZATION OF ORGANIC NITROGEN COMPOUNDS

Plants cannot utilize nitrogen as a gas, but can assimilate nitrogen when it is combined with the stable forms of either ammonium (NH_4^+) or as the nitrate ion (NO_3^-). For example, in western Tanzania (Robinson 1968), mineralizable-N correlated with maize yield ($r = 0.51$, $N = 44$). Mineralization describes the breakdown of organic forms of nitrogen and their conversion to mineral forms (NH_4^+, NO_2^-, NO_3^-), largely through the activities of micro-organisms. Mineralization proceeds in three stages as the final compounds of organic matter decomposition are produced. First, proteins are hydrolysed enzymatically to release the organic carbon required as an energy source by heterotrophic micro-

Chemical properties of the soil 183

organisms. As amines and amino acids are also released, this first stage is termed aminization

proteins → R − NH_2 + CO_2 + energy + other residues

where R represents a general class of organic compounds. The products of aminization are then rapidly utilized by micro-organisms or transformed in a second stage by other types of heterotrophic micro-organisms. The −NH_2 radical of amino compounds is converted to NH_3 by hydrolysis, the third hydrogen in the NH_3 molecule being derived from water. As ammoniacal compounds are released, the second stage is termed ammonification

R − NH_2 + H_2O → NH_3 + R − OH + energy

The NH_3 produced is rapidly converted to the NH_4^+ ion which can be absorbed by plants or adsorbed as an exchangeable ion on soil particle surfaces. Alternatively, two groups of obligate autotrophic bacteria take the mineralization of nitrogen through a third stage. They obtain carbon from CO_2 and energy from the oxidation of NH_4^+ to NO_3^- in a two-step process termed nitrification. The first step in nitrification ends with the formation of nitrite ions (NO_2^-)

$2NH_4^+ + 3O_2 \rightarrow 2NO_2^- + 2H_2O + 4H^+$

the work largely of the bacteria *Nitrosomonas*. Numerous heterotrophs including fungi and actinomycetes can also produce nitrite from reduced nitrogen compounds. The second step in nitrification which results in the final conversion to NO_3^- is

$2NO_2^- + O_2 \rightarrow 2NO_3^-$

with a second group of obligate autotrophic bacteria, *Nitrobacter*, being largely responsible. Again, some heterotrophs may also convert nitrite to nitrate.

For a plant to receive a continual supply of nitrogen during the growing season, mineralization must exceed immobilization. However, turnover rates are not high and available nitrogen released from soil organic matter is only about 2–4 per cent of the total per year. In fact, there are two contrasted rates of nitrogen cycling, with the bulk of annual plant growth requirements being met by rapidly cycling nitrogen derived from plant exudates and other easily decomposed constituents of plant litter. Other compounds release nitrogen only very slowly to the microbial pool and reside in the soil organic matter for long periods.

6.7.5 LOSSES OF NITROGEN FROM THE SOIL

The many possible transformations which produce gases, or create soluble forms of nitrogen which are subject to leaching, result in significant losses of nitrogen, including fertilizer nitrogen (N), from the soil-root zone.

(a) Denitrification

Some anaerobic organisms can obtain oxygen from the reduction of nitrates and nitrites, with the accompanying release of nitrogen gas and nitrous oxide. This biochemical reduction of both nitrite and nitrate is termed denitrification. In biological denitrification, the main sequence of products formed is NO_3^- (nitrate):

$$NO_3 \text{ (nitrate)} \rightarrow NO_2 \text{ (nitrite)} \rightarrow N_2O \text{ (nitrous oxide)} \rightarrow N_2 \text{ (nitrogen gas)}$$

(Cooper and Smith 1963). Circumstances which increase the biological demand for oxygen, such as the presence of quantities of actively decomposing organic matter, increases denitrification by depleting the oxygen supply.

(b) Leaching

NO_2^- (nitrite) and NO_3^- (nitrate) are highly soluble and are therefore readily lost in percolating water and in surface runoff when precipitation exceeds evapotranspiration and where the soil has reached field capacity. Such losses are much less from a cropped soil than from bare ground. There is direct assimilation of nitrate by the crop and evapotranspiration reduces the amount of water in the soil. Also, since the bulk of soil nitrogen resides in the organic-rich topsoil, loss of nitrogen by surface runoff and by soil erosion can be significant.

Since loss of soluble forms of nitrogen depends on the efficiency of the plant roots (MacLean 1977), profile depth is an important consideration. For instance, sampling below 180 cm in deep loessial soils in Nebraska revealed little more nitrate than is found in unfertilized soils. This indicates that there is little movement of nitrates below the crop-rooting zones in deep loessial soils. If penetration of crop roots is limited to shallow depths by high acidity, compact horizons, bedrock or other unfavourable growth factors in the subsoil (Herron et al. 1968), the leaching of soluble forms of nitrogen is correspondingly greater.

(c) Ammonia volatilization

Volatile losses of NH_3 (ammonia gas) may create soil nitrogen deficits or the loss of fertilizer N (see p.149), particularly at higher pH values. In an alkaline, damp soil ammonium salts react:

$$NH_4^+ + H_2O + OH^- \rightarrow NH_3 + 2H_2O.$$

6.7.6 RELEASE AND DIFFUSION OF PHOSPHORUS

The plant is thought to absorb phosphorus mainly in the forms $H_2PO_4^-$ and HPO_4^{2-} (di- and mono-hydrogen orthophosphate). $H_2PO_4^-$ is the more important form, particularly in acid soils. Under ideal conditions, as plants take in $H_2PO_4^-$ ions from the soil solution, other ions replace them from slowly soluble compounds in the soil. Phosphate continually dissolves from and is re-sorbed by the solid phase. The rate of replenishment of the soil solution with phosphate is important in determining the phosphate uptake of a crop and the concentration in solution is less important. Only if the rate of phosphate release from the solid phase is adequate can the root absorb sufficient phosphate to satisfy the metabolic requirements of the plant. A content of fertilizer P_2O_5 in the solution of 1 mg/litre provides completely for the growth of plants. The content of phosphorus in the soil solution is replenished quite rapidly. It has been estimated that the phosphorus content in the soil solution may be renewed from 100 to 500 times in a growing season. It is convenient to regard 'availability' as the ratio of available phosphate to total phosphate.

Compared with nitrogen, phosphorus mobility in soils is very slow indeed. An estimate of a diffusion gradient for P is for only 0.4 mm around the root in 24 hours (Olsen and Watanabe 1966). In consequence, whilst the rate of release of P in the total soil mass may be sufficiently rapid for purposes of plant nutrition, the solubility and diffusion rate of P may not be adequate within the limited volume of soil adjacent to plant roots. Thus, phosphate uptake over an extended period is maintained by root extension into fresh soil, rather than by more intensive uptake from the same rhizocylinders of soil.

6.7.7 MINERALIZATION AND LOSS OF SULPHUR

As in the cycling of nitrogen, the organic forms of sulphur must be mineralized by soil micro-organisms before sulphur can be used by plants. Reduced substances are oxidized just as ammoniacal compounds are formed when nitrogenous compounds are decomposed

organic sulphur → decay products → sulphates
e.g. H_2S, other sulphides

For example, sulphides are oxidized to sulphuric acid

$$H_2S + CO_2 \rightarrow H_2SO_4$$

and sulphates (SO_4^{--}) are rapidly available for plant use. Again, oxidation is predominantly a biochemical process involving numbers of obligate, autotrophic bacteria of the genus *Thiobacillus*. Like the nitrifiers, the thiobacilli obtain their carbon from CO_2 and their energy by oxidizing inorganic substances. Soils annually mineralizing about 6 ppm S or less respond to sulphur application. In addition, S once oxidized may be largely lost by leaching due to the anionic nature of sulphates and the solubility of most of its common salts (Chao et al. 1962). Sulphur may be lost in the annual bush fires of savannah areas as SO_2 volatilized would drift away in smoke (Enwezor 1976).

6.7.8 INFLUENCE OF SOIL TEXTURE

As fixation of nitrogen in the mineral fraction of the soil is that of the ammonium ion replacing interlayer cations in the expanded lattice of 2:1 clay minerals, the total amount of fixed ammonium in a soil is directly related to the amount and kind of clay present (fig. 6.7c). Vermiculite, the hydrous micas, and montmorillonite fix appreciable quantities of NH_4. Kaolinitic soils trap little ammonium. Sandy soils with low exchange capacities permit appreciable movement of ammonium nitrogen into the subsoil. The rapidity with which anions may move through the soil is well known. In sandy soils in North Carolina, nitrate may descend 1–5 mm for each 1 mm of rain. In addition, appreciable leaching of nitrate nitrogen may occur through certain clay soils. For instance, in clay pan soils in Missouri, nitrate was concentrated at 30 cm depth in the profile 1 month after treatment, moved through the pan 14–23 months later, and perhaps as much as 65 per cent of the amount applied was lost to underground aquifers. Thus, experiments in Georgia show that such losses occur only twice as rapidly in loamy sand soils compared with clay soils (Boswell and Anderson 1970).

In terms of mineralization, the silt content may be the most important fraction. The effect of a grass cover in inhibiting nitrification is marked in finer-textured soils, especially those high in silt content. Since denitrification is favoured when diffusion of oxygen is not rapid enough to replace the oxygen that is being consumed by the micro-organisms, finer-textured soils are conducive to nitrate reduction.

Total P increases with a decrease in particle size (Hanley et al. 1965). Also the release of phosphorus to both plants and to extracting agents depends on particle size. Sands may have substantial amounts of inorganic phosphorus, particularly the finer sizes if much is of primary origin (fig. 6.7d). Since apatites are their principal forms, phosphorus in sands is largely calcium-bound. Sands also contain significant amounts of soil organic P as films coating the mineral particles, but clays are the main site of phosphate fixation. Clays may adsorb phosphates if they have large anion exchange capacities like allophane which may fix over 1000 ppm of added P. Also, since larger P-fixing capacities are associated

Chemical properties of the soil 187

with higher contents of the amorphous sesquioxides, clays containing significant quantities of iron and aluminium oxides such as kaolinitic soils may fix 500–1000 ppm of added P. Compared with the calcium-bound, primary origin of P in sands, that in clay fractions is largely secondary, the result of accumulation of organic, inorganic and fertilizer P.

Soils with high clay contents have a low per cent saturation since they tend to fix P, whereas soils low in clay are more saturated. Thus, the inverse correlation between per cent clay and total P in clay was $r = -0.61$ for 24 soils representative of the major Irish soil groups (Hanley and Murphy 1970). Organic P is concentrated in clay separates because there it is associated with higher levels of organic carbon and nitrogen. Therefore, with decreased particle size, organic P as a percentage of total P tends to increase (Syers *et al.* 1969). Applied P is strongly held against leaching but in sandy soils some leaching of this nutrient occurs (Ozanne *et al.* 1961).

Leaching losses of sulphur are greatest on coarse-textured soils under high rainfall. Clay mineral type influences the retention of the sulphate anion as with phosphates. Thus kaolinitic clays retain more sulphate than montmorillonite due to the higher proportion of anion exchange sites on 1:1 type clays, together with the higher negative charge and its associated anion repulsion on the 2:1 type clays.

6.7.9 INFLUENCE OF ORGANIC MATTER

Maximum biological immobilization of nitrogen occurs when large quantities of decomposable crop residues are added to the soil (fig. 6.7e). Heterotrophic organisms then grow so rapidly that frequently all available soil nitrogen may be utilized by them. Where nitrogen is present in cationic form, NH_3 retention in organic soils may be correlated with per cent carbon. Phenolic hydroxyl groups, for example, may react readily with NH_3, whilst carboxyl groups may also react to a lesser extent.

P is not generally available when combined to form organic compounds, but is released when these compounds are mineralized. Thus, in recently cleared soils in tropical areas, P mineralization may be the most important factor in determining P availability (Adepetu and Corey 1976). Micro-organisms may immobilize a large amount of P when a source of nitrogen and energy is available. Part or most of organic P is incorporated into the soil humus, but certain additional phosphorus-containing compounds, such as nucleic acids and inositol phosphates, may occur in the soil independently from the humus fraction (McKercher and Anderson 1968).

There is no efficient mechanism on humus particles for holding exchangeable and available anions such as $H_2PO_4^-$, and so if organic matter coats clay mineral surfaces, P adsorption is reduced. On the other hand, decomposing organic anions form highly stable complex ions with Ca^{++}, Fe^{3+} and Al^{3+} and thereby release P for plant growth. To some extent sulphate can be held by the positive charge of soil organic matter due to its amphoteric properties, and may account for up to half the amount of sulphate retained. In highly weathered soils, the bulk of S reserves often occurs in the organic fraction. The C-S bonds of the humic acid fraction are a major source of mineralized sulphur (McLaren and Swift 1977).

6.7.10 INFLUENCE OF SOIL WATER

No comprehensive generalizations can be made about soil moisture and N transformations, as these differ for diffusion, nitrification and denitrification and for plant uptake of nitrogen. Fluctuations in soil moisture content affect the stability of nitrate markedly (fig. 6.9a) because of soil moisture control on the supply of oxygen diffusing through the soil and on the metabolism of micro-organisms. For example, the optimum moisture content for nitrate accumulation in the main agricultural soils of Colombia was 20–33 per cent of the maximum water-holding capacity (Blasco and Cornfield 1967).

Nitrate reduction after flooding can be very rapid, with rates of 55 ppm/day suggested. Redox potential is usually below 350 mV when denitrification starts. In fact, denitrification rates may be most rapid at water contents near saturation and drop at contents above or below this level. It is a major mechanism of nitrogen loss in waterlogged paddy soils particularly because the soil may be flooded and drained several times during one season. In addition, reduction reactions in a waterlogged soil produce ferrous and manganous ions which displace ammonium from the exchange complex to the soil solution where it is more likely to be removed.

Water must also be present if phosphate supply to the plant root is to be maintained (figs 6.8 and 6.9b). Porosity is an important control, with moisture in intercrumb pores providing the main and more rapidly moving supply. In the intracrumb pores a slower moving but larger store of phosphate in solution helps to maintain the concentration of phosphate in the intercrumb pores (Vaidyanathan and Talibudeen 1968). Drying increases retention of phosphate on the surface of kaolinite. Under flooded conditions, the inorganic form is the more important and iron phosphate becomes the dominant form of phosphate fixation perhaps owing to the greater surface area created by the reduction of ferric hydroxide to more reactive ferrous compounds (Khalid et al. 1977). There is also a marked increase in the availability of native and added phosphates in flooded

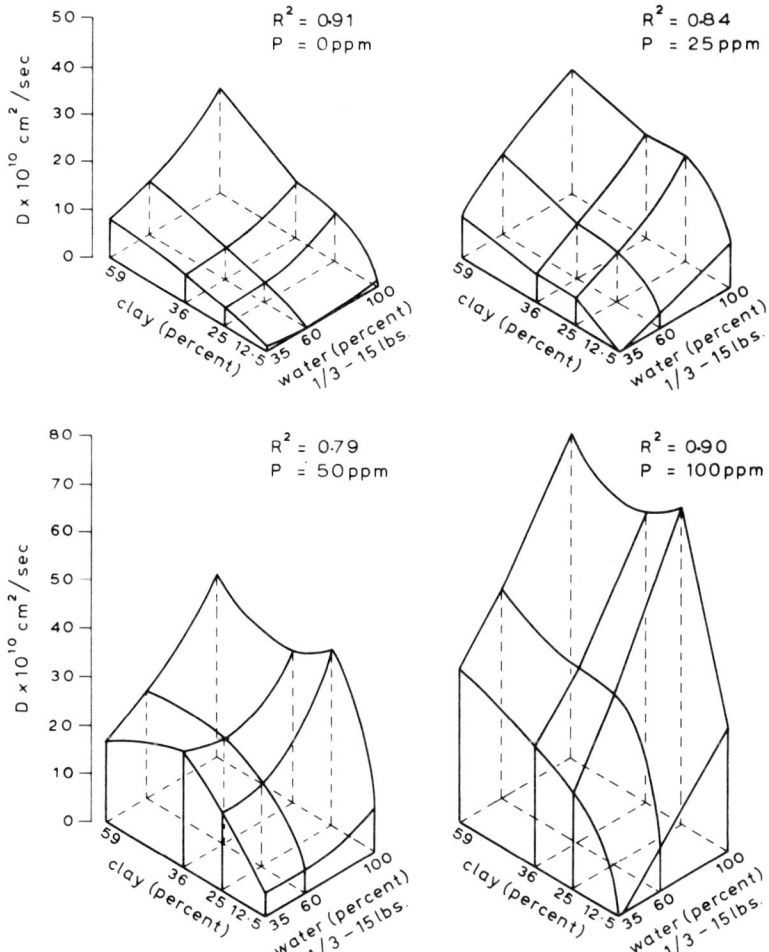

Figure 6.8 Increased diffusion of P with increase in water, clay, and amounts of applied P (Mahtab et al., 1971).

soils. In addition to the greater diffusion of $H_2PO_4^-$ ions in a larger volume of soil solution and the reduction to the more soluble ferrous forms, there is also dissolution of the previously oxidized layers surrounding phosphate particles.

The decreased oxidation of sulphur compounds at high moisture levels is due to decreased oxygen supply to the aerobic, sulphur-oxidizing thiobacilli. Like nitrates in wet soils, sulphates tend to be unstable if there is an energy source present, and in anaerobic environments containing decomposable organic matter, sulphate is readily reduced to sulphide

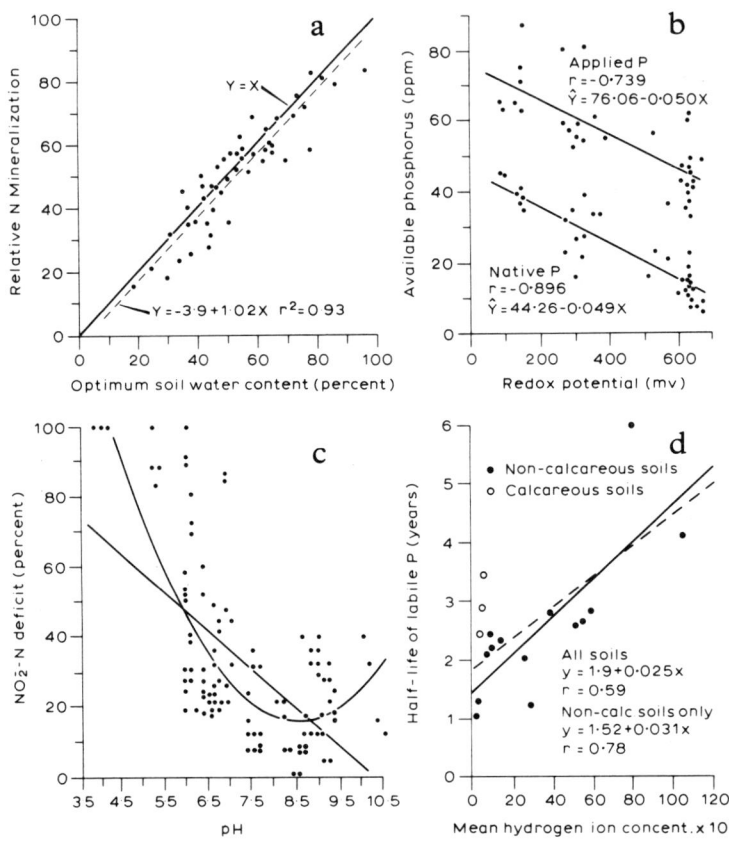

Figure 6.9 Influence of water content and soil pH on nitrate and phosphate behaviour in soils.
a Relative N mineralization in relation to water content of widely differing soils in the U.S.A. (Stanford and Epstein, 1974). Soil water is expressed as a percent of the optimum for the experiment, as observed at the end of a 2-week incubation at $35°C$.
b Influence of redox potential on available phosphorus (Savant and Ellis, 1964), the predominantly iron phosphate of an acid soil, the Cherokee silt loam from Kansas, becomes more soluble and more available under reducing conditions.
c Complexity of relationship between nitrite deficits and pH (Wullstein, 1969). In these non-cultivated soils in the Salt Lake Valley, Utah, the inverse relationship commonly observed between pH and nitrite deficits over the acid pH range, does not appear to extend to alkaline soils.
d Increase of half-life of available P with hydrogen-ion concentration for a wide range of British soils (Larsen *et al.*, 1965), possibly because a crystalline basic calcium phosphate forms at a rate which increases with pH.

Chemical properties of the soil 191

(S^{--}). A number of bacteria of the genera *Desulfovibro* and *Desulfotomaculum* utilize the combined oxygen in sulphate to oxidize organic materials. Submerged soils like rice paddy soils are usually well-supplied with iron in the ferrous form, with which the sulphide ion reacts instantly

$$Fe^{++} + S^{--} \rightarrow FeS \text{ (ferrous sulphide)}$$

If active iron is insufficient to precipitate the sulphide, free H_2S may be evolved. This inhibits enzymatic activity and is toxic to rice. However, the equilibrium between insoluble and soluble sulphides in submerged soils is more complex than the simple relation

$$FeS \rightleftharpoons H_2S$$

6.7.11 INFLUENCE OF SOIL REACTION

Soil reaction influences nitrogen transformations (fig. 6.9c) because microbial populations are susceptible to pH controls. For instance, in the British Isles, the proportion of poorly effective rhizobium strains increases as pH falls (Jones 1966). In addition, crop recoveries have also been low in many instances where the soil was so acid that nitrification of ammonium fertilizer was inhibited. In contrast, the genus *Beijerinckia* flourishes and fixes nitrogen in a strongly acid environment, possibly due to its high iron demand. Similarly, non-symbiotic nitrogen-fixing bacteria, *Clostridium pasteurianum* function in acid soils unfavourable to most species of Azotobacter. The net effect of these varying responses to soil acidity is that there appears to be a critical pH in the 5.7–6.3 range where nitrification rate exceeds ammonification rate, provided that moisture contents are within the aerobic range. In soils nearer neutrality in reaction, ammonium oxidation by *Nitrosomonas*, which requires calcium, and nitrite oxidation by *Nitrobacter* are influenced by pH, with little activity above pH 7.7. When nitrites are formed, or added to soil, they are stable if the pH is near 7 or above and gaseous N losses are small. Loss as ammonia is usually important only in alkaline soils.

pH influences phosphate availability, but its control is largely through inorganic processes (fig. 6.9d). When phosphates are added to soils highly saturated with bases, the free phosphate ions not taken up by the plant combine largely with calcium and magnesium. In contrast, acid mineral soils have a high P-fixing power. Phosphates in acid soils combine with the substantial supply of free iron and aluminium to form ferric or aluminium phosphates. Thus, in an Illinois silt loam under forest cover, the Al and Fe fractions of P decreased ($r = -0.82$ and $r = -0.93$, respectively, $N = 40$) as soil pH increases (Gilmore 1972).

The environmental requirements and tolerances of the five species of thiobacilli vary. In contrast to the narrow pH range near neutral requir-

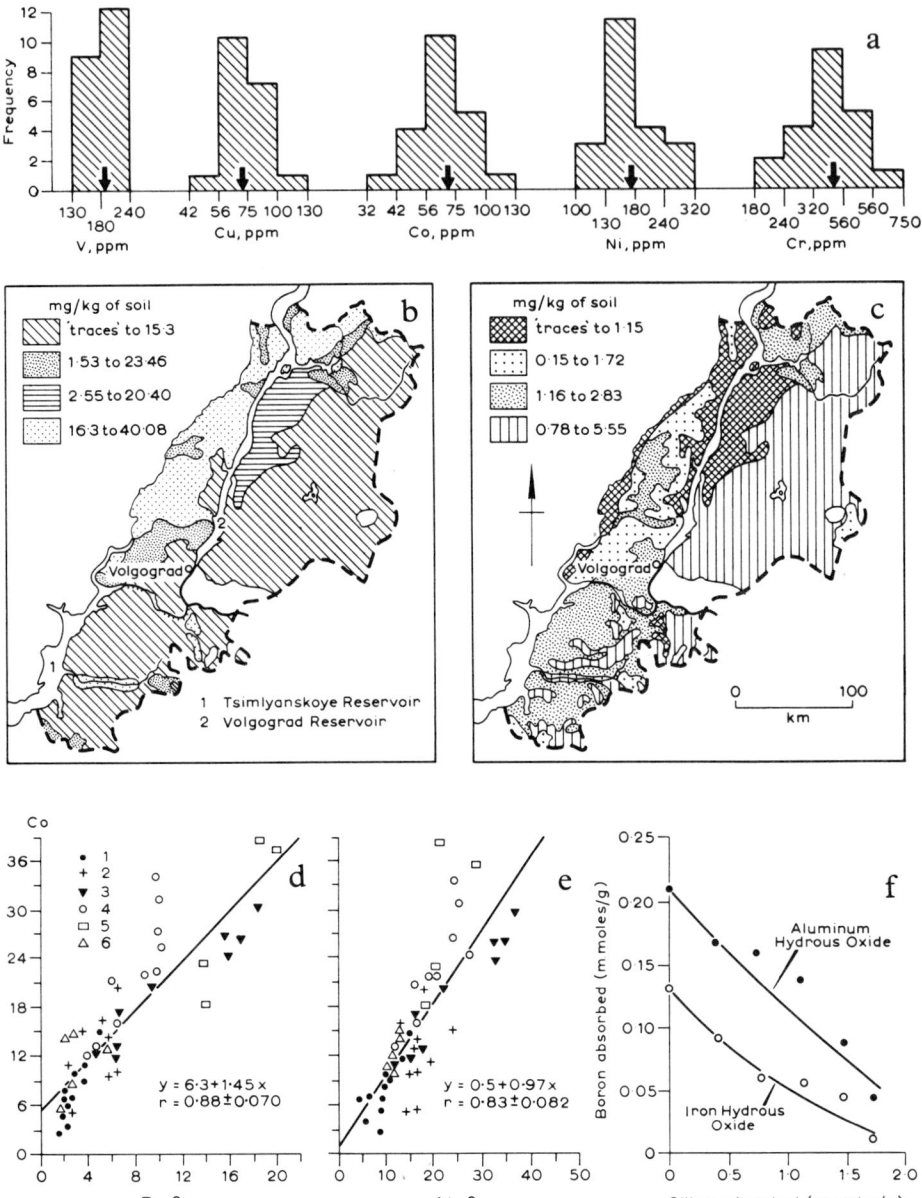

Figure 6.10 Variability of microelements and the importance of sesquioxides in their adsorption.

  a Frequency distribution of some microelements (Yaalon *et al.*, 1974), from several types of soil derived from basalts near Galilee. The mobility sequence in this Mediterranean climate is suggested as $Sr > Ba > Mn > Co, Ni, Cr, Cu, V > Ti$. Arithmetic means are indicated by arrows.

(Continued on facing page)

ed for nitrogen oxidation, sulphur is oxidized at all soil pH values. Retention of sulphate, however, is strongly pH-dependent because anion exchange increases as soil acidity favours exchangeable aluminium and iron increases. Thus leaching is least when soils are acid and the amounts of sulphur adsorbed above pH 6–7 are not significant (Harward and Reisenauer 1966). Sulphur processes may, in themselves, induce extremely low levels of pH. Sulphur oxidation is acidifying, since two hydrogen ions are produced for every sulphur atom which is oxidized, following drainage of submerged areas,

$$H_2S + 2O_2 \rightarrow H_2SO_4 \text{ (sulphuric acid)}$$

Where sulphur occurs in soils in pyrite minerals such as the sulphides of iron, copper, nickel and zinc, extreme pH conditions may again develop. Pyrite-rich soils are usually confined to former marine deposits, like the acid sulphate soils covering over 2 m ha in Vietnam. Free sulphuric acid is commonly present which dissolves clay minerals and produces large amounts of exchangeable aluminium in quantities toxic to most crops. Iron and manganese toxicities and phosphorus deficiencies are common. Under constantly flooded conditions, however, pH increases sufficiently to eliminate toxicities and paddy rice may be grown.

6.8 Micronutrients

6.8.1 MOBILITY AND ADSORPTION

Manganese (fig. 6.10b) may be immobilized as insoluble organic complexes, some of which may be of bacterial or fungal origin. In moving down the profile, after sesquioxides have been precipitated, Mn is oxidized and precipitated because of the sharp decrease in the amount of organic matter present in the soil solution. The oxidation and precipitation is

---

b and c. Maps showing geographical variability of manganese (b) and boron (c) content in the Chestnut and Light-Chestnut soils of Volgograd Oblast (Radov and Korchagina, 1967). The 0.78 – 5.55 mg/kg boron areas require less fertilizers than the other groups.

d and e. Association between cobalt and iron (d) and aluminium (e) in tropical soils of Burma (Obukhov, 1968). Differing tropical soils and environments are distinguished, according to the amount of iron and aluminium in the soil and the associated microelement content, dry savannas (1), tropical monsoon forests (2), tropical evergreen forests (3), sub-tropical Red Earths (4), lateritic soils (5), and dry forests and shrub thickets (6).

f Reduction in capacity of silicon-iron and silicon-aluminium oxide complexes to adsorb boron, depending on the amounts of Si adsorbed (McPhail et al., 1972). There is possibly direct competition between monosilicic acid and boric acid for adsorption sites.

aided by activity of the Mn bacteria and the catalysing effect of the MnO_2 already precipitated. Manganese moves in the form of organomineral compounds or as complexed colloids together with sesquioxides. Manganese also migrates in the ionic bivalent form, especially under anaerobic conditions, its oxidation and precipitation being prevented by reducing processes. The presence of mobile organic acids intensifies the reducing process. In calcareous soils, if favoured by carbon dioxide and organic acids, manganese may move as manganese bicarbonate. Much of the manganese in the soil is in the form of various oxides. These form a relatively insoluble reserve which may be brought into solution by reduction or through complex formation by root exudates. The active accumulation of Mn by plants results in a substantial amount of the total Mn present in the parent material being 'transferred' to the top organogenic horizons where it is converted into forms capable of migrating.

Molybdate is adsorbed by anion exchange with hydroxyl ions of hydrous ferric and aluminium oxides. Since ironstone nodules have been reported to contain 10 ppm Mo whereas aluminous nodules contained less then 1 ppm, a greater affinity of Mo for iron than for aluminium oxides is suggested. Molybdenum probably moves as an anion. By analogy with phophate transport and fixation by humic acid complexes, Mo could well be transported as a complexed Fe-molybdate (Taylor and Giles 1970).

Boron adsorption (fig. 6.10f) is a physical, non-metabolic process which can be described by the Langmuir adsorption equation for a wide range of boron concentrations. Boron, being poorly bound in the soil, is mobile and quite easily leached away. Cobalt in the soil is a component of the soil aluminosilicates (fig. 6.10d-e) and can also be found adsorbed on the surface of mineral and organic colloids. Cobalt, which is readily extracted from the soil by weak acids, is mobile and available to plants. Iodine is possibly adsorbed as are the anions phosphate and molybdate. In 23 soils from widely separated sites in the United Kingdom, iodine content, in the 2.7–36.9 mg/kg range, correlated with free aluminium oxide and ferric oxide, $r = 0.88$ and $r = 0.64$ respectively (Whitehead 1973).

6.8.2 AVAILABILITY

Plants take up divalent Mn from the soil solution. 'Available' manganese includes water-soluble and exchangeable manganese as well as some manganese bound to organic matter. Redox conditions in the soil are critical in the transition of manganese to an available state. When the redox potential is low, as for example in a rice field during inundation or when other flooded crops are grown, soils can usually provide adequate supplies of Mn to plants. Availability of Mo is largely controlled by the adsorption of molybdate from solution by hydrated ferric and aluminium oxides (Reisenauer *et al.* 1962). Availability is generally low in soils high in available manganese and iron oxides. The availability of sorbed Zn apparently decreases as the crystallinity of the sesquioxide increases.

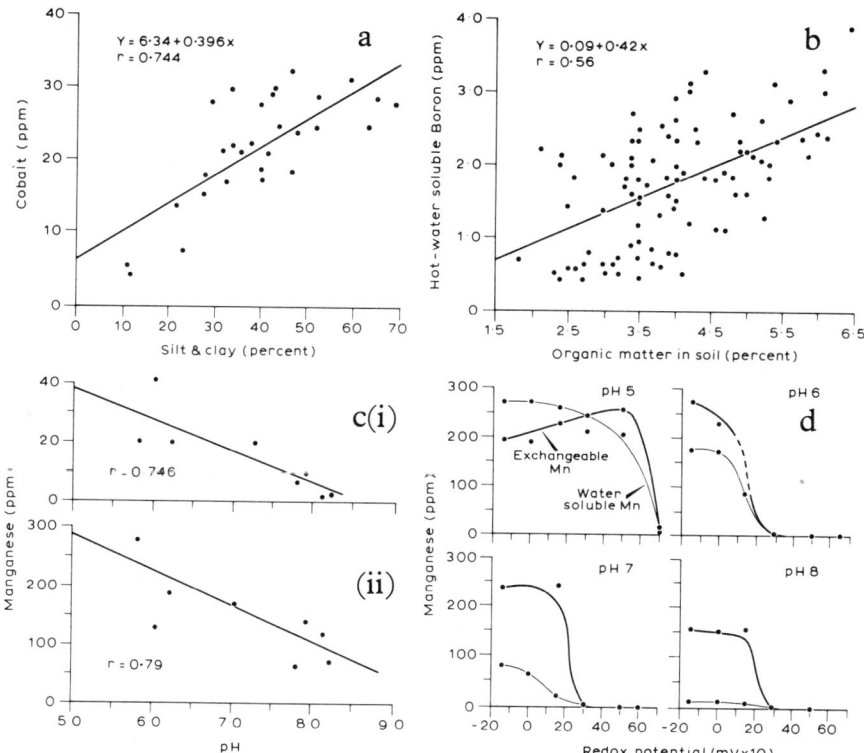

Figure 6.11 Several soil properties which influence microelement amounts and exchangeability.
a Influence of silt + clay content on total cobalt in the alluvial soils of the Indus-Ganges plains (Randhawa *et al.*, 1964).
b Influence of organic matter on boron content of soils in Nova Scotia and New Brunswick (Gupta, 1968).
c Influence of pH on the amounts of exchangeable (i) and active (ii) manganese (Randhawa *et al.*, 1961), in Punjab soils where both deficiencies and toxicity have been noted.
d Influence of redox potential and pH on manganese (Gotoh and Patrick, 1972), in the plough layer of the Crowley silt loam at the Rice Experiment Station in Louisiana.

6.8.3 INFLUENCE OF TEXTURE

The behaviour of most microelements is influenced by soil particle size and clay mineral type (fig. 6.11a). For instance, molybdate is sorbed by clay minerals like halloysite and kaolinite by anion exchange with hydroxyl ions. In Israel, total zinc contents of soils ranged widely from 40 to 142 ppm and increased with the clay content of the calcareous soils $r = 0.60$, $N = 21$. The soil clay fraction thus appears to be the main carrier of the total zinc (Navrot and Ravikovitch 1969). Silicate clays, in fact, adsorb zinc reversibly by cation exchange and irreversibly by lattice penetration.

6.8.4 ORGANIC MATTER

The presence of organic matter may promote the availability of a given nutrient by supplying complexing agents which reduce the fixation (fig. 6.11b). In Burma, organic matter has been observed to influence the content of acid-soluble Mn in non-calcareous soils, $r = 0.73$. Similarly, the correlation between available B and humus was $r = 0.27$ for calcareous soils and $r = 0.50$ for non-calcareous soils. Cobalt can form complex compounds with organic materials and peats may adsorb more Co than any of the clays. Copper is fixed in the soil in a form of low availability to plants since it combines with humic acid as an organometal complex. In consequence, copper deficiency is more frequent on organic soils than in comparable soils which have less organic matter. Total and extractable zinc increase with higher organic matter contents, again because zinc is also present in the soil solution largely as an organometal complex (Hodgson *et al.* 1966). Lead is rapidly immobilized in soils with high organic matter (Karamanos *et al.* 1976).

Water may be important where excess supply entails the disappearance of oxygen. An increase in manganous ion in the soil solution and on the exchange complex is, in fact, one of the first measurable effects of reducing conditions caused by flooding soils. Manganese in flooded soils is generally considered to exist in at least four forms, water soluble, exchangeable, reducible and residual manganese (Patrick and Turner 1968). The first two forms are largely manganous manganese, the third is composed of higher oxides of Mn, and the fourth is a minor constituent of soil minerals.

6.8.5 INFLUENCE OF pH

Soil pH has a marked effect on the activity of soil manganese, with a low pH readily bringing manganese into soil solution, even under aerobic conditions (fig. 6.11c,d). Mo deficiency is often associated with acid soils because it becomes less soluble as the pH falls. This relationship is the converse to that for most trace elements for which solubilities are higher in acid soils. Instead, the relationship between Mo and pH is analogous to that of phosphate when held in insoluble forms by iron and aluminium. Boron adsorption is distinctly different from that of other inorganic anions common to the soil system. Under acid conditions, adsorption of Cl, NO_3, and PO_4 is maximal. B adsorption is greatest at pH 8.5–9, A more acid medium favours the conversion of copper compounds to a soluble form (Sviridov *et al.* 1969).

6.8.6 THE INFLUENCE OF OTHER IONS

Soils having higher amounts of calcium carbonate fix higher amounts of nutrients. For instance, calcareous soils are generally considered as poor in 'available' zinc, owing to the ability of $CaCO_3$ to transform soil zinc into sparingly soluble compounds (Clarke and Graham 1968).

The chemistry of soil manganese is important to the nutrition of lowland rice (Tanaka and Navasero 1966). Hydrous manganese and iron oxides have been considered to be the principal solid phase components controlling the labile fraction of heavy metals in soils (Taylor and McKenzie 1966).

6.9 Sodium chloride and associated salts

The characteristics of sodium chloride and associated salts in soils are often little related to other soil properties. Instead, they are a function of climate, being mainly associated with hot, dry summers or arid climates, and with undrained depressions wherever a continental climate prevails. The distinctive features of salt-affected soils express seasonal changes in the balance between infiltration and evapotranspiration, as concentrations of highly soluble salts left when a predominantly upward flow of water in the soil evaporates.

All continents have salts in some of their soils, such as the vast areas in the northern Khartoum and Blue Nile provinces in the Sudan. In India, about 7.0 m ha are salt-affected, especially 2.5 m ha of the *usar* lands in the intensively cultivated arid and semi-arid regions of the Indo-Gangetic alluvial plain of Uttar Pradesh. There are many similar areas in the west and south of the U.S.A., including 60,000–80,000 ha in the Lower Rio Grande Valley in Texas. In Australia, the most alkaline continent, some 2.4 m km² are salt-affected. Locally, salt-affected soils occur wherever rainfall is seasonal and the balance between evapotranspiration-induced upward flow of soil water and the downward flow of infiltrating water and vapour becomes critical. In Saskatchewan, for instance, the control of climate is evident in the high salt concentrations in soils, where the average annual rainfall is 43 cm and evaporation is high, due to average wind velocities of 16 km/hour and long hours of bright sunshine favour high transpiration rates.

Salts are compounds formed from cations other then H^+ and anions other than OH^- which may alter the balance of the exchangeable cations on the soil colloids. The cations include Ca^{++}, Mg^{++}, and K^+ as well as Na^+ and the anions include chloride (Cl^-), sulphate (SO_4^{--}) and carbonate (CO_3^{--}).

6.9.1 CATIONS

Sodium is lost more readily in leaching waters than potassium or the divalent cations because of the low energy by which it is held by the soil colloids. Leaching is not simply a washing out of salts, but involves continuous internal reactions within the soil profile in which other exchangeable cations and the colloids are involved. Conversely, sodium moves upward equally readily by capillarity and, during dry spells, sodium is readily available to react on the soil exchange complex. Sodium is usually less than half the total of soluble cations and such proportions are not significantly adsorbed. However, the proportion adsorbed increases when the sodium percentage exceeds 50.

Upward movement depends on solubility, with highly soluble sodium moving upwards in greatest amounts. Such changes in the composition and concentration of dissolved cations alter the equilibrium between cations adsorbed on the clay surfaces and those in the soil solution. Calcium and magnesium salts move more slowly than sodium. As the electrolyte concentration increases, magnesium may precipitate as carbonate, followed by calcium precipitating as either carbonate or sulphate. Sodium has a lower affinity for clay than the other cations, and the proportion of sodium in solution tends to increase. Calcium and magnesium are the naturally dominant cations in arid-region soils. However, as groundwaters are raised by irrigation and as soluble salts from irrigation waters concentrate in the soil, the solubility limits of calcium sulphate, calcium carbonate and magnesium carbonate are exceeded, and calcium and magnesium are precipitated. Due to the dynamic equilibrium between soluble and adsorbed ions, the relative increase in the sodium percentage in the soil solution causes sodium to replace some of the calcium and magnesium originally present on the soil exchange complex.

A widely used criterion to describe the proportions of the major cations is the U.S. Salinity Laboratory's Exchangeable Sodium Percentage, ESP, where

$$\text{ESP} = \frac{\text{exchangeable sodium}}{\text{exchangeable (calcium + magnesium)}} \times 100$$

6.9.2 ANIONS

Chlorine is highly reactive and the chemical element Cl is not found free in soils. The main chloride compounds, sodium chloride (NaCl), calcium chloride ($CaCl_2$), and magnesium chloride ($MgCl_2$) are very soluble and are leached readily by rain and groundwaters or are drawn upwards by evaporation. Some crops are tolerant of high chloride concentrations, with clover and sugar cane being examples of crops which are very sensitive to chloride.

The chloride anion makes up 95 per cent of the total anions in well-drained areas in the Caspian region, with sodium chloride as the main salt component and calcium chloride or magnesium chloride also important. $CaCl_2$ and $MgCl_2$ are the dominant salts in the *sabakh* soils of southern Iran. In saline soils in Australia, chloride dominates the anions and sodium the cations to the extent that sodium chloride accounts for 50–80 per cent of the total soluble salts.

Sulphate (SO_4^{--}) behaviour in arid soils is predominantly the solution and precipitation of gypsum, calcium sulphate ($CaSO_4 \cdot 2H_2O$). Sulphates are also released during weathering, when oxidized mineral sulphides are precipitated as soluble and insoluble sulphate salts in semi-arid and arid regions. After sodium chloride, sodium sulphate and magnesium sulphate ($MgSO_4 \cdot 7H_2O$) are the most common soluble salts in soils. Soil moisture content, common ion effects, and ionic strength influence solubility. With greater amounts of water added, the net downward movement of sulphate is greater. The nature of the cations present influences sulphate movements. Leaching is greatest when monovalent cations such as sodium and potassium predominate, followed by the divalent cations, calcium and magnesium. The sulphates of calcium, magnesium, sodium, and potassium are frequently precipitated in soils in large quantities. Sulphate is weakly held, but is more strongly adsorbed than chloride or nitrate by soil clays and oxides. Sulphate leaches more slowly than chloride because the radius of the hydrated sulphate ion is 1.4 times that of the chloride ion. Also, if gypsum is present, sulphate ions continue to dissolve from gypsum crystals. Another factor which may result in a net retention of sulphate is the fact that it tends to move downwards more slowly than soil water. This is observed in Alberta where sodium sulphate ($NaSO_4$) and other soluble salts accumulate from groundwater as a dried salt crust which dissolves each time the soil is wetted (Khan and Webster 1968).

The carbonate ion (CO_3^{--}) is highly toxic to plants. Abundant carbonate ions in the soil are generally associated with high exchangeable sodium, and of all the salts which occur in soils, sodium carbonate is most harmful to both soils and plants. Their fertility, if any, is very low. In salt-affected soils in Europe, the percentage of sodium carbonate ($NaCO_3$) and bicarbonate may vary from a few tenths up to several per cent. Bicarbonate ions may be an appreciable proportion of anions in natural tropical and in irrigation waters. Soils having high organic matter contents release HCO_3^- and CO_3^- ions if anaerobic conditions prevail.

6.9.3 ENVIRONMENTAL INFLUENCES ON SALTS IN THE SOIL

Because of the high mobility of soluble salts, environmental influences are a more fundamental source of differences and distinctive characteristics in salt-affected soils than the solubilities of their ions and compounds. In addition to the control of climate and weather at all scales, the origin and evolution of the parent material is a major differentiating factor. In the Punjab alluvium, the main salt is sodium sulphate, inherited from the alluvium deposited under different climatic conditions in the past. In many areas, mineralized soil solutions have been inherited from a preceding stage of marine salt accumulation, including the 'quick-clays' of eastern Canada and southern Scandinavia. In Andalusia saline soils cover 160,000 ha in old tidal basins or estuaries. These *marismas* provide the pasturage on which most of Spain's fighting bulls are reared. In Antarctica, the sea is the probable source of salt crusts up to 10 cm thick, made up largely of chlorides, nitrates and sulphates of sodium, potassium, calcium and magnesium (Claridge and Campbell 1977).

Mineralized groundwaters influence soils decisively. In salt-affected soils in Europe, groundwater depths as shallow as 1 m are common. In Pakistan, a high salt level is maintained in B horizons by capillary rise from groundwater at depths of 3 m and more. The groundwater effect is very variable since flow may come from more than one source and direction may vary, depending on time of year. Interflow is also important (Conacher and Murray 1973) and direction of hillslope can be critical, as in Alberta where salt contents increase downslope and are highest near the base of slopes (Greenlee *et al.* 1968). Broad differences between soils at high elevations and those on the lower slopes can be explained by soil water flow which has removed salts from soils at higher elevations and increased sodium and magnesium sulphates and bicarbonates near the ground surface at the base of depressions (Maclean and Pawluk 1975).

6.9.4 SALINIZATION AND ALKALIZATION

The relative adsorption of different cations from solution occurs in proportion to their effective concentrations and strengths of adsorption. However, phenomena involving anion retention in soils are much less definite. The amount of anion retention is determined to a much greater extent by concentration, pH, and by the type of other ions in the solution.

In the U.S.A., the emphasis is on the cationic composition of salts, since cations are involved in exchange reactions with the soil, thus controlling its chemical and physical properties. In contrast, Russian scientists, in studying the composition of salts in soils, emphasize the significance of anionic composition. Thus, chlorine salinity and sulphate salinity are common terms in their soil science literature.

Salinization is essentially the presence of free sodium salts, mainly sodium chloride and sodium sulphate, dominating the soil. Other soluble salts that effectively contribute to soil salinity are mainly the chlorides and sulphates of calcium and magnesium, and the anions bicarbonate and carbonate may be present. If their concentration were equivalent to 1 mole/l of soil solution, an osmotic pressure equivalent to 24 atmospheres at 20° C is produced. Such osmotic pressure changes soil water behaviour, and reduces the availability of water to plants by adding to the soil moisture tension. The high osmotic pressure of the soils solution, rather than the presence of the salts *per se*, injures plants. Measuring osmotic pressure is tedious, but the analogy with an electric current provides an indirect estimate, if the conductivity of the soil solution is measured. Thus, the electrical conductivity of the saturation extract (EC_e), expressed in millimhos per cm at 25° C is recommended for estimating salinity hazard in relation to plant growth. Soils yielding saturation extracts with an EC_e greater than 4 millimhos are saline, according to the widely used specification of the U.S. Salinity Laboratory. At this level, the salt concentration in the soil extract is about half that which would exist at field capacity and about a quarter of that at permanent wilting point. In contrast, in most European countries, salinization is estimated directly from the soil rather from the conductivity of the soil water, with the determination of mineral residues in a 1:5 aqueous extract from salt-affected soils. Roughly 0.25 per cent of salinity measured in the 1:5 aqueous extract may be considered equivalent to the EC_e value of 4 mmhos/cm. The salinized soil profile is the Solonchak of Russian scientists, who recognize four types of salinization—carbonate and sulphate, sulphate and chloride, chloride with sulphate in small amounts, and preponderantly chloride. The original North American term was 'white alkali', a reflection of the efflorescences on the soil surface. Salinization is naturally favoured by the presence of impermeable subsoils and an occasional excess of water in the soil. Secondary salinization describes accumulations of soluble salts, mainly sodium chloride, as a result of irrigation, agricultural practices or by the clearance of native vegetation altering the natural cycle of moisture movement.

Alkalization describes the saturation of the soil exchange complex with sodium ions, identified when the ESP exceeds 15, and leading to the formation of a sodic soil. This process is the solonization of Russian soil scientists and involves both salinization and a change in the composition of the accumulated salts, with soils being affected by sodium salts capable of alkaline hydrolysis, such as $NaHCO_3$, $NaCO_3$ and $NaSiO_3$. Hydrolysis of sodium carbonate occurs as

$$2Na^+ + CO_3^{--} + 2H_2O \rightarrow 2Na^+ + 2OH^- + H_2CO_3$$

The resulting hydroxyl ions give pH values of 10 and above. A high pH indicates sodicity when exchangeable sodium data are lacking, but where local knowledge can be used to evaluate the significance of a given pH in terms of exchangeable sodium. At high pH and in the presence of the carbonate ion, calcium and magnesium are largely precipitated as calcium and magnesium carbonate. Such sodic soils therefore have few soluble salts, the ionic composition differs considerably from that of saline soils, and the ESP is high. An ESP of 15 is widely accepted as a lower limit, above which soils are adversely affected by the dominance of sodium in exchangeable form. The high pH values of 8.5–10 and the dominance of sodium ions brings organic matter into solution. One of the characteristic features of sodic soils is the redeposition of organic matter dissolved in the soil solution as a thin film on the surface of soil aggregates or on the ground surface itself. This characteristic suggested the term 'black alkali' for such soils, which are widely known by the Russian term Solonetz, and are distinguished by a prismatic or columnar structure developed in the B horizon. They form the most widespread group of salt-affected soils in Europe. As the salinity of the soil solution varies, non-saline-sodic soils are defined by an EC_e of less than 4 mmho/cm whereas saline-sodic soils have an EC_e greater than 4 mmho/cm. The combined processes of salinization and alkalization may create saline-sodic soils with sufficient quantities of both soluble salts and of adsorbed sodium to reduce the yield of most plants.

If desalinization occurs, following natural or artificial saturation of a sodic soil, degradation or 'solodization' describes the destruction of the adsorption complex as hydrogen replaces the adsorbed sodium. Solods may be leached completely of carbonates or they may contain carbonate concretions in the lower part of the illuvial horizon. The soil becomes acid, the clay remains dispersed and there is intense clay translocation. The topsoil may become sandy, bleached, and susceptible to wind erosion.

7 Soil mechanical properties

Soil mechanical properties are obviously an expression of the materials which make up the soil and of the water and air temperature changes within them (fig. 7.1a). However, it is logical to discuss soil mechanical properties after considering chemical characteristics of soils too, as it is the chemical processes and changes within soils which explain some of the widest contrasts in soil mechanical properties.

Inevitably, there is an inverse, functional relationship between porosity and bulk density as complementary measures of the same soil condition. Here the interest is in the relative bulk of the soil through which stresses are transmitted rather than the voids through which fluids and gases pass.

7.1 Bulk density

7.1.1 SOIL PROPERTIES AFFECTING BULK DENSITY

Bulk density is one of the main characteristics which describe the relative proportions of solid and void in a soil. Its expression in any one soil is related to texture. For example, the average value for a field of Panoche clay loam, a deep alluvial profile in the Central Valley of California, was 1.356 g/cm^3 (Nielsen *et al.* 1972). There is a tendency for bulk density to increase as texture becomes coarser. Such variations with texture are illustrated by a range of Chernozem soils, where the bulk density of coarse silt plough horizons averaged 1.44 g/cm^3 compared with 1.03 g/cm^3 for clay (Pokotilo 1967).

Organic matter is the major factor affecting bulk density (fig. 7.1b,c), particularly in uncultivated soils. Even in arable and grassland soils

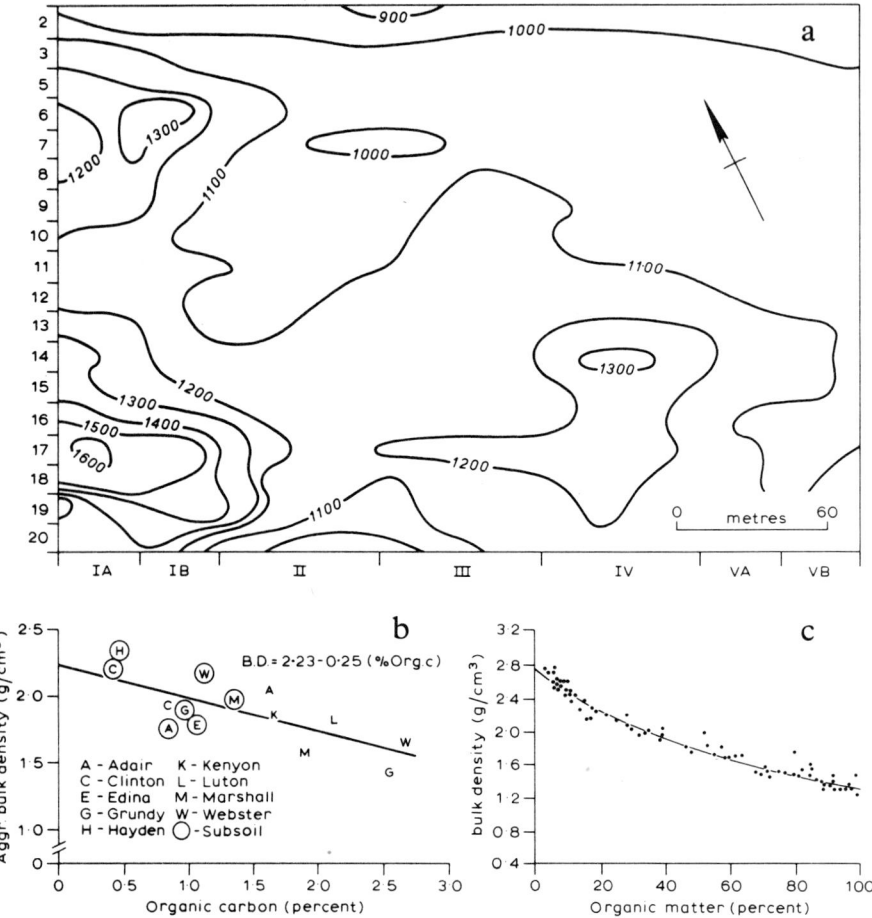

Figure 7.1 Soil bulk density and the significance of organic matter at both high and low values.
a Isodynes indicating patches of heavier soil on the Broadbalk plot (Avery *et al.*, 1969). Isodynes indicate lines of equal soil resistance measured in lb drawbar pull.
b Inverse relationship between organic carbon and soil bulk density (Tabatabai and Hanway, 1968), showing greater bulk density in subsoils of a wide range of soils in Iowa.
c Progressive reduction in soil true density up to 100 per cent organic matter (Adams, 1973). This relationship, observed in mudstone soils in Wales, is due to low true density of fibrous, poorly humified organic matter.

50 per cent of the variation in bulk density can be accounted for by variations in the organic content of the soil (Williams 1971). The control of organic matter on bulk density is particularly marked in peaty soils, as in the High Arctic terrain on Devon Island, N.W.T., where bulk

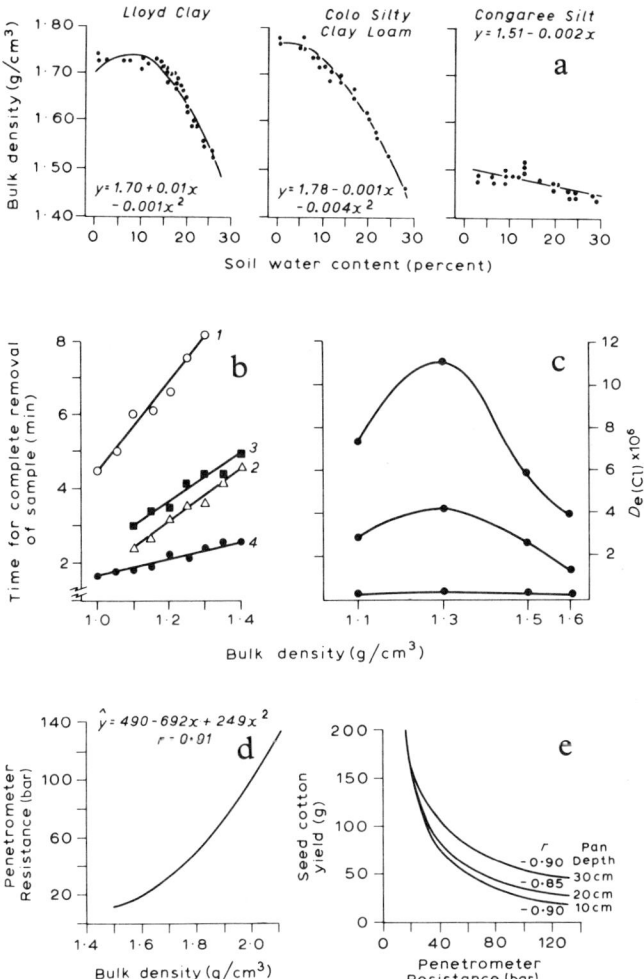

Figure 7.2 Soil bulk density, as affected by texture and water content, and its influence on other soil properties.
a Influence of texture and soil water on bulk density (Camp and Gill, 1969). The Lloyd clay and the Congaree silt are southern Piedmont soils and the Colo Silty Clay is found in Iowa.
b Influence of soil bulk density on erodibility of soils in the Yergeni Upland (Kuznetsov, 1967). The soils are the A_1 and B_1 horizons of a clay loam (1 and 2), A_1 horizon of a coarse loam (3), and the B_2 horizon of a medium loam (4).
c Optimum bulk density for diffusion in a silt loam (Warncke and Barber, 1972), Cl diffusion being examined at three moisture levels of 30, 20, and 10 per cent.
d and e. Relation between soil bulk density and penetrometer resistance after cotton harvest (d) and on seed cotton seedling yield (e) (Lowry *et al.*, 1970), on the Norfolk loamy sand soil of southern U.S.A.

density and humin content are strongly correlated ($r = 0.777$, $N = 23$) as are bulk density and ash content ($r = 0.671$) (Bunting and Hathout 1971). Degree of decomposition is the major factor.

Bulk density declines at increased moisture contents (fig. 7.2a). This results from the expansion of mineral particles when colloids swell and from the vertical movement of the soil if the moisture freezes. Similarly, the water-swollen fibres in peat deposits are much less dense than those of artificially dried materials. The chemical character of the soil solution is also significant in determining bulk densities.

7.1.2 SIGNIFICANCE OF BULK DENSITY

Changes in bulk density strongly influence permeability, drainage rate, trafficability and penetration by plant roots and burrowing animals. A change of 5 per cent in the bulk density of a sandy soil may alter the hydraulic conductivity by 1–3 fold. Such physical changes influence the diffusion of ions through the soil. At low bulk densities and with low moisture contents, compaction may increase the continuity of the liquid phases and hence increase the thickness of the water films around soil particles. This more continuous aqueous system in soil pores makes the diffusion path less tortuous, and reduces the interaction of the ion with the soil. However, at some critical bulk density value, the increased contact between solids in a unit volume begins to obstruct diffusion paths (fig. 7.2c). Such a critical value would vary according to soil texture and ion concentration.

From the point of view of plant growth, the mechanical impedance of soil bulk density has a widespread influence on root penetration and growth (fig. 7.2d). In sandy soils, growing roots penetrate bulk densities of 1.6–1.8 g/cm^3 only with difficulty. For example, critical values for pine root penetration in sands of the Middle Don are 1.5–1.8 g/cm^3 (Zyuz 1968). Critical densities at which impeded root growth is reported range from 1.3 to 1.8 g/cm^3 (Schuurman 1965), with 1.94 g/cm^3 as an upper limit. For example, cotton root weight and depth of penetration decrease gradually from soil bulk densities of 1.3 to 1.5 g/cm^3 (fig. 7.2d). There is a sharp decrease in root development with higher bulk densities. In rice soils, bulk density may be far more significant than texture. In monsoon Asia, a near-functional relationship between bulk density and available water capacity ($r = -0.90$, $N = 215$) has been observed (Kyuma et al. 1977) and with only low or insignificant correlations with texture ($r = 0.10$ and $r = -0.11$ for clay and for total sand, respectively). In leached residual laterite profiles, the leaching may have reduced bulk density in undisturbed soils. In such profiles in Brazil, voids of up to 60 per cent of the total volume have been noted, and serious engineering problems can arise if such high void ratios are not anticipated.

7.2 Soil strength

Although bulk density is a major control on the strength of a soil, other closely related properties are also recognized.

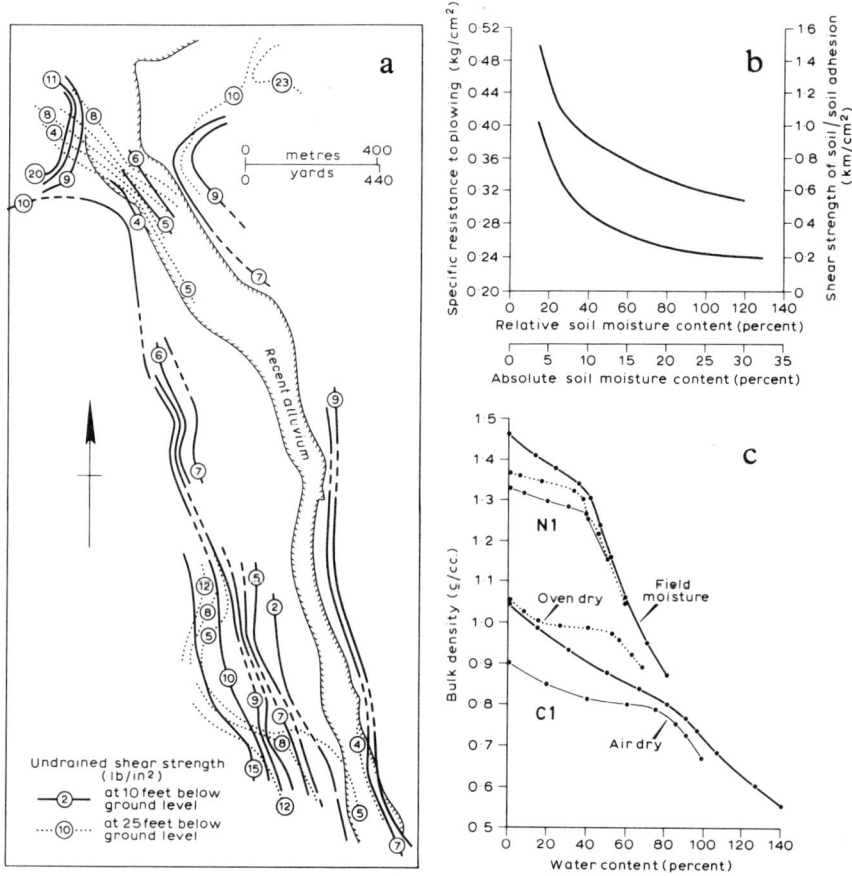

Figure 7.3 Properties affecting strength and stability of soils.
a Iso-strength lines (undrained shear strength) for Team valley sediments, Gateshead (Taylor, 1972). Such superficial deposits, associated with buried channels, create similar slope instability problems from the R. Tyne to the R. Tees in north-east England.
b Soil shear strength (above) and adhesion (below) in Sod-Podzolic soils near Moscow, and their specific resistance to ploughing (Bakhtin et al., 1969).
c High water content at shrinkage limit, one of the unusual physical properties of allophane-containing soils (Warkentin and Maeda, 1974). Soil N1 is from Hokodate and C1 is from Dominica. The Dominican soil has a high shrinkage limit of 105 per cent, compared with 40 per cent for the Japanese soil, 30 per cent for kaolinitic clays and 15–20 per cent for swelling clay.

7.2.1 COHESION

As a soil dries out, bulk density increases and even the loss of small amounts of water may increase significantly the strength or cohesion of soils (fig. 7.3). Particles pack more closely and non-capillary porosity decreases. The number of inter-particle contacts increases. With continued drying, water forms a membrane retreating into the micro-pores of peds and their cohesive strength is largely dependent on soil suction. The maximum bond is developed when soils are wet with an extremely restricted amount of water. With small amounts less or more than the optimum amount of water, the strength of the bond is greatly reduced. The number of points of contact between aggregates is minimal in dry soils and the maximum cohesion is about field capacity because of the increased thickness of the water films which, when they thicken at higher moisture contents, then decrease the number of contact points. Cohesion does not depend entirely on contact and thickness of water films. Electro-negative charges, polarized water molecules and dissolved ions all complicate the character of the bonding strength. In the case of the 'subplasticity' property of some Australian soils, cohesive strength is sufficient for aggregates to display properties associated with sands and gravels (McIntyre 1976).

Cohesion is commonly measured as the soil's resistance to penetration, shear, or crushing. Resistance measurements by penetrometer may correlate closely with rates of plant root elongation (Taylor and Ratliff 1969). Cohesion also affects the draft and performance of tillage implements, and erosion by surface runoff is greatest where cohesion between soil aggregates is low. Soil erodibility is directly associated with the mechanical strength of soil clods, and the terms cohesion and wind-resistance can be used synonymously to express the same concept (Shiyatyy et al. 1972). In northern Kazakhstan, resistance to wind erosion is highly correlated with percentage fine clay ($r = 0.89, N = 34$). In engineering, the range of soil moisture conditions is critical, since all unsaturated soils have a large cohesive component of strength. Saturation may follow intense tropical downpours and in Hong Kong, where cuttings in rhyolite soils are stable during the dry season, softening follows the infiltration of the rainy season and collapse is common (Lumb 1965).

7.2.2 SHEAR STRENGTH

Soil will deform under stress, its mechanical strength being described by the shear strength at failure. Shear strength describes that major soil-engineering property which provides the support for artificial structures and maintains the stability of a hillside (fig. 7.3a). It is defined as the force required to initiate sliding between two soil surfaces of unit area.

In cohesive, non-granular materials the shear strength, S, may be expressed as S = C, where C equals cohesion, usually expressed adequately by the Voids Ratio. With coarser particles present in the soil, a second variable is considered, ϕ, the angle of internal friction. This term expresses the friction between adjacent coarse particles and the interlocking action which prevents the movement of individual grains. The angle of internal friction is thus related to the repose angle of a pile of granular material. It is greater for angular particles and for poorly sorted materials. The shear strength of a pile of loose, coarse material may therefore be expressed as

$$S = N \tan \phi$$

where N equals the normal stress. For most engineering soils, the shear strength depends on both cohesive strength and on the angle of internal friction, described by the generalized Coulomb equation as

$$S = C + N \tan \phi$$

A feature which, in practice, the failure criterion must accommodate is the prefailure increase in strength as the average effective stress increases. In the field, deformation by shearing is very closely linked with the amount of water adsorbed by the soil (fig. 7.3b).

To variable degrees, a soil loses shear strength on being remoulded, either by artificial or by natural disturbance. Sensitivity is defined as the ratio of the unconfined compressive strength of an undisturbed clayey soil to the strength of the same clay once disturbed and remoulded. Ratios of more than 4 indicate sensitive materials. The quickclays of Scandinavia and eastern Canada lose virtually all their strength on being disturbed. Thus, the Leda clay with a sensitivity ratio of 150 is very prone to rapid landsliding.

7.2.3 ADHESION

Adhesion is a soil property which is only recognized when an implement is drawn through the soil. It is attributable partly to molecular forces. Adhesion increases as the soil texture becomes finer. Adhesion also increases with greater contents of soil organic matter which improves the microstructure of the soil. In general, the adhesion between soil and a foreign object can be attributed essentially to the water film between the joined surfaces (fig. 7.4). Within the 'sticky' range, the value of adhesion has been found to equal the product of the area of the water film joining the two surfaces and the tension within it. Electrostatic forces are also involved and adhesion is related to the electrolytic character of the soil solution. For example, in soils in the Tatar A.S.S.R. adhesion

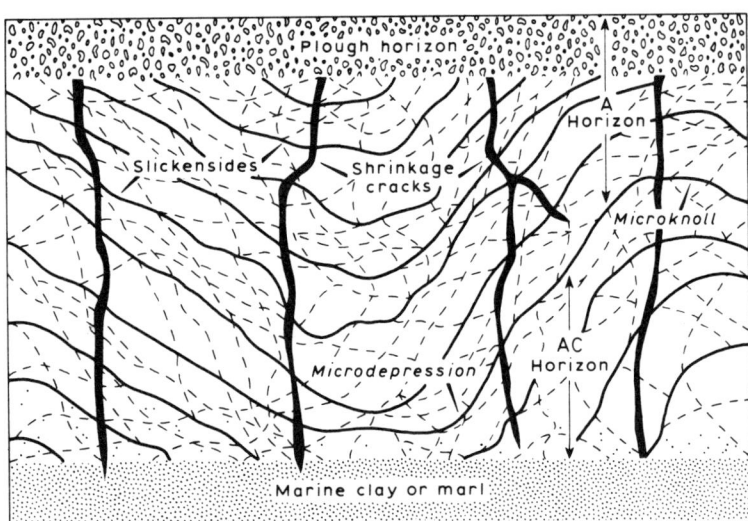

Figure 7.4 Contrast between the properties of adhesion and cohesion.
a Non-soil material adhering by water films to relatively dry, coarse soil (left) and wet, fine soil (right) (Fountaine, 1954).
b Cohesion holding natural soil structures of varying size together during periodic shrinkage (Ritchie *et al.*, 1972). The example is the Houston Black clay, a typical Vertisol covering 10 m/ha in Texas.

in sodium-saturated soils was $30.7\,\text{g/cm}^2$ compared with $17.4\,\text{g/cm}^2$ in calcium-saturated soil (Koloskova and Burlakov 1969).

Soil 'stickiness' which arises at moisture contents when the cohesion within soil crumbs and clods is less than its adhesion to a foreign object is one of the farmer's most troublesome soil physical properties. An excessively sticky soil makes cultivation difficult and prevents the harvesting of root crops in a clean condition.

7.3 Swelling and dispersion

The soil engineer may encounter the severe risk of structural damage caused by swelling soils. The seat of swelling forces is the physico-

Soil mechanical properties 211

chemical characteristics of expanding-lattice clay minerals. As such clays occur in sub-humid areas, adsorbed ions, particularly sodium, have a pronounced effect on swelling properties. Engineering problems relate to the magnitude of the swelling pressures, the volume increase that may take place, and the reduction in shear strength due to the swelling pressures in the soil mass.

7.3.1 DISPERSION

The disruptive effect of the sodium ion on soil physical conditions explains the behaviour of Solonetz soils (Harker et al. 1977). The effect of a sodium-rich soil solution is exerted primarily on the micro-aggregate characteristics. Upon hydration, the radius of the sodium ion increases nearly eight-fold, from 0.98 Å to 7.8 Å, a size too large to fit within clay crystals or even within the adsorbed water layer on mineral particle surfaces. Such exchangeable ions with their shells of water are not held very tightly by the electronegative charge on the clay particle surfaces and move away to positions of equilibrium in the soil solution. As these highly hydrated sodium ions are only loosely held, they do not neutralize the electronegative charge of clay particles which repel each other and remain dispersed. Dispersion will not take place until the quantity of exchangeable sodium reaches a certain percentage of the total CEC since calcium ions are held in the adsorbed layer whilst sodium ions occur mostly in solution. Change in structure or dispersion is not gradual but occurs quickly when the ESP reaches the critical limit. The clay particles disperse because a pressure builds up due to the replacement of Ca^{++} ions with two Na^+ ions at a given exchange site.

Other ions and soil properties may be involved in dispersion. For instance, the sodic Houston Black Clay (fig. 7.4b) tends to lose structural stability less than the Sudan Gezira clay, due to higher organic matter, less sand and perhaps less native sodium (Mukhtar et al. 1974). Exchangeable K and NH_4 may exert a dispersing action on clays intermediate between that of Ca and Na, but they do not necessarily adversely affect physical properties. After flooding in paddy fields, hydration of reducible iron and manganese oxides, silicates and organic matter facilitates swelling.

7.3.2 AMOUNTS OF SWELLING AND WATER ADSORPTION

Swelling of a rice field Vertisol has been measured in New South Wales, near Coleambally (Talsma and van der Lelij 1976), with a maximum vertical movement of 25 mm involving swelling of soil material between 5 and 80 cm depth (fig. 7.5). Elsewhere, swelling has been attributed to moisture adsorption alone, as at the Western Galilee Experimental

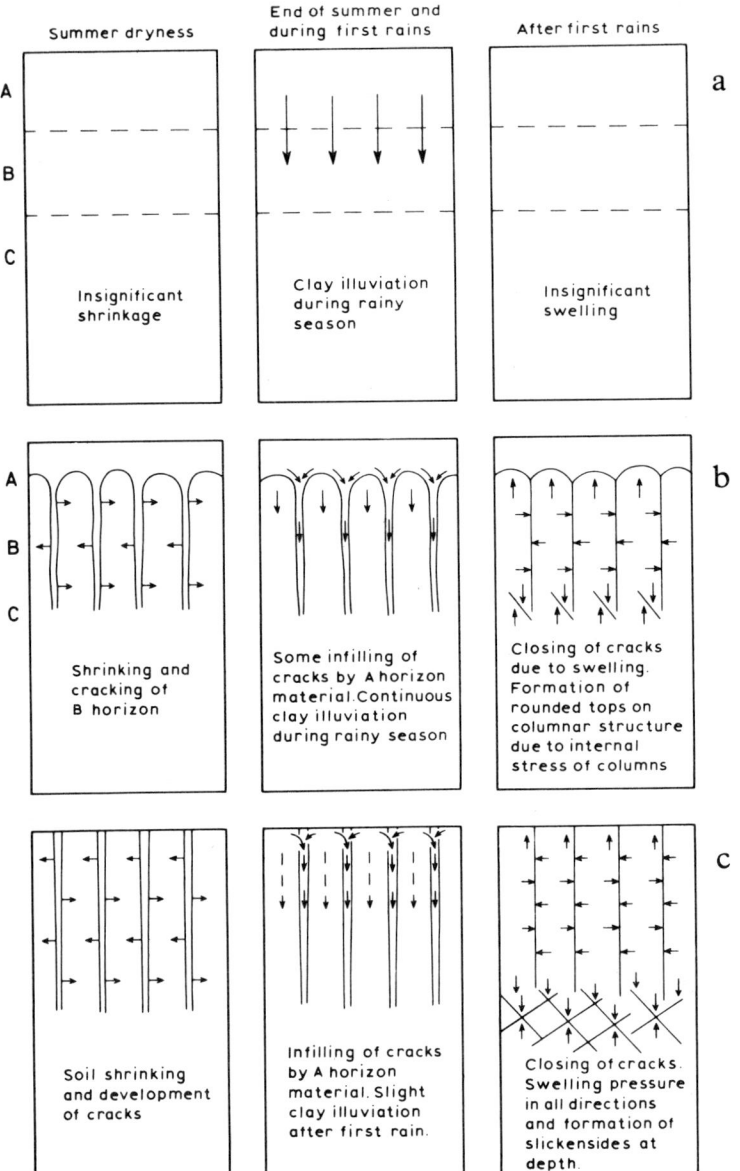

Figure 7.5 Influence of texture and clay mineralogy on shrinking and swelling (Dan *et al.*, 1968), relating to soils in the Sharon region of Israel.
a Sandy clay loam, tending towards kaolinization.
b Sandy loam with pronounced clay movement into the B_2 horizon.
c Non-leached, montmorillonite-dominant swamp soil.

Farm, where vertical rises of 0.3 mm/day were recorded (Yaalon and Kalmar 1972).

7.3.3 HYDRAULIC CONDUCTIVITY

Water flow can be slowed by the chemical dispersion and swelling of soils as this results in the plugging of the water-conducting pores and fissures by the dispersed or swollen materials. In sandy loams in southern Alberta, the hydraulic conductivity of Ca-treated soils is 10.2–15 mm/hour. The greatest decrease in hydraulic conductivity with increased ESP was between ESP values of 10 and 15, with rates less than 0.2 mm/hour when the ESP exceeded 20–30 (van Schaik 1967). Potassium ions also may have deleterious effects on hydraulic conductivity, chiefly attributable to clay swelling and the subsequent restriction of conducting pores. In acid soils, NH_4 oxidizes slowly and accumulates. At pH 4 or below, soil aggregation deteriorates slightly and permeability is greatly reduced due to the high exchangeable NH_4.

7.4 Shrinkage and fissuring

The shrinking (fig. 7.4b) and swelling of clays in compact soils leads to a change in volume of the aggregates. Wherever there is cohesion within an aggregate or clod, this soil structural unit is compressed or expanded. Thus, a specific characteristic of the Very Compact soils of Cuba is their capacity for severe contraction on drying. Volumes change by 30 per cent, porosity declines from 47–56 per cent to 33 per cent, and bulk density of the clods increases from 1.2–1.5 to 1.7–1.8 g/cm³ (Val'kov 1968). In fact, cracking influences soil development to such a degree that it is recognized at the highest level of soil classification. Vertical fissuring is the main field characteristic of Vertisols as defined in both the FAO and the U.S.D.A. Soil Taxonomy. Desiccation fissures, however, are present in a much wider range of soils. In desert soils in Mesopotamia, desiccation cracks in silt loams, coated with humus and clay, persist below a depth of 70 cm. Similar structures develop in the Netherlands as shrinkage follows the gradual drying out of recently reclaimed polders. In South Dakota, giant desiccation cracks have been located where surface collapse in holes, 1–3 m wide and equally deep, have appeared. They are most common where alfalfa has been grown for a period of 5–10 years (White 1970).

Under natural soils in the Gezira, shrinkage is about 15 per cent at the surface, declining to 3 per cent at the 0.5 m depth and then decreasing slowly to nil at 130 cm depth (Zein el Abedine and Robinson 1971). Crack widths show fewer differences than depths because, once soil moisture reaches the hygroscopic range, surface shrinkage stops. Cracks will continue to deepen, however, as long as depletion of moisture continues from the wetter layers at depth. The cracks that are produced by shrinkage can be semi-permanent. Even with the cracks becoming plugged with surface materials, deep fissures remain open and the cracking pattern will reform along these planes of weakness.

The effects of clay content on the width of cracks has been demonstrated for the Gezira where soils with 55–67 per cent clay, compared with soils with 71–76 per cent clay, have average crack widths of 3.4–3.7 cm and 4.8–5.1 cm, respectively. Soil characteristics affected by cracking include the distribution of organic matter and the processes of humifying plant residues. Upper horizons are deprived of organic matter whilst subsoils are enriched in it.

From the agricultural point of view, cracks are important as pathways for downward movement of water during the growing season, especially in drier years. However, all types of problems can arise in fissured soils. It is difficult to ensure uniform distribution of irrigation water. Cracks, as in the Tuares soils in Algeria, favour the initiation and rapid extension of water erosion. Cracking may sever plant roots and expose the broken ends to infection. Where cracking is bad it is difficult to pasture cattle in safety and farmers using machinery are at risk.

7.5 Consistency or Atterberg limits

The greater the amount of moisture in the soil the more it may behave like a liquid, due to reduced interaction between adjacent particles (fig. 1.2b). The physical state of water held directly on the surfaces of clay particles differs from that of liquid water. Plasticity develops when all exposed surfaces have a water film with its definite configuration, together with a little more which develops little or no definite configuration. These water molecules are free to behave as liquid water and greatly weaken inter-particle bonds. The three Atterberg limits are the shrinkage, plastic, and liquid limits and they define four states of mechanical behaviour in engineering soils which integrate the effects of several soil properties. They are also applicable in describing the plastic behaviour of agricultural soils (Towner 1974) and the consistency or 'feel' of the soil when kneaded between finger and thumb by the field surveyor. The plastic limit appears to be closely related to the upper moisture limit for ploughing.

Greater amounts of organic matter in a soil tend to increase the plastic and liquid limits, probably due to their effect in reducing compaction and to the strong attraction of water to humus. For a wide range of soils in south-east Scotland, readily oxidizable organic matter, which had an inverse correlation with bulk density ($r = -0.81$, $N = 58$), correlated closely with plastic limit ($r = 0.71$, $N = 51$) and with liquid limit ($r = 0.57$, $N = 56$) (Soane et al. 1972). In sandy halloysite and chlorite soils in Japan, the limits increased almost linearly with an increase in organic carbon (Kubota 1971). Also, the ions adsorbed or present in the diffusion layer around the soil colloids influence the percentage moisture at which critical limits are reached. CEC expresses well the attraction to water of the soil colloids and correlations between CEC and both plastic and liquid limits have been reported (Farrar and Coleman 1967) for a range of British clay soils ($r = 0.90$, $r = 0.81$, respectively, $N = 19$). In drier areas, sodium clays require less water to develop plasticity than do calcium-clays.

There are two distinctive features of consistency limits in tropical soils and lowland rice soils. Soils with high water contents lose their rigidity if mechanical force is applied. An infinite number of shear planes are developed within the soil. This artificial liquifying of aggregates is termed puddling, the detrimental and unintended result of working soils when too wet for the growth of crops other than rice. However, it is the deliberate and painstaking purpose of lowland rice agriculture to puddle soils. Aggregate destruction eliminates non-capillary pore space, the increased capillary porosity and increased moisture retention thereby retarding evaporation and percolation losses. The mean diameter of water-stable aggregates has been noted to decrease from 1.7 mm to 0.36 mm due to puddling. Structure is reformed as the rice crop grows and by cracks a few millimeters apart which form when the soil dries (Chaudhary and Ghildyal 1969). Like ploughing for dry-foot crops, the aggregate disruption of puddling accelerates organic matter decomposition. The availability of nutrient elements, particularly P, may increase with puddling, but this is probably an indirect effect, due to the greater continuity in the soil water system (Sanchez 1973). Puddling of sodium-saturated clays is easy, and montmorillonite clays puddle more readily than kaolinitic clays or sandy soils containing iron and aluminium oxides.

When siliceous particles are completely covered with hydroxides of iron and aluminium, water saturation is rarely attained. The low moisture absorption, usually associated with kaolinitic clays and sands is completely masked by the several hydroxide layers of such lateritic soils. The red clay soil in Kenya, although mainly kaolinitic, has a moisture content *in situ* as high as 30 per cent, and a plastic limit of 35 and liquid limit of 69 (Coleman et al. 1964). Also, because laterite soils may represent various stages of weathering, the Atterberg limits may not indicate their engineering properties adequately if the degree of decomposition is not defined (Lohnes et al. 1971). The desirable engineering properties of low plasticity, high-bearing strength and permeability may disappear after reworking in the presence of water, as this removes the free iron oxide, and structure breaks down irreversibly.

7.6 Compaction

Bulk density increases (fig. 7.7a), with compaction occurring primarily at the expense of the largest pores when the soil is compressed by an external load, as this is the zone in which the soil has least mechanical stability. In most engineering constructions involving earth foundations, fills or subgrades, compaction is the vital soil property. Compaction is now a major limitation on the productivity of many soils, particularly where chemical and biological imbalances are readily corrected with chemical additives.

The narrow front wheels of a tractor impose a pressure of up to 40 kg/cm² on soils (Ishii and Tokunga 1972). This compactive effect is essentially the same relationship of soil mechanics (fig. 7.6). As it is effective below the depth to which normal tillage implements work, compacted subsoil pans may develop, as in the southern part of the U.S. Corn Belt. Compac-

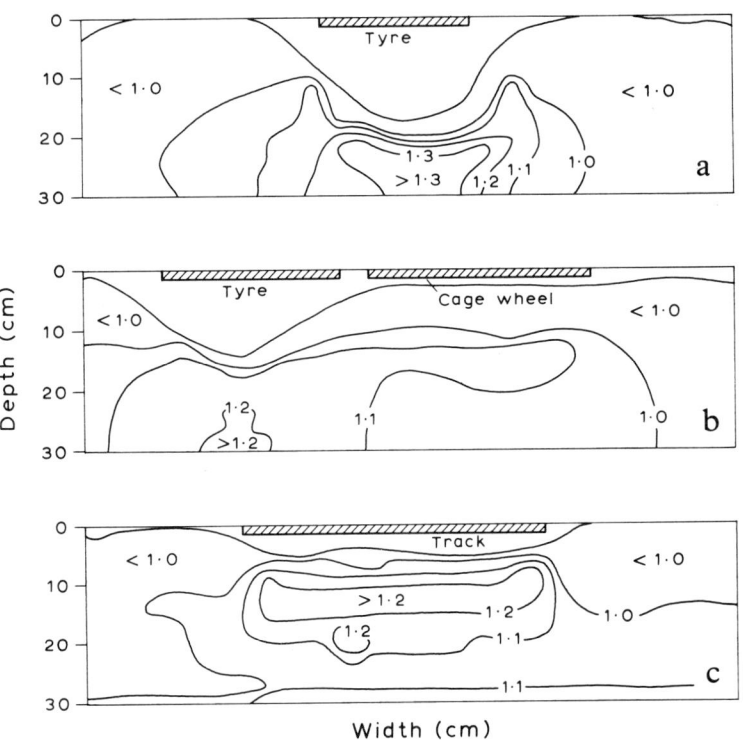

Figure 7.6 Bulk density of a test bed of sandy loam after compaction by passage of wheeled traffic (Soane, 1973). The soil below the tyre bears the greatest part of the axle load, which is not eliminated by the cage wheel. The tracked vehicle eliminates wheelslip but does not eliminate some compactive effects.

tion by trucks at harvest can almost double the energy required for subsequent ploughing. Tyre traffic may reduce air voids to the critical 10 per cent or less and hydraulic conductivity may be severally restricted. For instance, compaction of a fine sandy loam from 1.31 g/cm^3 to 1.5 g/cm^3 changed infiltration from 40 mm/hour to 1 mm/hour (Bateman 1963).

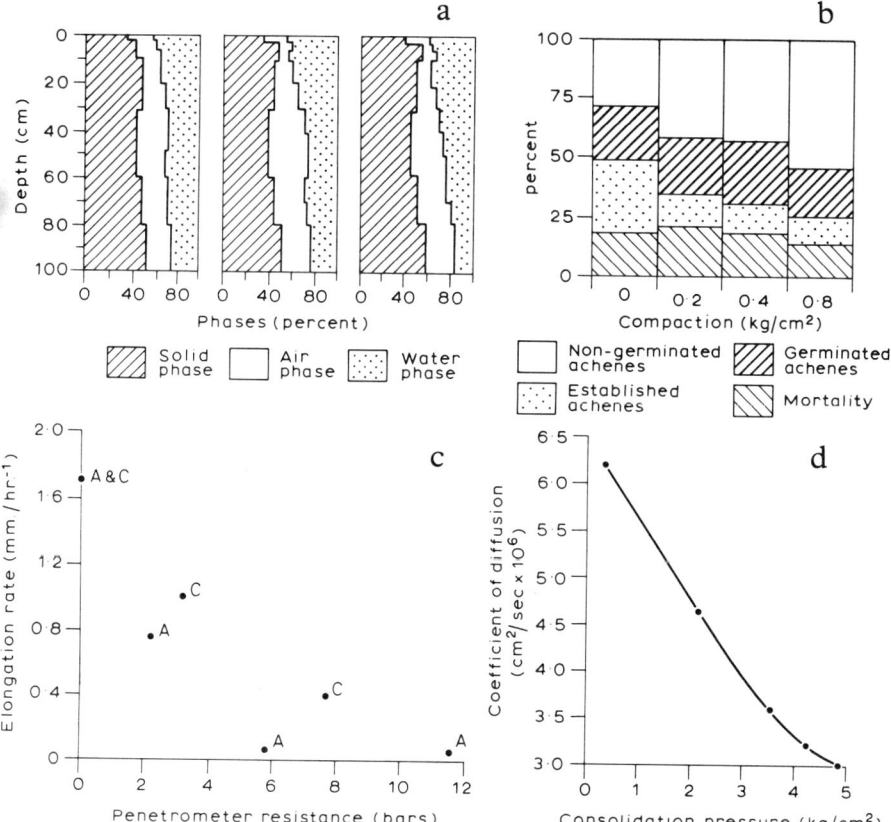

Figure 7.7 Various effects of soil compaction.
a Effect of compaction of a Chernozem (left) in the Bashkir region (Ishem'yakov and Taychinov, 1966), with a 250 g/cm^2 load (centre) and with a 300 g/cm^2 load (right).
b Effect of compaction on germination, establishment, and mortality of Compositae (Sheldon, 1974), recorded 11 weeks after sowing and at compactions of 0, 200, 400, and 800 g/cm^2.
c Reduction in root elongation due to increases in soil strength (Taylor *et al.*, 1967). Soils used are the Amarillo fine sandy loam from San Joaquin Co., California, and the Columbia loam from the Texas High Plains.
d Effect of consolidation on the coefficient of air diffusion through saturated kaolin (Sides and Barden, 1970), due to constriction and increased viscosity of adsorbed pore water as the clay surface is approached.

An increase of bulk density by tyre traffic of only 0.1 g/cm^3 is sufficient to flatten sugar cane roots in Hawaii.

To some extent, compaction is necessary for seed germination and for the soil to support plants physically. Yield responses therefore have a parabolic relationship with soil bulk density, if a range of compactive forces have operated on a soil. For agricultural soils in Estonia, the best compaction for optimal growth and development is no more than 1.15 g/cm^3 for sugar beets, 1.15–1.20 g/cm^3 for spring wheat, and 1.25 g/cm^3 for millet and corn. There are also several natural processes which compact the soil. Thus, compaction is most marked where clays swell and shrink in response to moisture changes. Ubiquitous micro-compaction effects are induced by burrowing animals and by plant root pressures, their effect being confined to the immediate neighbourhood of the burrow or plant root (fig. 4.2d). Due to particle size distribution or to insufficient aggregation, pore sizes may be less than the diameter of a growing root tip (fig. 7.7c) and roots may continue to elongate by deforming the surrounding soil.

7.7 Brittleness

Brittleness is a condition in which a coherent portion of soil withstands increasing pressure without deformation until the material shatters abruptly (Grossman and Carlisle 1969). Compact subsoils of high bulk density which are hard when dry and remain firm when moist display this property. The term fragipan identifies this distinctive B horizon, formerly called hardpans or siltpans and now designated by the suffix *x*. The most strongly expressed pans occur in the transition stages between moderately well- and poorly drained soils. The upper surface of fragipans has been reported at depths of 55 cm in south Wales (Crampton 1963), 35–55 cm in Scotland, and at 30–90 cm in Pennsylvania. The Scottish fragipans may have a pore space of only 20 per cent, compared with 50 per cent in the soil horizons above. The Erie fragipan has a bulk density of 1.85–1.9 g/cm^3 compared with 1.7 g/cm^3 for the soil as a whole. Progressive increases in bulk density from well- to poorly drained soils have been noted.

Fragipans present management problems for agriculture and they complicate some non-agricultural uses. Saturated conditions above the fragipan during much of the year harm roots of perennials such as trees and vines (Bradford and Blancher 1977). A restricted air and water supply due to the limited rooting volume is probably the principal factor causing reduced yields of cotton in the southern U.S.A. Roots cannot usually penetrate the dense fragipan. Slow subsurface flows complicate the engineering uses of such soils. On the other hand, such horizons are locally important in subarctic Quebec as the base of many dirt roads (Moore 1976).

Fragipans also present considerable problems in interpreting their distinctive physical characteristic. In the Condroz area of Belgium, for instance, differences between the fragipan horizons and the rest of the soil are not sufficiently significant to indicate that any particular cementing agent is present. Brittleness is attributed simply to compaction, with fragments held together by their compressed clay constituents. The scarcity of fragipans in calcareous soils and the influence of carbonates on silica solubility could be related. In fact, more recently, Indiana loessial fragipans were found to have more than 0.2 per cent SiO_2 whereas non-pan horizons had less than 0.2 per cent SiO_2 (Harlan et al. 1977). It is suggested that such traces of silica, after moving down the profile in solution, accumulate with soil water in a wet zone. If this zone is within the rooting depth of trees and thus becomes seasonally dry, silica would concentrate, precipitate and bind particles together as a brittle pan.

7.8 Cementation of iron concretions

The segregation of iron and manganese and their cementation into concretions is common in the upper horizons of several soils. In cool temperate environments where thick organic surface layers and gleying characteristics indicate periodic waterlogging, concretions are common, containing all the constituents released during soil weathering. Characteristic features of concretions, or ortsteins, are the inclusion of skeletal silt particles, chiefly quartz, and their cementation with an ochreous-brown substance. Both iron and aluminium, in conjunction with organic matter, can be the cement. Black manganiferous material is also cemented to soil fabrics (Taylor et al. 1964). Typically, soils with concretions in their eluvial horizons contain soft diffuse nodules in the B horizon. Towards the soil surface, nodules increase in abundance and hardness, suggesting that frequent wetting and drying is important in hardening.

The initial form of concretions is a water-stable aggregate, perhaps located near a living root or an accumulation of organic matter. A relatively narrow interval of redox conditions is critical, particularly for the migration and accumulation of manganese. Coatings of even traces of MnO_2 appear to function as condensation nuclei for catalytic acceleration of the precipitation of coagulations of MnO_2 and $Fe(OH)_3$ if these are in suspension. Thus small ortsteins enlarge, become compact and coalesce to form larger nodules. Iron, in particular, is involved, due partly to the closeness of the ionic radii of Mn^{++} (0.80Å) and Mn^{3+} (0.66Å) to those of Fe^{++} (0.76Å) and Fe^{3+} (0.64Å) respectively, and to similar chemical properties, including reversible oxidation-reduction solubility. Some concretions have a distinct concentric structure (Ojanuga and Lee 1973) whilst others may be formed by progressive weathering of small soil peds rather than by accretion. Even small fragments of bedrock may be the initial form, like the basalt fragments in some alluvial British Columbian soils (Clark and Brydon 1963).

Concretions make up 1–10 per cent or more of the weight of podzols and as much as 18–20 per cent of brown earths. Some tropical soil horizons consist almost entirely of concretions. In the Aripo profile in Trinidad, mottled areas occupy at least two-thirds of the volume of a clayey subsoil which harden on exposure into numerous nodules (Ahmad and Jones 1969). The size of concretions is also noteworthy, increasing from 0.25–1 mm in non-gleyed podzols to 1–2 mm in gleyed podzols, frequently 2.5 mm and reaching a maximum of 1 cm. Similar sizes are observed in some tropical soils. In West Africa, rock fragments less than 2 mm diameter do not usually harden to concretions.

The total chemical composition of ferromanganiferous concretions varies quite widely, depending on the degree of waterlogging and with differing depths in a profile. Elemental iron may account for 8–33 per cent of ortsteins, with about 85–95 per cent of the total Fe_2O_3 being extractable (Rusanov et al. 1975). Perhaps it is co-precipitation with organic compounds which inhibits crystallization, as in savannas where concretions are often confined to the feeding roots of plants, and the entire space around roots may overflow with concretions. On the surface of tropical soils, over 60 per cent of concretions 3–4 mm in size or more may be iron. Aluminium contents fluctuate irregularly between values of about 3 and 17 per cent, much being present in the silicate nucleus of the concretions which may contain much clay. Only iron and manganese (0.1–20 per cent) invariably concentrate in concretions in cooler environments, conspicuously so as size of concretion increases, thus differing from concentration of alumina as a characteristic feature of the tropical laterization process. Silica content, however, is lower in concretions than in the soil and becomes progressively less as concretion size increases (Dobrovol'skiy and Tereshina 1970).

7.9 Induration

7.9.1 LATERITE

After prolonged physico-chemical weathering in the tropics, during which even clay minerals are destroyed and silica leached, the sesquioxide-rich residues and other secondary materials may become partially or completely dehydrated during subsequent desiccation. Laterite results from the alternating mobilization of iron, which moves readily under reducing conditions, followed by its precipitation and concentration as hydrated oxides under the ensuing oxidizing conditions. The dehydration of colloidal hydrated iron oxides involves loss of water and the concentration and crystallization of the amorphous iron colloids into denser crystalline iron minerals, starting with limonite, then goethite and finally haematite. As hydrated oxides are precipitated close to concretions, a concentration gradient is set up. Iron moves with this gradient and the concretions increase in size. The growth of concretions and the linings of pores and channels may increase until a continuous, indurated crust is formed. Thin-sections (fig. 4.5c p.108) suggest that it is the reorganization and crystallization of the interlocking goethite crystals and the near-continuous crystalline phase of the iron oxides which causes induration (Alexander and Cady 1962). Aluminium is also taken into solution in the presence of organic matter and under acidic conditions, but iron and aluminium tend to segregate during deposition and secondary alteration. Comparatively small per cents of iron may be involved but the soft clay enclosed within the ferruginous framework has much less iron. The higher the sesquioxide content, the greater the degree of induration, hardness being a function of the iron content. Although aluminium oxides may crystallize and harden and hard, aluminous laterite crusts are known, iron is believed to play the key role in the induration process.

Laterites vary from scarcely cohesive masses to the 'armour plate' form, the *carapace* of French workers. Three main types can be distinguished. First, laterites may be little more than volumes of iron-impregnated clayey concretions giving undisturbed tropical soils a porous, granular structure, with no induration beyond the cementation of the concretions. Second, as the clay surrounding the concretions is removed, the iron-rich concretions harden and once in contact become bonded together in a massive pisolithic laterite, which may incorporate transported materials too. Third, there is the indurated vesicular skeleton of sesquioxides which appear to be continuous and homogeneously impregnated through a soft matrix of clay and detrital materials.

The mode of accumulation of iron may result in two ways. First, residual sesquioxide-enriched materials may form an absolute accumulation after a prolonged period of humid tropical weathering and the removal of other constituents. Absolute accumulations are most common on fine-grained basic rocks and are commonly concretionary (Alexander and Cady 1962). Second, relative accumulations are controlled by relief and groundwater influence in the seasonally dry savanna climate, with sesquioxides being added laterally from adjacent, higher areas. Sheets are often related to the present-day drainage and may coincide with the level at which the dry season groundwater emerges. If the horizon rich in sesquioxide nodules remains covered by some meters of soil, it may remain sufficiently soft to be penetrated by plant roots. Only indurated horizons are observed at or close to the land surface.

Induration of laterites is an important property in the land-use of tropical soils (fig. 1.3a). Crops grown above thick and fused horizons suffer from either waterlogging because the indurated horizon is impervious, or drought following the rapid loss of trapped moisture by evapotranspiration. Forest is less tall and less luxuriant and attempts to afforest man-induced savanna underlain by indurated laterite usually fail because of limited rooting depth. As laterites and their degree of induration are so widely variable, available data on their geotechnical properties shows that these soils range in performance from excellent to poor for engineering purposes (Grant and Aitchinson 1970).

7.9.2 SILCRETE

Appreciable amounts of silica may be released during the formation of the clay minerals kaolinite and halloysite (fig. 1.3a). The silica is derived from the weathering mantle by solution and transformed by precipitation from soil or groundwater solutions. Thus, silcrete in Australia (fig. 1.3a), for example, is predominantly an absolute rather than a relative accumulation (Stephens 1971). The evaporation of the natural solution, rich in silicic acid leads to the precipitation of silica in the form of amorphous hydroxide. This takes place in the absence of humic acids and under intense insolation in a dry climate. The precipitated silica is rapidly degraded to form secondary quartz, either directly through crystal growth or through the dehydration of silica gels being followed by transformation into quartz. Accumulations of silica are most often in the form of opalescent crusts, films, powders, and layers. Silica may occur as coatings on the ped exteriors, especially on vertical surfaces. Where more strongly developed, silica appears as extremely thin chalcedony-like stringers throughout the matrix and in clay skins (Flach *et al.* 1969). Even small amounts of silica cement clay into aggregates and the clay ceases to shrink and swell. Ultimately, there is interstitial precipitation and variable replacement of invaded materials, commonly waterworn sands and gravels. This becomes a hard indurated mass, resembling conglomerate or quartzite.

7.9.3 CALCRETE

The most visually prominent and striking feature of most soils in arid and semi-arid regions is the light-coloured, carbonate-enriched layer that tends to develop at the base of the illuvial horizon (fig. 1.6a). This horizon of prominent carbonate accumulation has been termed the K horizon (after German, *Kalk*) and has a diagnostic fabric termed K-fabric (Gile et al. 1965). Consistency ranges from soft to extremely hard, but most K-horizons are cemented to some degree. For instance, in the Seychelles, a hard impervious calcrete horizon, known locally as 'platin', is found just above or at a small distance above the watertable. This fine-grained conglomerate of calcium carbonate is impenetrable to deep coconut roots going down to the watertable and is often an obstacle to successful plantations. In Bermuda the phenomenon of hardening of exposure is observed and utilized in constructions.

Calcrete or 'caliche' are largely confined to areas with rainfall less than 500 mm (Goudie 1973). The accumulation of a calcrete horizon is due to the inflow of dissolved carbonates into the soil in solutions rising by capillarity from calcareous groundwaters, rather than to illuvial processes. Silica contents in arid zone soils are highest in indurated calcrete layers, which probably contributes appreciably to the induration of the carbonate layer. Aluminium silicates may make up 10–15 per cent of the total calcrete. Calcareous nodules or concretions are commonly associated with or are an integral part of calcrete. On the lower terraces of the Snake River in Idaho, for example, platy calcareous nodules or concretions 6 mm or less in size are common, whilst on the older, upper terraces, firmly cemented and strongly indurated plates 45–75 cm in diameter occur. A classification of calcrete types is particularly useful because of their obvious influence on hydraulic conductivities and land-use (Weatherby and Oades 1975).

8 Soil colour

8.1 The influence of colour on other soil characteristics

Soil colour is the property which exerts least influence of the other soil properties, but is important because it is largely a synthesis and expression of all other soil properties.

8.2 Description of soil colour

Colour is one of the most immediately obvious field characteristics of cultivated soils. Initially, adjectives describing soil colour were local or non-standardized, like the 'foxy red' of the boulder clay soils in East Yorkshire, and in sub-humid parts of the Soviet Union the 'Chestnut' soil remains a standard type. Now, the Munsell code, coloured chips, arrayed according to three variables, expresses a colour objectively as a point in a three-dimensional space (fig. 8.1). First, the 'hue' notation of an initial letter or letters of a colour, and a number scale, express the relation of a soil colour to red and yellow, or to blue or purple. In fact, tinges of purple and olive are unusual in soils, and soils are most commonly brown, with a 10YR hue. The hue grades from chart to chart, with the second variable, value, ranged vertically on each chart, indicating the colour's lightness. Although value notations in theory range from 0 for black to 10 for white, most soils fall between very dark grey, with a value of about 3, and light grey, which has a value of 7. The third variable, chroma, describes the strength of the colour's pigment. Thus to the right on each chart, more vividly coloured soils increase in chroma. To the left, greyer colours approach a neutral grey, for which the Munsell chroma notation is 0. Soil chromas rarely exceed 8. Thus, for a given soil, appropriate numbers for value and chroma, separated by a dash, are added to the hue notation,

Soil colour 225

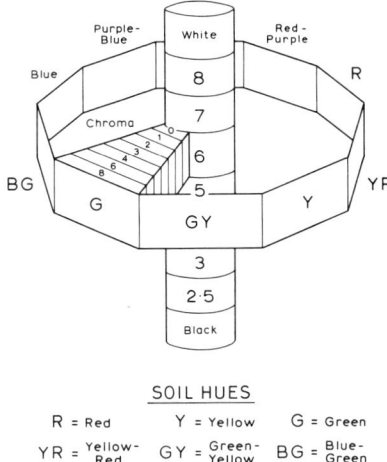

Figure 8.1 Three-dimensional representation of coding of soil colour according to the Munsell notation.

and a standard Munsell code for a brown soil might be 10YR 5/3 (i.e. hue, value/chroma). A 'reddish grey' soil might be designated 5YR 5/2, whereas a 'yellowish red' soil could be 5YR 5/8. As the individual Munsell parameters are interdependent, a 'colour code' (hue × value) has been suggested (Soileau and McCracken 1967).

8.3 Sources of pigments in soil colours

Iron is usually the most important mineral colouring agent in soils, especially in clays. In ferruginous soils the two common forms of free iron oxide produce red to yellow-brown colours, haematite (Fe_2O_3) being red and goethite (FeOOH) yellow to light yellowish brown. For soils developed on basic igneous rocks in the Malwa Plateau, India, for instance, there is a significant correlation between ferrous iron contents and the soil hue notation ($r = 0.83$, $N = 23$) and more than 95 per cent of variations observed in soil hue could be assessed by ferrous iron content together with that of titanium (Krishna Murti and Satyanarayana 1971). In southern Brazil, soils with dark red colours in the B horizon (2.5YR, 10R 3/6 or 3/4) usually had an iron content of more than 9 per cent, but when iron was less than 9 per cent, yellow or orange red colours occurred. With iron contents above 18 per cent, an associated increase in manganese added dusky red, purple colourations (Bennema *et al.* 1970). The hydration of aluminium compounds, as well as those of iron, can produce yellow colours. Calcium carbonate can also be significant when present in large amounts and colours may become whitish-grey.

Organic matter may influence soils colours to an over-riding degree. The more humified the organic matter, the darker the colour as humins and humic acids contrast with the lighter-coloured fulvic acids. Where organic colloids are saturated with iron, red colours predominate, whilst the black colour of Chernozems is the result of calcium saturating the exchange complex. Therefore, organic matter is not necessarily directly related with the relative blackness of a soil, although the correlation between organic matter and value for the Malwa soils ($r = -0.52, N = 23$) exemplifies the common instance of diminishing contents of organic matter permitting increased lightness of soil values.

8.4 Influence of other soil properties on soil colour

The texture of a soil can influence its colour. In a sandy soil, with little colloidal organic matter, colours more intense than greyish-white of residual quartz depend on the amount of ferric iron oxidized on the grain surfaces. Coarse-grained particles, with their much smaller surface areas, require much less pigment to impart a certain colour. Clay may tend to darken a soil, with a similar effect on value as noted for organic matter. In fact, it is common for the combined effect of clay–organic matter complexes to determine the general lightness, or value, of a soil colour, but only certain types of clays can form a dark clay–organic complex. In particular, the clay must have a high exchange capacity to permit a thorough dispersion of organic colloidal matter on its surface. In several parts of India, black soils usually comprise 40–60 per cent clay, most of which is montmorillonite, whereas the red soils have a much lower clay fraction and kaolinitic soil types with iron oxides predominate (Singh 1954).

Soil structure may influence soil colour indirectly, depending on the amount of light reflected from a soil surface. Coarse aggregates form a very complex surface, trapping and extinguishing light in the large number of inter-aggregate spaces. Thus, the darkness of Chernozems is partly due to the reflection of only 10–15 per cent of the light energy incident on them.

Chemical composition of a soil may also be an indirect influence on soil colour. In particular, sodium in clays favours the adsorption of large quantities of organic matter. The intensity of the dark colours developed is also due to dispersion of the clay by sodium which facilitates a thorough mixing of organic matter throughout the soil. Where sodium and potassium carbonates are dominant, surface horizons tend to be black or brown, due to their dissolving organic matter, and contrast markedly with the white incrustations of other saline soils which are largely chlorides or sulphates.

Despite the expression of many soil properties in a range of colours, invariable correlations cannot be established owing to varying environmental influences, particularly those affecting air and moisture conditions in the soil. For example, the red colour of soils is better developed in more arid environments which induce the irreversible precipitation of ferric oxide on the surface of the coarser mineral particles or its crystallization into separate minerals like haematite or ilmenite. In contrast, this link between iron content and redder colours does not develop in cooler, damp environments, where colours are yellow or brown, even if the iron contents are high. Here iron occurs in a hydrated form, combining with water, not oxygen, and in most poorly drained soils with the shortage of oxygen, reducing conditions predominate and ferrous iron compounds form producing grey soils with gleyed, dark blue, bluish or greenish-grey subsoils. The grey colour, becoming more intense in compounds with organic acids, can be due to the presence of goethite. Where very poor drainage encourages the accumulation of organic matter and peat, inevitably brown and black colours develop. Where substantial volumes of water move through a soil light-grey or whitish colours characterize the leaching regime as the mineral grains are freed of colloidal films of humus or of oxides and hydroxides of iron and aluminium. In tropical deep-weathering profiles, the grey horizons of the pallid zone may result either from the removal of hydroxides or from reduction.

Many soil colours reflect neither very dry, oxidizing conditions nor damp, reducing environments, but a seasonal oscillation or balance between the two. The dark colour of certain A horizons of tropical soils, for instance, is due to the thorough mixing of small organic contents (0.5–1.5 per cent) with clay. The coating of the clay surfaces with these traces or organic matter appears to result from temporary anaerobic conditions in the wet season.

8.5 Significance of soil colour

Clearly, the observation of a given soil colour can be useful in the description, mapping, and classification of soils and may elucidate problems associated with soil genesis or soil evaluation. Coloured mineral compounds, such as iron oxides, are clearly visible in the field and are particularly useful guides to the relative permeability of a given profile and in establishing a range of soil drainage classes. Despite the misapprehension, particularly in tropical areas, that black soils were invariably diagnostic of reserves of organic matter and hence of high fertility, the frequent correlation between organic matter and soil colour remains useful for many areas. Where a rock has a distinctive, valuable mineral, this may be apparent in the soil colour. In eastern South Australia, for instance, a correlation between copper content and soil chroma ($r = 0.53$, $N = 43$) could indicate the copper level in these soils (Russell and Moore 1967). Also, in studies of soil genesis, changes in colour down a profile can indicate processes of soil formation. In studying buried soils, colour changes are the most immediately obvious characteristic. More commonly, soil colouration increases in illuvial horizons owing to enrichment with iron and aluminium compounds. In soil classification, differences in colour can be critical, as between terra rossas and rendzinas.

Although the mineral responsible for certain soil colours is not always known and whilst quite different processes can produce similar colours, it is clear why soil colour is nonetheless widely regarded as a critical, comprehensive soil characteristic.

8.6 Soil colour and man in contact with the earth

In all continents, soil colour has always caught the eye of the primary producers tilling the land. In Latin America, Yucatan agriculturalists still use the Maya words Eck-Lum and Kankab to describe black earth and reddish earth, respectively. In the valleys of northern California, the atmosphere of the life of pioneer settlers was evoked by John Steinbeck in *To a God Unknown* with repeated references to the greyness of the soil.

In the Soviet Union, the singer-poet D. Dzhabayev spent his hundred years studying the changing shades of nature in his native Kazakhstan. He wrote:

> And the poor people went, cursing their fate,
> In the hungry years, to the salt marshes,
> Where the salt, seeping out, shone white in the distance,
> Like bitter tears of the earth . . . .

In another recently settled continent, a D. H. Lawrence bush-style hero set foot in western Australia and, after crossing the sandy coastal plain exclaimed:

> "The soil is red!"
> "Clay! That's clay! No more sand, except in patches, all the way to Albany. This is Guildford where the roses grow."

Bibliography

Chapter 1 Geography and soils

1.1 PHYSICAL GEOGRAPHY AND SOILS

ANDERSON, G. D. and HERLOCKER, D. J. (1973) Soil factors affecting the distribution of the vegetation types and their utilization by wild animals in Ngorongoro Crater, Tanzania, *J. Ecol.* 61, 627–51.

ARNETT, R. R. (1976) Some pedological features affecting the permeability of hillside soils in Caydale, Yorkshire, *Earth Surface Processes*, 1, 3–16.

BETTENAY, E., BLACKMORE, A. V. and HINGSTON, F. J. (1964) Aspects of the hydrological cycle and related salinity in the Bekla valley, Western Australia, *Aust. J. Soil Res.* 2, 187–210.

BORCHARDT, G. A., HOLE, F. D. and JACKSON, M. L. (1968) Genesis of layer silicates in representative soils in a glacial landscape of southeastern Wisconsin, *Soil Sci. Soc. Amer. Proc.* 32, 399–403.

BRIDGE, J. S. and JARVIS, J. (1976) Flow and sedimentary processes in the meandering river South Esk, Glen Clova, Scotland, *Earth Surface Processes* 1, 30–36.

DANIELS, R. B., GAMBLE, E. E. and CADY, J. G. (1971) The relationship between geomorphology and soil morphology and genesis, *Adv. Agron.* 23, 51–88.

DENNESS, B. et al. (1975) Investigation of a coastal landslip at Charmouth, Dorset, *Quart. J. Eng. Geol.* 8, 119–40.

EL—GHONEMY, A. A. et al. (1977) Mineral element composition of the natural vegetation along a transect at Mareotis, Egypt, *Soil Sci.* 124, 16–26.

GATES, D. H., STODDART, L. A. and COOK, C. W. (1956) Soil as a factor influencing plant distributions on salt-deserts of Utah, *Ecol. Monog.* 26, 155–75.

GERSMEHL, P. J. (1976) An alternative biogeography, *Ann. Ass. Amer. Geogrs.* 66, 223–41.

GILE, L. H. and HAWLEY, J. W. (1966) Periodic sedimentation and soil formation on an alluvial–fan piedmont in southern New Mexico, *Soil Sci. Soc. Amer. Proc.* 30, 261–68.

GILE, L. H. and HAWLEY, J. W. (1968) Age and comparative development of desert soils at the Gardner Spring radiocarbon site, New Mexico, *Soil Sci. Soc. Amer. Proc.* 32, 709–16.

GRAY, D. M. (ed.) (1970) *Handbook on the principles of hydrology* (Nat. Res. Council of Canada, Ottawa).

HANKS, R. J. and BOWERS, S. A. (1962) Numerical solution of the moisture flow equation for infiltration into layered soils, *Soil Sci. Soc. Amer. Proc.* 26, 530–34.

HUNT, P. A., MITCHELL, P. B. and PATON, T. R. (1977) "Laterite profiles" and "lateritic ironstones" on the Hawkesbury Sandstone, Australia, *Geoderma*, 19, 105–21.

LEPSCH, I. F., BUOL, S. W. and DANIELS, R. B. (1977) Soil–landscape relationships in the occidental plateau of Sao Paulo State, Brazil: I. Geomorphic surfaces and soil mapping units, *Soil Sci. Soc. Amer. Proc.* 41, 104–9.

MARTIN, A. F. and STEEL, R. W. (eds) (1954). *The Oxford Region* (Oxford University Press).

MITCHELL, J. E., WEST, N. E. and MILLER, R. W. (1966) Soil physical properties in relation to plant community patterns in the shadscale zone of northwestern Utah, *Ecology* 47, 627–30.

MOORE, C. W. E. (1959) Interaction of species and soil in relation to the distribution of Eucalypts, *Ecology* 40, 734–35.

ODUM, E. P. (1971) *Fundamentals of ecology*, 3rd edn. (Saunders, Philadelphia) 574 pp.

PARKER, B. S., MYERS, K. and CASKEY, R. L. (1976) An attempt at rabbit control by warren ripping in semi-arid western New South Wales, *J. appl. Ecol.* 13, 353–67.

REID, I. (1973) The influence of slope orientation upon the soil moisture regime, and its hydrogeomorphological significance, *J. Hydrol.* 19, pp. 309–21.

STEPHENS, C. G. (1971) Laterite and silcrete in Australia: a study of the genetic relationships of laterite and silcrete and their companion materials, and their collective significance in the formation of the weathered mantle, soils, relief and drainage of the Australian continent, *Geoderma* 5, 5–52.

VALENTINE, K. W. G. and DALRYMPLE, J. B. (1976) The identification of a buried paleosol developed in place at Pitstone, Buckinghamshire, *J. Soil Sci.* 27, 541–53.

1.2 SOILS AND HUMAN GEOGRAPHY

ALLAWAY, W. H. (1968) Agronomic controls over the environmental cycling of trace elements, *Adv. Agron.* 20, 235–74.

ARMSTRONG, W. et al. (1976) The relationship between soil aeration, stability and growth of Sitka spruce (*Picea sitchensis* (Bong.) Carr.) on upland peaty gleys, *J. appl. Ecol.* 13, 585–91.

BAILEY, G. W. and WHITE, J. L. (1970) Factors influencing the adsorption, desorption and movement of pesticides in soil, *Residue Rev.* 32, 29–92.

BERROW, M. L. and WEBBER, J. (1972) Trace elements in sewage sludges, *J. Sci. Food Agric.* 23, 93–100.

BINGAMAN, D. E. and KOHNKE, H. (1970). Evaluating sands for athletic turf, *Agron. J.* 62, 464–67.

BLEVINS, R. L., BAILEY, H. H. and BALLARD, G. E. (1970) The effect of acid mine water on floodplain soils in the western Kentucky coal fields, *Soil Sci.* 110, 191–96.

BOUMA, J. (1974) New concepts in soil survey interpretations for onsite disposal of septic tank effluent, *Soil Sci. Soc. Amer. Proc.* 36, 941–46.

BRIDGES, E. M. (1969) Eroded soils of the Lower Swansea Valley, *J. Soil Sci.* 20, 236–45.

CLARE, K. E. and BEAVEN, P. J. (1962) *Soils and other roadmaking materials in Nigeria* (H.M.S.O., London) 54 pp.

CLARKE, G. R. (1951) The evaluation of soils and the definition of quality classes from studies of the physical properties of the soil profile in the field, *J. Soil Sci.* 2, 50–60.

COPPOCK, J. T. (1976) Projections for British agriculture, *Geog. J.* 142, 1–11.

CORCORAN, P. et al. (1977) Soil corrosiveness in south Oxfordshire, *J. Soil Sci.* 28, 473–84.

GIBSON, R. E. and MORGENSTERN, N. (1962) A note on the stability of cuttings in normally consolidated clays, *Geotechnique*, 12, 212–16.

GIDDENS, J. (1976) Spent motor oil effects on soil and crops, *J. Environ. Qual.* 5, 179–81.

GORHAM, E. (1974) The relationship between standing crop in sedge meadows and summer temperature, *J. Ecol.* 62, 487–91.

GROSS, E. R. and RUST, R. H. (1972) Estimation of corn and soybean yields utilizing multiple curvilinear regression methods, *Soil Sci. Soc. Amer. Proc.* 36, 316–20.

HAANS, J. C. F. M. and WESTERVELD, G. J. W. (1970) The application of soil survey in the Netherlands, *Geoderma* 4, 279–309.

HAIG, I. T. (1929) Colloidal content and related soil factors as indicators of site quality, *Yale Univ. Sch. Forestry Bull.* 24, 29 pp.

HARRIS, D. R. (1971) The ecology of swidden cultivation in the Upper Orinoco Rain forest, Venezuela, *Geog. Rev.* 61, 475–95.

HERSHAFT, A. (1972) Solid waste treatment technology, *Env. Sci. Technol.* 6, 412–21.

HUTTON, E. M. (1970) Tropical pastures, *Adv. Agron.* 22, 1–73.

JACKSON, W. A., LEONARD, R. A. and WILKINSON, S. R. (1975) Land disposal of broiler litter–changes in soil potassium, calcium and magnesium, *J. Environ. Qual.* 4, 202–6.

JOHNSON, B. L. C. (1972) Recent developments in rice breeding and some implications for tropical Asia, *Geography* 57, 307–20.

JOHNSON, M. S. (1976) Plant growth in fluorspar mine tailing, *J. Soil Water Conserv.* 31, 17–20.

JUMIKIS, A. R. (1962) *Soil mechanics* (Von Nostrand, Princeton, New Jersey).

KIRILLINA, A. V. (1968) Change in the water-physical properties of Chernozems during irrigation, *Soviet Soil Sci.* 503–8.

KORTE, N. E. *et al.* (1976) Trace element movement in soils: influence of soil physical and chemical properties, *Soil Sci.* 122, 350–59.

LEAF, A. L. (1956) Growth of forest plantations on different soils of Finland, *For. Sci.* 2, 121–26.

LUTZ, H. J. (1945) Soil conditions of picnic grounds in public forest parks, *J. Forestry* 43, 121–27.

MCKEE, N. H. and SHOULDERS, E. (1974) Slashpine biomass response to site preparation and soil properties, *Soil Sci. Soc. Amer. Proc.* 38, 144–48.

MEAD, W. R. (1953), *Farming in Finland* (Athlone Press, London).

MILLER, W. L. *et al.* (1976) Formation of soil acidity in carbonaceous soil materials exposed by highway excavations in East Texas, *Soil Sci.* 121, 162–69.

MORGAN, G. W. (1974) Crop productivity as affected by depths of topsoil spread for reclaiming bauxite-mined lands in Jamaica, *Trop. Agric.* 51, 332–46.

MUNRO, J. M. M. and DAVIES, D. A. (1973) Potential pasture production in the uplands of Wales. 2. Climatic limitations on production, *J. Br. Grassl. Soc.* 28, 161–69.

NISHITA, H. *et al.* (1956) Fixation and extractability of fission products contaminating various soils and clays: I. Sr^{90}, Y^{91}, Ru^{106}, Cs^{137} and Ce^{144}, *Soil Sci.* 81, 312–26.

NORTHEY, R. D. (1966) Correlation of engineering and pedological soil classification in New Zealand, *N.Z.J. Sci.* 9, 809–33.

OLSON, G. W. and PULESTON, D. E. (1972) Soils and the Maya, *Americas* 24, 33–39.

OLSON, T. C. (1977) Restoring the productivity of a glacial till soil after topsoil removal, *J. Soil Water Conserv.* 32, 130–32.

PAAUW, F. van der (1972) Quantification of the effects of weather conditions prior to the growing season on crop yields, *Plant Soil* 37, 375–88.

PARKES, J. G. M. and DAY, J. C. (1975) The hazard of sensitive clays. A case study of the Ottawa-Hull area, *Geog. Rev.* 65, 198–213.

PLAMBECK, H. (1976) Mainland China: its people and agriculture, *J. Soil Water Conserv.* 31, 248–51.

PULFORD, I. D. and DUNCAN, H. J. (1975) The influence of pyrite oxidation products on the adsorption of phosphate by coal-mine waste, *J. Soil Sci.* 26, 74–80.

RADLEY, J. and SIMMS, C. (1967) Wind erosion in East Yorkshire, *Nature* 216, 20–1.

ROBINSON, D. N. (1968) Soil erosion by wind in Lincolnshire, March 1968, *East Midld Geogr.* 4, 351–62.

ROBINSON, G. H. (1971) Exchangeable sodium and yields of cotton on certain clay soils of Sudan, *J. Soil Sci.* 22, 328–35.

SEMPLE, E. C. (1928) Ancient Mediterranean agriculture, *Agr. Hist.* 2, 61–98.

SHORTRIDGE, J. R. (1976) The collapse of frontier farming in Alaska, *Ann. Ass. Amer. Geogrs.* 66, 583–604.

SKAGGS, R. H. (1975) Drought in the United States, 1931–40, *Ann. Ass. Amer. Geogrs.* 65, 391–402.

SMITH, J. H. (1976) Treatment of potato processing waste waters on agricultural land, *J. Environ. Qual.* 5, 113–16.

SWINER, G. L., NELSON, L. E. and SMITH, W. H. (1966) The characterization of dry matter and nitrogen accumulation by Loblolly Pine (*Pinus taeda* L.), *Soil Sci. Soc. Amer. Proc.* 30, 114–19.

TANAKA, A. and YOSHIDA, S. (1970) Nutritional disorders of the rice plant in Asia, *Int. Rice Res. Inst. Tech. Bull.* 10.

TERZAGHI, K. (1955) Influence of geological factors on the engineering properties of sediments, *Econ. Geol.* 50th Ann. Vol., 557–618.

THOMSON, I., THORNTON, I. and WEBB, J. S. (1972) Molybdenum in black shales and the incidence of bovine hypocuprosis, *J. Sci. Food Agric.* 23, 879–91.

UNANYANTS, T. P. (1970) Development of mineral fertilizer production in the USSR, *Soviet Soil Sci.* 189–91.

VINK, A. P. A. (1963) Soil survey as related to agricultural productivity, *J. Soil Sci.* 14, 88–101.

WALTON, K. (1950) The distribution of population in Aberdeenshire, 1696, *Scot. geog. Mag.* 66, 17–26.

WILLIAMS, C. H. and DAVID, D. J. (1976) Effects of pasture improvement with subterranean clover and superphosphate on the availability of trace metals to plants, *J. Aust. Soil Res.* 14, 85–93.

WILLIAMS, G. D. V., JOYNT, M. I. and MCCORMICK, P. A. (1975) Regression analyses of Canadian Prairie crop-district cereal yield 1961–1972 in relation to weather, soil and trend, *Can. J. Soil Sci.* 55, 43–53.

WILSON, J. R. and HAYDOCK, K. P. (1971) The comparative response of tropical and temperate grasses to varying levels of nitrogen and phosphorus nutrition, *Aust. J. agric. Res.* 22, 573–87.

1.3 SOILS AND GEOGRAPHICAL METHOD

ANDERSON, K. E. and FURLEY, P. A. (1975) An assessment of the relationship between the surface properties of chalk soils and slope form using principal components analysis, *J. Soil Sci.* 26, 130–43.

AVERY, B. W. *et al.* (1972) The soil of Barnfield, *Rothamsted Exp. Sta. Rep. for 1971* 2, 5–37.

BOUWER, H. (1969) Returning wastes to the land, a new role for agriculture, *J. Soil Water Conserv.* 24, 164–68.

BRIDGES, E. M. and DOORNKAMP, J. C. (1963) Morphological mapping and the study of soil patterns, *Geography* 48, 175–81.

BURINGH, P., STEUR, G. G. L. and VINK, A. P. A. (1962) Some techniques and method of soil survey in the Netherlands, *Neth. J. Agric. Sci.* 10, 157–72.

BURNHAM, C. P. (1973) Discovering about soils, *Geography* 58, 43–46.

BURROUGH, P. A., BECKETT, P. H. T. and JARVIS, M. G. (1971) The relation between cost and utility in soil survey (I–III), *J. Soil Sci.* 22, 359–94.

CAMPBELL, J. B. (1977) Variation of selected properties across a soil boundary, *Soil Sci. Soc. Amer. Proc.* 41, 578–82.

CATT, J. A., KING, D. W. and WEIR, A. H. (1975) The Soils of Woburn Experimental Farm I Great Hill, Road Piece and Butt Close, *Rothamsted Exp. Sta. Rep. for 1974* 2, 5–28.

COLLINS, J. F. (1976) Soil heterogeneity and profile development in a stratified gravel deposit in Ireland, *Geoderma* 15, 143–56.

CONACHER, A. J. and DALRYMPLE, J. B. (1977) The nine-unit landsurface model: an approach to pedogeomorphic research, *Geoderma* 18, 1–154.

COURTNEY, F. M. and WEBSTER, R. (1973) A taxonomic study of the Sherborne soil mapping unit, *Trans. Inst. Brit. Geogrs.* 58, 113–24.

CRUICKSHANK, J. G. (1976) Soil and land valuation in New England, New South Wales, *Aust. Geogr.* 13, 249–55.

DAVIES, B. E. (1976) Mercury content of soils in western Britain with special reference to contamination from base metal mining, *Geoderma* 16, 183–92.

ENGELSTAD, O. P. and RUSSEL, D. A. (1975) Fertilizers for use under tropical conditions, *Adv. Agron.* 27, 175–208.

FERRIANS, O. J., KACHADOORIAN, R. and GREENE, C. W. (1969) Permafrost and related engineering problems in Alaska, *U.S. Geol. Surv. Prof. Paper* 678, 37 pp.

FRIDLAND, V. M. (1967) Structure of soil mantle in the principal soil zones and subzones of Western USSR, *Soviet Soil Sci.* 589–97.

GERASIMOV, I. P. (1973) A genetic approach to the subdivision of tropical soils, regolith and their products of redeposition, *Soviet Geog.* 14, 165–77.

GERSMEHL, P. J. (1977) Soil taxonomy and mapping, *Ann. Ass. Amer. Geogrs.* 67, 419–28.

GERSPER, P. L. and HOLOWAYCHUK, N. (1970) Effects of stemflow water on a Miami soil under a beech tree: II. Chemical properties, *Soil Sci. Soc. Amer. Proc.* 34, 786–94.

HERATH, J. W. and GRIMSHAW, R. W. (1971) A general evaluation of the frequency distribution of clay and associated minerals in the alluvial soils of Ceylon *Geoderma* 5, 119–30.

JONES, R. L. and BEAVERS, A. H. (1966) Weathering in surface horizons of Illinois soils, *Soil Sci. Soc. Amer. Proc.* 30, 621–24.

KYUMA, K., HATTORI, T. and KAWAGUCHI, K. (1974) Rice cultivation and its environmental conditions in the Mediterranean countries. II. Soils, their fertility and mineralogy, *Soil Sci. Plant Nutrn (Tokyo)* 20, 225–40.

LLOYD, P. S., GRIME, J. P. and RORISON, I. H. (1971) The grassland vegetation of the Sheffield region. I. General Features, *J. Ecol.* 59, 863–86.

LOPEZ-HERNANDEZ, I. D. and BURNHAM, C. P. (1974) The covariance of phosphate sorption with other soil properties in some British and tropical soils, *J. Soil Sci.* 25, 196–200.

LUND, L. J., PAGE, A. L. and NELSON, C. O. (1976) Nitrogen and phosphorus levels in soils beneath sewage disposal ponds, *J. Environ. Qual.* 5, 26–30.

MCCORMACK, D. E. and WILDING, L. P. (1969) Variation of soil properties within mapping units of soils with contrasting substrata in northwestern Ohio, *Soil Sci. Soc. Amer. Proc.* 33, 587–93.

MCNEAL, B. L. et al. (1966) Effect of rice culture on the reclamation of sodic soils, *Agron. J.* 58, 238–40.

MAHILUM, B. C., FOX, R. L. and SILVA, J. A. (1970) Residual effects of liming volcanic ash soils in the humid tropics, *Soil Sci.* 109, 102–9.

MINAMI, K. and ARAKI, K. (1975) Distribution of trace element in arable soil affected by automobile exhausts, *Soil Sci. Plant Nutrn (Tokyo)* 21, 185–88.

ODUM, E. P. (1975) *Ecology: the link between the natural and the social sciences*, 2nd edn. (Holt, Rinehart Winston, London) 244 pp.

OLLIER, C. D. and THOMASSON, A. J. (1957) Asymmetrical valleys of the Chiltern Hills, *Geog. J.* 123, 71–80.

PAVLIK, H. F. and HOLE, F. D. (1977) Soilscape analysis of slightly contrasting terrains in southeastern Wisconsin, *Soil Sci. Soc. Amer. Proc.* 41, 407–13.

RAGG, J. M. (1977) The recording and organisation of soil field data for computer areal mapping, *Geoderma* 19, 81–89.

RUBTSOVA, L. P. (1967) Changes in Grey Forest soils when cultivated, *Soviet Soil Sci.* 308–19.

RUSSEL, D. A. and WILLIAMS, G. G. (1977) History of chemical fertilizer development, *Soil Sci. Soc. Amer. Proc.* 41, 260–65.

SOKOLOV, I. A. and TARGUL'YAU, V. O. (1976) Taiga soils of Transbaykalia in relation to the originalness of soils in the permafrost-taiga region, *Soviet Soil Sci.* 405–16.

STEPANOV, I. N. (1966) Soil erosion in Turkmenia, *Soviet Soil Sci.* 196–200.

STEWART, D. K. R. and CHISHOLM, D. (1971) Long-term persistence of BHC, DDT, and chlordane in a sandy loam soil, *Can. J. Soil Sci.* 51, 379–83.

VORONKOV, N. A. (1967) Spatial variation in the moisture content of sandy soils under pine stands, *Soviet Soil Sci.* 1337–43.

WATSON, J. P. (1962) The soil below a termite mound, *J. Soil Sci.* 13, 46–51.

WEBSTER, R. (1968) Fundamental objections to the 7th Approximation, *J. Soil Sci.* 19, 354–65.

WEERT, R. VAN DER (1974) Influence of mechanical forest clearing on soil conditions and the resulting effects on root growth, *Trop. Agric.* 51, 325–31.

WHITTLESEY, D. (1936) Major agricultural regions of the earth, *Ann. Ass. Amer. Geogrs* 26, 199–242.

WILDING, L. P., SCHAFER, G. M. and JONES, R. B. (1964) Morley and Blount soils: a statistical summary of certain physical and chemical properties of some selected profiles from Ohio, *Soil Sci. Soc. Amer. Proc.* 28, 674–79.

WILLIAMS, A. R. and MORGAN, R. P. C. (1976) Geomorphological mapping applied to soil erosion evaluation, *J. Soil Water Conserv.* 31, 164–68.

YAALON, D. H. and YARON, B. (1966) Framework for man-made soil changes an outline of metapedogenesis, *Soil Sci.* 102, 272–77.

YOUNG, A. (1973) Soil survey procedures in land development planning, *Geog. J.* 139, 53–64.

Chapter 2 The mineral fraction of the soil

2.1 THE CLASSIFICATION OF PARTICLE SIZES

FOLK, R. L. (1966) A review of grain-size parameters, *Sedimentology* 6, 73–94.
LIEBEROTH, L. (1968) Mapping of soils in the German Democratic Republic, *Soviet Soil Sci.* 1350–63.
MARSHALL, T. J. (1947) Mechanical composition of soil in relation to field descriptions of Texture, *C. S. I. R. Bull. No. 224* 20 pp.
PETERSEN, G. W., CUNNINGHAM, R. L. and MATELSKI, R. P. (1968) Moisture characteristics of Pennsylvania soils: I. Moisture retention as related to texture, *Soil Sci. Soc. Amer. Proc.* 32, 271–75.
TANNER, W. F. (1969) The particle size scale, *J. Sed. Petrol.* 39, 809–12.
WHITESIDE, E. P. *et al.* (1967) Considerations relative to a common particle size scale of earthy materials, *Soil Sci. Soc. Amer. Proc.* 31, 579–84.
WISCHMEIER, W. H. and MANNERING, J. V. (1969) Relation of soil properties to its erodibility, *Soil Sci. Soc. Amer. Proc.* 33, 131–37.

2.2 THE COARSE FRACTION

CLARKE, G. R. and BECKETT, P. H. T. (1965) The textures of soils in the Oxford region, *J. Soil Sci.* 16, 318–21.
JESSUP, R. W. (1960) The stony tableland soils of the south-eastern portion of the Australian arid zone and their evolutionary history, *J. Soil Sci.* 11, 188–96.
KASK, R. P. (1965) Qualitative evaluation of agricultural land in the Estonia SSR, *Soviet Soil Sci.* 954–63.
NIKIFOROFF, C. C. (1948) Stony soils and their classification, *Soil Sci.* 66, 347–63.
RUTHERFORD, G. K. (1971) The properties, genesis and geomorphological relationships of a sequence of soils on a limestone plain in southeast Ontario, Canada, *Geoderma* 5, 179–96.
WILDE, S. A. and KRAUSE, H. H. (1960) Soil-forest types of the Yukon and Tanana valleys in subarctic Alaska, *J. Soil Sci.* 11, 266–79.
WILDE, S. A., VOIGT, G. K. and PIERCE, R. S. (1954) The relationship of soils and forest growth in the Algoma district of Ontario, Canada, *J. Soil Sci.* 5, 22–38.

2.3 SAND

AL-RAWI, A. H., JACKSON, M. L. and HOLE, F. D. (1969) Mineralogy of some arid and semiarid land soils of Iraq, *Soil Sci.* 107, 480–86.

HUIZING, H. G. J. (1971) A reconnaissance study of the mineralogy of sand fractions from East Pakistan sediments and soils, *Geoderma* 6, 109–33.

NYE, P. H. (1955) Some soil-forming processes in the humid tropics. II. The development of the upper-slope member of the catena, *J. Soil Sci.* 6, 51–62.

SAWHNEY, B. L. and VOIGT, G. K. (1969) Chemical and biological weathering in vermiculite from Transvaal, *Soil Sci. Soc. Amer. Proc.* 33, 625–29.

SMITH, J. (1957) A mineralogical study of weathering and soil formation from olivine basalt in Northern Ireland, *J. Soil Sci.* 8, 225–39.

SMITHSON, F. (1953) The micro-mineralogy of North Wales soils, *J. Soil Sci.* 4, 194–210.

SOMARSIRI, S. and HUANG, P. M. (1971) The nature of K-feldspars of a chernozemic soil in the Canadian Prairies, *Soil Sci. Soc. Amer. Proc.* 35, 810–15.

STORRIER, R. R. and MUIR, A. (1962) The characteristics and genesis of a ferritic brown earth, *J. Soil Sci.* 13, 259–70.

VAKULIN, A. A. (1966) Mineralogic composition of the sands of the Lower Volga region, *Soviet Soil Sci.* 785–89.

WALKER, T. W. (1974) Phosphorus as an index of soil development, *Trans. 10th Int. Congr. Soil Sci.* 6, 451–57.

WILLIAMS, J. D. H. et al. (1969). Apatite transformations during soil development, *Agrochimica* 13, 491–501.

2.4 SILT

ALRAWI, G. J. and SYS, C. (1967) A comparative study between Euphrates and Tigris sediments in the Mesopotamian flood plain, *Pedologie* 17, 187–211.

ALTAIE, F. H., SYS, C. and STOOPS, G. (1969). Soil groups of Iraq: their classification and characterization, *Pedologie* 9, 65–148.

NABHAN, H. M., SYS, C. and STOOPS, G. (1969) Mineralogical study of the suspended matter in the Nile water, *Pedologie*, 19, 34–48.

PITTY, A. F. (1968) Particle size of the Saharan dust which fell in Britain in July 1968, *Nature* 220, 364–65.

SNEDDON, J. I., LAVKULICH, L. M. and FARSTAD, L. (1972) The morphology and genesis of some alpine soils in British Columbia, Canada: I. Morphology, classification, and genesis, *Soil Sci. Soc. Amer. Proc.* 36, 100–110.

WEIR, A. H., CATT, J. A. and MADGETT, P. A. (1971) Postglacial soil formation in the loess of Pegwell Bay, Kent (England), *Geoderma* 5, 131–49.

2.5 CLAYS

ALTSCHULER, Z. S., DWORNIK, E. J. and KRAMER, H. (1963) Transformation of montmorillonite to kaolinite during weathering, *Science* 141, 148–52.

BALL, D. F. (1966) Chlorite clay minerals in Ordovician pumice-tuff and derived soils in Snowdonia, North Wales, *Clay Min.* 6, 195–209.

BRAMAO, L. et al. (1952) Criteria for the characterization of kaolinite, halloysite, and a related mineral in clays and soils, *Soil Sci.* 73, 273–87.

DROSTE, J. B. (1956) Alteration of clay minerals by weathering in Wisconsin tills, *Bull. geol. Soc. Amer.* 67, 911–18.

ESWARAN, H. and HENG, Y. Y. (1976) The weathering of biotite in a profile on genesis in Malaysia, *Geoderma* 16, 9–20.

FIELDES, M. (1955) Clay mineralogy of New Zealand soils. Part II: Allophane and related mineral colloids, *N. Z. J. Sci. Technol.* 37B, 336–50.

FINCK, A. and OCHTMAN, L. H. J. (1961) Problems of soil evaluation in the Sudan, *J. Soil Sci.* 12, 87–95.

GAUDETTE, H. E., EADES, J. L. and GRIM, R. E. (1966) The nature of illite, *Clays Clay Min.* 13, 33–48.

GJEMS, O. (1967) Studies on clay minerals and clay mineral formation in soil profiles in Scandinavia, *Medd. Norske Skogsforsves.* 21, 303–415.

GRIM, R. E. (1968) *Clay mineralogy*, 2nd edn. (McGraw-Hill, New York) 596 pp.

GUPTA, R. N. (1968) Clay mineralogy of the Indian Gangetic Alluvium of Uttar Pradesh, *J. Indian Soc. Soil Sci.* 16, 115–27.

HANNA, F. S. and BECKMANN, H. (1975) Clay minerals of some soils of the Nile Valley in Egypt, *Geoderma* 14, 159–70.

HARDER, H. (1972) The role of magnesium in the formation of smectite minerals, *Chem. Geol.* 10, 31–39.

HENDRICK, S. B. and FRY, W. H. (1930) The results of X-ray and mineralogical examination of soil colloids, *Soil Sci.* 29, 457–76.

HOUTEN, F. B. VAN (1953) Clay minerals in sedimentary rocks and derived soils, *Amer. J. Sci.* 251, 61–82.

HOWER, J. and MOWATT, T. C. (1966) The mineralogy of illites and mixed-layer illite/montmorillonites, *Amer. Mineral.* 51, 825–54.

HUANG, P. M. et al. (1977) Sesquioxidic components of selected Taiwan soils, *Geoderma* 18, 251–63.

LINARES, J. and HUERTAS, F. (1971) Kaolinite: synthesis at room temperature, *Science* 171, 896–97.

MARSHALL, C. E. (1964) *The physical chemistry and mineralogy of soils* (John Wiley, New York) 388 pp.

MERWE, G. R. VAN DER and WEBER, H. W. (1965) The clay minerals of

South African soils developed from dolerite under different climatic conditions, *S. Afr. J. Agric. Sci.* 8, 11–142.
REESMAN, A. L. and KELLER, W. D. (1967) Chemical composition of illite, *J. Sed. Petrol.* 37, 592–96.
ROSS, C. S. and HENDRICKS, S. B. (1945) Minerals of the montmorillonite group, *U.S. Geol. Surv. Prof. Paper* 205B, 79 pp.
ROSS, C. S. and KERR, P. F. (1931) The kaolin minerals, *U.S. Geol. Surv. Prof. Paper* 165E, 151–76.
SCHULTZ, L. G. et al. (1971) Mixed layer kaolinite-montmorillonite from the Yucatan Peninsula, Mexico, *Clays Clay Min.* 19, 137–50.
SILLS, I. D. (1975) Internal surface area and porosity of kaolinitic glaebules, *Aust. J. Soil Res.* 13, 235–38.

Chapter 3 Soil organic matter

3.1 SUPPLY OF ORGANIC MATTER TO THE SOIL

ADACHI, T. (1971) On the area and humus content of Ando soils in Japan, *J. Sci. Soil Manure (Japan)* 42, 305–22.
BAL, L. (1970) Morphological investigation in two moder-humus profiles and the role of the soil fauna in their genesis, *Geoderma* 4, 5–36.
BATES, J. A. R. (1960) Studies on a Nigerian forest soil. I. The distribution of organic matter in the profile and various soil fractions, *J. Soil Sci.* 11, 246–56.
BEDRNA, Z. and MIČIAN, L. (1967) Soil geographic patterns in central and southeast Europe, *Soviet Soil Sci.* 1465–71.
BENACCHIO, S. S., BAUMGARDNER, M. F. and MOTT, G. O. (1970) Residual effect of grain-pasture feeding systems on the fertility of the soil under a pasture sward, *Soil Sci. Soc. Amer. Proc.* 34, 621–24.
BIRCH, H. F. and FRIEND, M. T. (1956) The organic-matter and nitrogen status of East African soils, *J. Soil Sci.* 7, 156–67.
BORNEMISSZA, G. F. (1971) A new variant of the paracoptic resting type in the Australian dung beetle, *Onthophagus compositus, Pedobiologia* 11, 1–10.
BRAY, J. R. and GORHAM, E. (1964) Litter production in the forests of the world, *Adv. Ecol. Res.* 2, 101–57.
CARLISLE, A. (1965) Carbohydrates in the precipitation beneath a sessile oak *Quercus petraea* (Mattushka) Liebl. canopy, *Plant Soil* 24, 399–400.
DESPHANDE, S. B., FEHRENBACHER, J. B. and BEAVERS, A. H. (1971) Mollisols of Tarai region of Uttar Pradesh, northern India, I. Morphology and mineralogy, *Geoderma* 6, 179–93.
FORREST, G. I. (1971) Structure and production of North Pennine blanket bog vegetation, *J. Ecol.* 59, 453–79.

GAYEL', A. G. and SHTINA, E. A. (1974) Algae on the sands of arid regions and their role in soil formation, *Soviet Soil Sci.* 311–19.

GERKIYAL, A. M. (1974) Accumulation of organic substances and nutrients in soil with the after-harvest and root residues of fallow crops, *Soviet Soil Sci.* 42–46.

HANDLEY, W. R. C. (1954) Mull and mor formation in relation to forest soils, *For. Comm. Bull.* No. 23, 115 pp.

JONES, M. J. (1973) The organic matter content of the savanna soils of West Africa, *J. Soil Sci.* 24, 42–53.

KRUPSKIY, N. K., KUZ'MICHEV, V. P. and DEREVYANKO, R. G. (1970) Humus content in Ukrainian soils, *Soviet Soil Sci.* 278–88.

MOIR, W. H. and GRIER, H. (1969) Weight and nitrogen, phosphorus, potassium, and calcium content of forest floor humus of lodgepole pine stands in Colorado, *Soil Sci. Soc. Amer. Proc.* 33, 137–40.

MUSTAFAYEV, K. M. et al. (1976) Biomass and bioenergy of the plant communities of some soil groups in the northeastern part of the Lower Caucasus, *Soviet Soil Sci.* 540–47.

NYE, P. (1961) Organic matter and nutrient cycles under moist tropical forest, *Plant Soil* 13, 333–46.

RUBILIN, Y. V. and DOLOTOV, V. A. (1967) Effect of cultivation on the amounts and composition of humus in Grey Forest soils, *Soviet Soil Sci.* 733–38.

RUNDNEVA, Y. N., TONKONGOV, V. D. and DOROKHOVA, K. Y. (1966) Ash elements and nitrogen cycle in the green-moss–spruce stands of the northern taiga in the basin of the Mezen' River, *Soviet Soil Sci.* 254–65.

SHCHERBAKOV, M. F. (1970) Effect of cultivation on the physical properties of soil and grass yields after accelerated regrassing of a dry meadow, *Soviet Soil Sci.* 589–95.

SOMMERS, L. E. (1977) Chemical composition of sewage sludges and analysis of their potential use as fertilizers, *J. Environ. Qual.* 6, 225–32.

STUTZBACH, S. J., LEAF, A. L. and LEONARD, R. E., (1972) Variation in forest floor under a red pine plantation, *Soil Sci.* 114, 24–28.

TEDROW, J. C. F. (1968) Pedogenic gradients of the polar regions, *J. Soil Sci.* 19, 197–204.

VANNIER, G. (1971) Les fourmis, prédateurs permanents de certain types de Collomboles, *Rev. Ecol. Biol. Sol.* 8, 119–32.

3.2 BREAKDOWN OF ORGANIC MATTER IN THE SOIL BY MACROFAUNA

BAKER, K. F. and SNYDER, W. C. (eds) (1965) *Ecology of soil-borne plant pathogens* (University of California, Berkeley).

BOCOCK, K. L. (1964) Changes in the amounts of dry matter, nitrogen, carbon, and energy in decomposing woodland leaf litter in relation to the activities of the soil fauna, *J. Ecol.* 52, 273–84.

BURGES, A. and RAW, F. (eds) (1967) *Soil biology* (Academic Press, London) 532 pp.

CURRY, J. P. (1969) The decomposition of organic matter in soil—I. The role of the fauna in decaying grassland herbage, *Soil Biol. Biochem.* 1, 53–58.

DICKINSON, C. H. and PUGH, G. J. F. (eds) (1974) *Biology of plant litter decomposition. Vol. 2. The organisms* (Academic Press, London) 775 pp.

GRAY, T. R. G. and WILLIAMS, S. T. (1971) *Soil micro-organisms* (Oliver and Boyd, Edinburgh) 240 pp.

HARLEY, J. L. (1971) Fungi in ecosystems, *J. Ecol.* 59, 653–68.

KÜHNELT, W. (1976) *Soil biology*, 2nd edn. (Faber and Faber, London) 483 pp.

LEE, K. E. and WOOD, T. G. (1971) *Termites and soils* (Academic Press, London) 251 pp.

LINDQUIST, B. (1941) Investigations on the significance of some Scandinavian earthworms in decomposition of leaf litter and structure of soil, *Svensk. Skags. U. Foren. Tidsler* 37, 179–242.

MACAULEY, B. J. (1975) Biodegradation of litter in *Eucalyptus pauciflora* communities–1. Techniques for comparing the effects of fungi and insects, *Soil Biol. Biochem.* 7, 341–44.

MURPHY, P. W. (1953) The biology of forest soils with special reference to the mesofauna or meiofauna, *J. Soil Sci.* 4, 155–93.

NYE, P. H. (1955) Some soil-forming processes in the humid tropics. IV. The action of the soil fauna, *J. Soil Sci.* 6, 73–83.

YEATES, G. W. (1977) Soil nematodes in New Zealand pastures, *Soil Sci.* 123, 415–22.

3.3 BIOCHEMICAL COMPOUNDS IN SOIL ORGANIC MATTER

CHESHIRE, M. V., GREAVES, M. P. and MUNDIE, C. M. (1974) Decomposition of soil polysaccharide, J. Soil Sci. 25, 483–98.

CHESHIRE, M. V., MUNDIE, C. M. and SHEPHERD, H. (1969) Transformation of ^{14}C glucose and starch in soil, *Soil Biol. Biochem.* 1, 117–30.

DUBACH, P. and MEHTA, N. C. (1963) The chemistry of soil humic substances *Soils Fert.* 26, 293–300.

GUPTA, V. C. and SOWDEN, F. J. (1964) Isolation and characterisation of cellulose from soil organic matter, *Soil Sci.* 97, 328–33.

KISS, S., DRĂGAN-BULARDA, M. and RĂDULESCU, D. (1975) Biological significance of enzymes accumulated in soil, *Adv. Agron.* 27, 25–87.

LOWE, L. E. (1968) Soluble polysaccharide fractions in selected Alberta soils, *Can. J. Soil Sci.* 48, 215–17.

MANSKAYA, S. M. and KODINA, V. I. (1968) Aromatic structure of lignins and their role in the formation of humic acids, *Soviet Soil Sci.* 1107–7.

OADES, J. M. (1967) Carbohydrates in some Australian soils, *Aust. J. Soil Res.* 5, 103–15.

ORLOV, D. S. and SADOVNIKOVA, L. K. (1975) Content and distribution of carbohydrates in the major soil groups of the U.S.S.R. *Soviet Soil Sci.* 440–49.

RAO, A. V. et al. (1975) Phenols in rice soils, *Soil Biol. Biochem.* 7, 227–29.

SCHNITZER, M. and DESJARDIN, J. G. (1965) Carboxyl and phenolic hydroxyl groups in some organic soils and their relation to the degree of humification, *Can. J. Soil Sci.* 45, 257–64.

SWINCER, G. D., OADES, J. M. and GREENLAND, D. J. (1969) The extraction, characterization, and significance of soil polysaccharides, *Adv. Agron.* 21, 195–235.

WHITEHEAD, D. C., BUCHAN, H. and HARTLEY, R. D. (1975) Components of soil organic matter under grass and arable cropping, *Soil Biol. Biochem.* 7, 65–71.

3.4 HUMUS

ANDERSON, D. W., PAUL, E. A. and ST.ARNAUD. R. J. (1974) Extraction and characterization of humus of Saskatchewan soils, with particular reference to clay associated humus, *Can. J. Soil Sci.* 54, 317–24.

ASEYEVA, I. V. and VELIKZHANINA, G. A. (1966) Biosynthesis of free amino acids by microorganisms in the soil, *Soviet Soil Sci.* 63–67.

BAL, L. (1970) Morphological investigation in two moder-humus profiles and the role of the soil fauna in their genesis, *Geoderma* 4, 5–36.

BEL'CHIKOVA, N. P. (1966) Characteristics of humic substances of taiga soils in Central Siberia which have developed on basic rocks, *Soviet Soil Sci.* 1136–45.

DUCHAUFOUR, P. (1976) Dynamics of organic matter in soils of temperate regions: its action on pedogenesis, *Geoderma* 15, 31–40.

FELBECK, G. T. (1965) Structural chemistry of soil humic substances, *Adv. Agron.* 17, 327–68.

FLAIG, W. (1964) Effects of micro-organisms in the transformation of lignin to humic substances, *Geochim. Cosmoch. Acta* 28, 1523–35.

GODLIN, M. M. and SON'KO, M. P. (1970) Humus of Ordinary Steppe Chernozems in the Ukraine, *Soviet Soil Sci.* 8–18.

HUNTJENS, J. L. M. (1972) Amino acid composition of humic acid-like polymers produced by streptomycetes and of humic acids from pasture and arable land, *Soil Biol. Biochem.* 4, 339–45.

JENKINSON, D. S. (1965) Studies on the decomposition of plant material in soil. I. Losses of carbon from ^{14}C labelled ryegrass incubated with soil in the field, *J. Soil Sci.* 16, 104–15.

KONONOVA, M. M. (1961) *Soil organic matter* (Pergamon, London) 450 pp.

LINEHAN, D. J. (1977) A comparison of the polycarboxylic acids extracted water from an agricultural top soil with those extracted by alkali, *J. Soil Sci.* 28, 369–78.

LOWE, L. E. (1969) Distribution and properties of organic fractions in selected Alberta soils, *Can. J. Soil Sci.* 49, 129–41.

MCGILL, W. B., SHIELDS, J. A. and PAUL, E. A. (1975) Relation between carbon and nitrogen turnover in soil organic fractions of microbial origins, *Soil Biol. Biochem.* 7, 57–63.

MARTIN, J. P. and HAIDER, K. (1971) Microbial activity in relation to soil humus formation, *Soil Sci.* 111, 54–63.

PONOMAREVA, V. V. and PLOTNIKOVA, T. A. (1975) Comparison of the humus profiles of typical Chernozem, Dark Gray Forest soil, and Dark Chestnut soil, *Soviet Soil Sci.* 404–13.

SCHNITZER, M. and KODAMA, H. (1975) An electron microscope examination of fulvic acid, *Geoderma* 13, 279–87.

SØRENSEN, L. H. (1975) The influence of clay on the rate of decay of amino acid metabolites synthesised in soils during decomposition of cellulose, *Soil Biol. Biochem.* 7, 171–77.

TAN, K. H. *et al.* (1972) Humic-fulvic acid content in soils as related to ley clipping management and fertilization, *Soil Sci. Soc. Amer. Proc.* 36, 565–67.

TITOVA, N. A. (1968) Nature of humus in the dry steppe soils of the central Yergeni and northwestern part of the Caspian lowland, *Soviet Soil Sci.* 377–90.

WHITEHEAD, D. C. and TINSLEY, J. (1963) The biochemistry of humus formation, *J. Sci. Fd Agric.* 14, 849–57.

WILLIAMS, S. T., MCNEILLY, T. and WELLINGTON, E. M. H. (1977) The decomposition of vegetation growing on metal mine waste, *Soil Biol. Biochem.* 9, 271–75.

3.5 PEAT

BASCOMB, C. L., BANFIELD, C. F. and BURTON, R. G. O. (1977) Characteristics of peaty materials from organic soils (Histosols) in England and Wales, *Geoderma* 19, 131–47.

BOELTER, D. H. (1965) Hydraulic conductivity of peats, *Soil Sci.* 100, 227–31.

FARNHAM, R. S. and FINNEY, H. R. (1965) Classification and properties of organic soils, *Adv. Agron.* 17, 115–62.

KATS, N. J. (1966) Bogs of the subantarctic and cold-temperate zone of the Southern hemisphere, *Soviet Soil Sci.* 180–88.

KONDO, R. (1974) Humus composition of peat land plant remains, *Soil Sci. Plant Nutrn* 20, 17–32.

LEVIN, I. and SHOHAM, D. (1972) Nitrate formation in peat soils of the reclaimed Hula swamp in Israel, *4th Intern. Peat Cong. Proc.* III, 47–57.

MARTEL, Y. A. and PAUL, E. A. (1974) The use of radiocarbon dating of organic matter in the study of soil genesis, *Soil Sci. Soc. Amer. Proc.* 38, 501–6.

OKRUSZKO, H. (1975) Classification and description of the peat soils of Poland, *Soviet Soil Sci.* 458–67.

O'TOOLE, M. A. (1968) Grassland research on blanket peat in Ireland, *J. Br. Grassl. Soc.* 23, 43–50.

RICHARDSON, S. J. and SMITH, J. (1977) Peat wastage in the East Anglian Fens, *J. Soil Sci.* 28, 485–89.

SCHNITZER, M. (1967) Humic-fulvic acid relationship in organic soils and humification of the organic matter of these soils, *Can. J. Soil Sci.* 47, 245–50.

SCHOTHORST, C. J. (1977) Subsidence of low moor peat soils in the western Netherlands, *Geoderma* 17, 265–91.

SJORS, H. (1961) Surface patterns in Boreal peatland, *Endeavour* 20, 217–24.

TAYLOR, J. A. (1964) Distribution and development of the world's peat deposits, *Nature* 201, 454–56.

WALMSLEY, M. E. and LAVKULICH, L. M. (1975) Chemical, physical, and land-use investigations of organic terrain, *Can. J. Soil Sci.* 55, 331–42.

WALSH, T. and BARRY, T. A. (1958) The chemical composition of some Irish peats, *Proc. Roy. Ir. Acad. B* 59, 325–28.

WARD, W. H., PENMAN, A. and GIBSON, R. E. (1955) Stability of a bank on a thin peat layer, *Geotechnique* 5, 154–63.

Chapter 4 Soil structure and porosity

4.1 SOIL STRUCTURE

ALLISON, F. E. (1968) Soil aggregation–some facts and fallacies as seen by a microbiologist, *Soil Sci.* 106, 136–43.

ASPIRAS, R. B. *et al.* (1971) Chemical and physical stability of microbially stabilized aggregates, *Soil Sci. Soc. Amer. Proc.* 35, 283–86.

BLEVINS, R. L., HOLOWAYCHUK, N. and WILDING, L. P. (1970) Micromorphology of soil fabric at tree root-soil interface, *Soil Sci. Soc. Amer. Proc.* 34, 460–65.

BOUMA, J. and HOLE, F. D. (1971) Soil structure and hydraulic conductivity of adjacent virgin and cultivated pedons at two sites: a Typic Argiudoll (Silt loam) and a Typic Eutrochrept (clay), *Soil Sci. Soc. Amer. Proc.* 35, 316–19.

BRAMMER, H. (1971) Coatings in seasonally flooded soils, *Geoderma* 6, 5–16.

BREWER, R. (1960) Cutans: their definition, recognition and interpretation, *J. Soil Sci.* 11, 280–92.

BREWER, R. and SLEEMAN, J. R. (1960) Soil structure and fabric: their definition and description, *J. Soil Sci.* 11, 172–85.

BREWER, R. and SLEEMAN, J. R. (1963) Pedotubules: their definition, classification and interpretation, *J. Soil Sci.* 14, 156–66.

BREWER, R. and SLEEMAN, J. R. (1964) Glaebules: their definition, classification and interpretation, *J. Soil Sci.* 15, 66–78.

BUOL, S. W. and HOLE, F. D. (1961) Clay skin genesis in Wisconsin soils, *Soil Sci. Soc. Amer. Proc.* 25, 371–79.

EDWARDS, A. P. and BREMNER, J. M. (1967) Microaggregates in soils, *J. Soil Sci.* 18, 64–73.

EMERSON, W. W. (1959) The structure of soil crumbs, *J. Soil Sci.* 10, 235–44.

EMERSON, W. W. (1967) A classification of soil aggregates based on their coherence in water, *Aust. J. Soil Res.* 5, 47–57.

ESWARAN, H. (1968) Point-count analysis as applied to soil micromorphology, *Pedologie* 18, 238–52.

HAMBLIN, A. P. and GREENLAND, D. J. (1977) Effect of organic constituents and complexed metal ions on aggregate stability of some East Anglian soils, *J. Soil Sci.* 28, 410–16.

HARRIS, R. F., CHESTERS, G. and ALLEN, O. N. (1966) Dynamics of soil aggregation, *Adv. Agron.* 18, 107–69.

JONGERIUS, A. (1970) Some morphological aspects of regrouping phenomena in Dutch soils, *Geoderma* 4, 311–31.

KHALIFA, E. M. and BOUL, S. W. (1969) Studies of clay skins in a Cecil (Typic Hapludult) soil: II. Effect on plant growth and nutrient uptake, *Soil Sci. Soc. Amer. Proc.* 33, 102–5.

KUZNETSOVA, I. V. (1966) Evaluation of the role of the different soil constituents in creating water-stable soil structure, *Soviet Soil Sci.* 1030–37.

KUZNETSOVA, I. V. (1967) Mechanical stability of soil structure, *Soviet Soil Sci.* 1073–80.

LOW, A. J. (1973) Soil structure and crop yield, *J. Soil Sci.* 24, 249–59.

LYNN, W. C. and GROSSMAN, R. B. (1970) Observations of certain soil fabrics with the scanning electron microscope, *Soil Sci. Soc. Amer. Proc.* 34, 645–48.

MILLER, F. P., WILDING, L. P. and HOLOWAYCHUK, N. (1971) Canfield silt loam, a Fragiudalf: II. Micromorphology, physical and chemical properties, *Soil Sci. Soc. Amer. Proc.* 35, 324–31.

MOLDENHAUER, W. C., WISCHMEIER, W. H. and PARKER, D. T. (1967) The influence of crop management on runoff erosion, and soil properties of a Marshall silty clay loam, *Soil Sci. Soc. Amer. Proc.* 31, 541–46.

PETTAPIECE, W. W. and ZWARICH, M. A. (1970) Micropedological study of a chernozemic to Grey Wooded sequence of soils in Manitoba, *J. Soil Sci.* 21, 138–45.

ROSE, C. W. (1962) Some effects of rainfall, radient drying, and soil factors on infiltration under rainfall into soils, *J. Soil Sci.* 13, 286–98.

RUSSELL, E. W. (1971) Soil structure: its maintenance and improvement, *J. Soil Sci.* 22, 137–51.

4.2 POROSITY

AUGUSTINUS, P. G. E. F. and SLAGER, S. (1971) Soil formation in swamp soils of the coastal fringe of Surinam, *Geoderma* 6, 203–11.

AYLMORE, L. A. G. and QUIRK, J. P. (1967) The micropore size distribution of clay mineral systems, *J. Soil Sci.* 18, 1–17.

CARY, J. W. and HAYDEN, C. W. (1973) An index for pore size distribution, *Geoderma* 9, 249–56.

CURRIE, J. A. (1966) The volume and porosity of soil crumbs, *J. Soil Sci.* 17, 24–35.

GREEN, R. D. and ASKEW, G. P. (1965) Observations on the biological development of macropores in soils of Romney Marsh, *J. Soil Sci.* 16, 342–49.

LEENHER, L. de (1971) The influence of weather, crop and sampling depth on the measurement of pore size distribution in the arable layer of some cultivated silt soils, *Soil Sci.* 112, 89–99.

ROMANS, J. C. C. (1959) Some measurements of air space in Scottish soils, *J. Soil Sci.* 10, 201–14.

SOKOLOVSKAYA, N. A. (1966) Distribution of pores by size and some water properties of aggregate fractions in a fine clay loam Thick Chernozem, *Soviet Soil Sci.* 682–87.

WILKINSON, G. E. (1975) Effect of grass fallow rotations on the infiltration of water into a savanna zone soil of Northern Nigeria, *Trop. Agric.* 52, 97–103.

ZEIN, EL ABEDINE, A., ROBINSON, G. H. and TYEGO, J. (1969) A study of certain physical properties of a Vertisol in the Gezira area, Republic of Sudan, *Soil Sci.* 108, 359–66.

Chapter 5 Physical properties of the soil

5.2 SOIL WATER

ABATUROV, B. D. and KARPACHEVSKIY, L. O. (1966) Effect of moles on the water. Physical properties of Sod-Podzolic soils, *Soviet Soil Sci.* 667–74.

BARTELLI, L. J. and PETERS, D. B. (1959) Integrating soil moisture characteristics with classification units of some Illinois soils, *Soil Sci. Soc. Amer. Proc.* 23, 149–51.

BOUMA, J. et al. (1971) Field measurement of unsaturated hydraulic conductivity by infiltration through artificial crusts, *Soil Sci. Soc. Amer. Proc.* 35, 362–64.

FAIRBOURN, M. L. and GARDNER, H. R. (1972) Vertical mulch effects on soil water storage, *Soil Sci. Soc. Amer. Proc.* 36, 823–27.

GREENSLADE, P. J. M. (1974) Some relations of the meat ant, *Iridomyomex purpurens* (Hymenoptera: Formicidae) with soil in South Australia, *Soil Biol. Biochem.* 6, 7–14.

GRIN, A. M. and NAZAROV, G. V. (1967) Investigation of soil permeability of different land uses in the steppe zone of the European USSR, *Soviet Soil Sci.* 1669–74.

GUSENKOV, Y. P., KASITSYN, V. N. and YEVTYUSHKIN, V. N. (1966) Characteristics of the soil-meliorative grouping of land in the northeastern part of the Tselinnoy Kray for irrigation purposes, *Soviet Soil Sci.* 7–13.

HILLEL, D., KRENTOS, V. D. and STYLIANON, Y. (1972) Procedure and test of an internal drainage method for measuring soil hydraulic characteristics *in situ*, *Soil Sci.* 114, 395–400.

HOLLIS, J. M., JONES, R. J. A. and PALMER, R. C. (1977) The effects of organic matter and particle size on the water retention properties of some soils in the West Midlands of England, *Geoderma* 17, 225–38.

KEMPER, W. D. and NOONAN, L. (1970) Runoff as affected by salt treatments and soil texture, *Soil Sci. Soc. Amer. Proc.* 34, 126–30.

LEBEDEVA, I. I. (1969) Nature of the illuvial horizon of light-grey forest soils on moraine loams, *Soviet Soil Sci.* 10–19.

LOW, A. J. (1954) The study of soil structure in the field and the laboratory, *J. Soil Sci.* 5, 57–74.

MacLEAN, A. H. and YAGER, T. U. (1972) Available water capacities of Zambian soils in relation to pressure plate measurements and particle size analysis, *Soil Sci.* 113, 23–29.

MASLOV, B. S. (1967) Influence of bulk density of sand on relationship between suction and moisture content, *Soviet Soil Sci.* 616–20.

MICHURIN, B. N. and LYTAYEV, I. A. (1967) Relationship between moisture content, moisture tension, and specific surface area in soil, *Soviet Soil Sci.* 1093–1103.

NAZAROV, G. V. (1970) Water permeability of soils in the European USSR in the zonal aspect, *Soviet Soil Sci.* 441–43.

PACHIKINA, L. I. and SHAROSHKINA, N. B. (1970) Composition, properties, and formation of Compact Meadow soils in the ancient Ural River delta, *Soviet Soil Sci.* 385–96.

PETERSEN, G. W., CUNNINGHAM, R. L. and MATELSKI, R. P. (1968) Moisture characteristics of Pennsylvania soils: II. Soil factors affecting moisture retention within a textural class—silt loam, *Soil Sci. Soc. Amer. Proc.* 32, 866–70.

PHILIPSON, W. R. and DROSDOFF, M. (1972) Relationships among physical and chemical properties of representative soils of the tropics from Puerto Rico, *Soil Sci. Soc. Amer. Proc.* 36, 815–19.

PIDGEON, J. D. (1972) The measurement and prediction of available water capacity of ferralitic soils in Uganda, *J. Soil Sci.* 23, 431–41.

RAO, K. S. and RAMACHARLU, P. T. (1959) pF-water relationship in typical Indian soils, *Soil Sci.* 87, 174–78.

RICHARDS, L. A. and WEAVER, L. R. (1944) Moisture retention by some irrigated soils as related to moisture tension, *J. Agr. Res.* 69, 215–35.

RICHARDS, S. J. and MARSH, A. W. (1961) Irrigation based on soil suction measurements, *Soil Sci. Soc. Amer. Proc.* 25, 65–68.

RIVERS, E. D. and SHIPP, R. F. (1972) Available water capacity of sandy and gravelly North Dakota soils, *Soil Sci.* 113, 74–80.

ROSE, G. W. (1962) Some effects of rainfall, radiant drying, and soil factors on infiltration under rainfall into soils, *J. Soil Sci.* 13, 286–98.

SALTER, P. J., BERRY, J. and WILLIAMS, J. B. (1966) The influence of texture on the moisture characteristics of soils, III. Quantitative relationships between particle size composition and available water capacity, *J. Soil Sci.* 17, 93–98.

SALTER, P. J. and WILLIAMS, J. B. (1963) The effect of farmyard manure on the moisture characteristics of a sandy loam soil, *J. Soil Sci.* 14, 73–81.

SALTER, P. J. and WILLIAMS, J. B. (1967) The influence of texture on the moisture characteristics of soils, *J. Soil Sci.* 18, 174–81.

SOKOLOVSKAYA, N. A. (1967) Role of soil structure in the movement of water to the consumption zone, *Soviet Soil Sci.* 1343–49.

THOMASSON, A. J. and ROBSON, J. D. (1967) The moisture régimes of soils developed on Keuper Marl, *J. Soil Sci.* 18, 329–40.

UNGER, P. W. (1970) Water relations of a profile-modified slowly permeable soil, *Soil Sci. Soc. Amer. Proc.* 34, 492–95.

VASHENIN, I. G., DOL GOPOLOVA, R. V. and SNETKOYA, A. P. (1969) Microvariation of characteristics and properties of soils within a soil profile, *Soviet Soil Sci.* 141–56.

VERSTRAETEN, L. M. J., FEYEN, J. and LIVENS, J. (1971) Hygroscopicity as a valuable complement in soil analysis, 2. Correlation between maximum hygroscopicity and physico-chemical characteristics of Ap-horizons with different textures, *Geoderma* 6, 263–74.

VORONIN, A. D. (1974) The energy state of soil moisture as related to soil fabric, *Geoderma* 12, 183–89.

WHITESIDE, E. P. *et al.* (1967) Considerations relative to a common particle size scale of earthy materials, *Soil Sci. Soc. Amer. Proc.* 31, 579–84.

WILKINSON, G. E. and AINA, P. O. (1976) Infiltration of water into two Nigerian soils under secondary forest and subsequent arable cropping, *Geoderma* 15, 51–59.

WILLIAMS, W. A. and DONEEN, L. D. (1960) Field infiltration studies with green manure and crop residues in irrigation soils, *Soil Sci. Soc. Amer. Proc.* 24, 58–61.

WISCHMEIER, W. H. (1966) Relation of field-plot runoff to management and physical factors, *Soil Sci. Soc. Amer. Proc.* 30, 272–77.

WOOD, H. B. (1977) Hydrologic differences between selected forested and agricultural soils in Hawaii, *Soil Sci. Soc. Amer. Proc.* 41, 132–36.

ZEIN, EL ABEDINE, A., ROBINSON, G. H. and TYEGO, J. (1969) A study of certain physical properties of a Vertisol in the Gezira area, Republic of Sudan, *Soil Sci.* 108, 359–66.

5.3 SOIL TEMPERATURE

CARY, J. W. and MAYLAND, H. F. (1972) Salt and water movement in unsaturated frozen soil, *Soil Sci. Soc. Amer. Proc.* 36, 549–55.

CHO, D. Y. and PONNAMPERUMA, F. N. (1971) Influence of soil temperature on the chemical kinetics of flooded soils and the growth of rice, *Soil Sci.* 112, 184–94.

DIMO, V. N. (1967) Soil and air temperature in the soil-climatic regions of the USSR, *Soviet Soil Sci.* 1641–54.

FERGUSEN, H., BROWN, P. L. and DICKEY, D. D. (1964) Water movement and loss under frozen soil conditions, *Soil Sci. Soc. Amer. Proc.* 28, 700–2.

FLUKER, B. J. (1958) Soil temperature, *Soil Sci.* 86, 35–46.

HART, G. (1961) Snow and frost conditions in New Hampshire under hardwoods and pines and in the open, *J. Forestry* 61, 287–89.

HOECKSTRA, P. (1966) Moisture movement in soil under temperature gradients with cold-site temperature below freezing, *Water Resour. Res.* 2, 241–50.

KULIK, N. F. (1967) Thermal-gradient transport of water in sandy soils, *Soviet Soil Sci.* 1507–20.

LAL, R. (1974) Soil temperature, soil moisture and maize yields from mulched and unmulched tropical soils, *Plant Soil* 40, 129–43.

LEONARD, R. E. *et al.* (1971) Annual soil moisture-temperature patterns as influenced by irrigation, *Soil Sci.* 111, 220–27.

LETTAU, B. (1971) Determination of the thermal diffusivity in the upper

layers of a natural ground cover, *Soil Sci.* 112, 173–77.
MCDOLE, R. E. and FOSBERG, M. A. (1974) Soil temperature in selected southeastern Idaho soils. II. Relation to soil and site characteristics, *Soil Sci. Soc. Amer. Proc.* 38, 486–91.
MATTHEWS, B. (1967) Automatic measurement of frost-heave: results from Malham and Rodley (Yorkshire), *Geoderma* 1, 107–15.
NECHAYEVA, Y. G. (1967) Characteristics of the dynamics of some properties of forest soils in the Far East during winter, *Soviet Soil Sci.* 1210–16.
PHILIP, J. R. and VRIES, D. A. de (1957) Moisture movement in porous materials under temperature gradients, *Trans. Amer. Geogphys. Un.* 38, 222–32.
RAHI, G. S. and JENSEN, R. D. (1975) Effect of temperature on soil-water diffusivity, *Geoderma* 14, 115–24.
SARTZ, R. S. (1970) Natural freezing and thawing in a silt and a sand, *Soil Sci.* 109, 319–23.
SELIM, H. M. and KIRKHAM, D. (1970) Soil temperature and water content changes during drying as influenced by cracks: a laboratory experiment, *Soil Sci. Soc. Amer. Proc.* 34, 565–69.
THORNE, D. B. and DUNCAN, D. P. (1972) Effects of snow removal, litter removal and soil compaction on soil freezing and thawing in a Minnesota oak stand, *Soil Sci. Soc. Amer. Proc.* 36, 153–57.
VASHCHENKO, I. M. and GAYEL, A. G. (1967) Distribution of the roots of fruit trees in relation to thermal characteristics of coarse-textured steppe soils, *Soviet Soil Sci.* 1764–67.
WILLIS, W. O., WIERENGA, P. J. and VREDENBURG, R. T. (1977) Fall soil water: effect on summer soil temperature, *Soil Sci. Soc. Amer. Proc.* 41, 615–17.
YOUNG, A. and STEPHEN, I. (1965) Rock weathering and soil formation on high-altitude plateaux of Malawi, *J. Soil Sci.* 16, 322–33.

5.4 SOIL AIR

ABRAMOVA, M. M. (1968) Evaporation of soil water under drought conditions, *Soviet Soil Sci.* 1151–58.
ARNOLD, P. W. (1954) Losses of nitrous oxide from soil, *J. Soil Sci.* 5, 116–28.
CURRIE, J. A. (1961) Gaseous diffusion in the aeration of aggregated soils, *Soil Sci.* 92, 40–45.
DOWDELL, R. J. and SMITH, K. A. (1974) Field studies of the soil atmosphere. II. Occurrence of nitrous oxide, *J. Soil Sci.* 25, 231–38.
GRABLE, A. R. (1966) Soil aeration and plant growth, *Adv. Agron.* 18, 57–106.
MACGREGOR, A. N. (1972) Impact of wetting on microbial respiration in

desert soil, *Soil Sci. Soc. Amer. Proc.* 36, 851–52.
ONCHUKOV, D. N., OSTAPCHIK, V. P. and CHANYY, V. G. (1972) Movement of water vapour in the soil under isothermal conditions, *Soviet Soil Sci.* 345–51.
PERRIER, E. R. and PRAKASH, O. (1977) Heat and vapour movement during infiltration into dry soils, *Soil Sci.* 124, 73–76.
RADFORD, P. J. and GREENWOOD, D. J. (1970) The simulation of gaseous diffusion in soils, *J. Soil Sci.* 21, 304–13.
REPNEVSKAYA, M. A. (1967) Liberation of CO_2 from soil in the pine stands of the Kola Peninsula, *Soviet Soil Sci.* 1067–72.
RICHTER, J. (1972) Evidence for significance of other-than-normal diffusion transport in soil gas exchange, *Geoderma* 8, 95–101.
ROSS, D. J. (1965) A seasonal study of oxygen uptake of some pasture soils and activities of enzymes hydrolysing sucrose and starch, *J. Soil Sci.* 16, 73–85.
SMITH, K. A. (1977) Soil aeration, *Soil Sci.* 123, 284–91.
TANAKA, A. and NAVASERO, S. A. (1967) Carbon dioxide and organic acids in relation to the growth of rice, *Soil Sci. Plant Nutrn (Tokyo)* 13, 25–30.
VERETENNIKOV, A. V. (1968) Carbon dioxide content of soil water in the waterlogged forests of Archangel Oblast, *Soviet Soil Sci.* 1417–22.
VUGAKOV, P. S. and POPOVA, Y. P. (1968) Carbon dioxide regime in soils of the Krasnoyarsk Forest Steppe, *Soviet Soil Sci.* 795–800.
WANNER, H. (1970) Soil respiration, litterfall and productivity of tropical rain forest, *J. Ecol.* 58, 543–47.

Chapter 6 Chemical properties of the soil

6.1 OXIDATION–REDUCTION POTENTIALS

AOMINE, S. (1961) A review of research on redox potentials of paddy soils in Japan, *Soil Sci.* 94, 6–13.
ARMSTRONG, W. A. (1967) The relationship between oxidation–reduction potentials and oxygen-diffusion levels in some waterlogged organic soils, *J. Soil Sci.* 18, 27–34.
BOHN, H. L. (1971) Redox potentials, *Soil Sci.* 112, 39–45.
FLUHLER, H., STOLZY, L. H. and ARDAKANI, M. S. (1976) A statistical approach to define soil aeration in respect to denitrification, *Soil Sci.* 122, 115–23.
JEFFREY, J. W. O. (1961) Measuring the state of reduction of a waterlogged soil, *J. Soil Sci.* 12, 317–25.
KAURICHEV, I. S. and SHISHOVA, V. S. (1967) Oxidation-reduction

conditions of coarse-textured soils of the Meshchera Lowland, *Soviet Soil Sci.* 636–43.
KAURICHEV, I. S. and TARARINA, L. F. (1972) Oxidation-reduction conditions inside and outside Gray Forest soil aggregates, *Soviet Soil Sci.* 547–50.
PATRICK, W. H. and MAHAPATRA, I. C. (1968) Transformation and availability to rice of nitrogen and phosphorus in Waterlogged soils. *Adv. Agron.* 20, 323–59.
PEARSALL, W. H. (1938) The soil complex in relation to plant communities. I. Oxidation–reduction potentials in soil, *J. Ecol.* 26, 180–93.
SANCHEZ, P. A. (1972) Nitrogen fertilization and management in tropical rice, *North Carolina Agr. Exp. Sta. Tech. Bull.* 213.
YAMANE, I. and SATO, K. (1968) Initial rapid drop of oxidation–reduction in submerged air-dried soils, *Soil Sci. Plant Nutrn (Tokyo)* 14, 68–69.

6.2 ION EXCHANGE

BURFORD, J. R. et al. (1964) Influence of organic materials on the determination of the specific surface areas of soils, *J. Soil Sci.* 15, 192–201.
DAVIDTZ, J. C. and SUMNER, M. E. (1965) Blocked charges on clay minerals in sub-tropical soils, *J. Soil Sci.* 16, 270–74.
GILLMAN, G. P. and BELL, L. C. (1976) Surface charge characteristics of six weathered soils from tropical North Queensland, *Aust. J. Soil Res.* 14, 351–60.
HINGSTON, F. J., POSNER, A. M. and QUIRK, J. P. (1972) Anion adsorption by goethite and gibbsite. I. The role of the proton in determining adsorption envelopes, *J. Soil Sci.* 23, 177–92.
KORNBLYUM, E. A. (1967) Changes in clay minerals accompanying soil development on the Volga-Akbtuba floodplain, *Soviet Soil Sci.* 1527–40.
MCALEESE, D. M. and MCCONAGHY, S. (1957) Studies on the basaltic soils of Northern Ireland, *J. Soil Sci.* 8, 135–40.
MARTINI, J. A. (1970) Allocation of cation exchange capacity to soil fractions in seven surface soils from Panama and the application of a cation exchange factor as a weathering index, *Soil Sci.* 109, 324–31.
MEKARU, T. and VEHARA, G. (1972) Anion adsorption in ferruginous tropical soils, *Soil Sci. Soc. Amer. Proc.* 36, 296–300.
MUTWEWINGABO, B. et al. (1975) Etude comparative de sols des Laurentides, Quebec, *Can. J. Soil Sci.* 55, 363–79.
PERKINS, H. F., TIWARI, S. C. and TAN, K. H. (1971) Characterization and classification of a highly weathered soil of the Southern Piedmont, *Soil Sci.* 111, 119–23.
PRITCHARD, D. T. (1971) Aluminum distribution in soils in relation

to surface area and cation exchange capacity, *Geoderma* 5, 255–60.
SAWHNEY, B. L. and NORRISH, K. (1971) pH dependent cation exchange capacity: minerals and soils of tropical regions, *Soil Sci.* 112, 213–15.
SCHOFIELD, R. K. (1949) Effect of pH on electric charges carried by clay particles, *J. Soil Sci.* 1, 1–8.
SINGH, S. and BHANDARI, G. S. (1965) Studies on humic acids of seven typical soils of Rajasthan, *J. Soil Sci.* 16, 183–91.
STEPHENS, F. R. (1969) Source of cation exchange capacity and water retention in southeast Alaskan Spodosols, *Soil Sci.* 108, 429–31.
WESTERN, S. (1972) The classification of arid zone soils I. An approach to the classification of arid zone soils using depositional features, *J. Soil Sci.* 23, 266–78.
WILDING, L. P. and RUTLEDGE, E. M. (1966) Cation-exchange capacity as a function of organic matter, total clay, and various clay fractions in a soil toposequence, *Soil Sci. Soc. Amer. Proc.* 30, 782–85.
WILSON, M. J. and LOGAN, J. (1976) Exchange properties and mineralogy of some soils derived from lavas of Lower Old Red Sandstone (Devonian) age. I. Exchangeable cations, *Geoderma* 15, 273–88.
WRIGHT, W. R. and FOSS, J. E. (1972) Contributions of clay and organic matter to the cation exchange capacity of Maryland soils, *Soil Sci. Soc. Amer. Proc.* 36, 115–18.

6.3 SOIL ACIDITY

COLEMAN, N. T., KAMPRATH, E. J. and WEED, S. B. (1958) Liming, *Adv. Agron.* 10, 475–522.
COLEMAN, N. T. and THOMAS, G. W. (1967) The basic chemistry of soil acidity, *Agron. Monog.* 12, 1–41.
GROENEWOUD, H. VAN (1961) Variation in pH and buffering capacity of the organic layer of grey wooded soils, *Soil Sci.* 92, 100–5.
JACKSON, M. L. (1963) Aluminum bonding in soils: a unifying principle in soil science, *Soil Sci. Soc. Amer. Proc.* 27, 1–10.
LARSEN, G. and CHILINGER, G. V. (eds.) (1967) *Diagenesis of sediments* (Elsevier, New York).
PERKINS, H. F., TIWARI, S. C. and TAN, K. H. (1971) Characterization and classification of a highly weathered soil of the Southern Piedmont, *Soil Sci.*, III, 119–23.
PIONKE, H. B. and COREY, R. B. (1967) Relationship between acidic aluminium and soil pH, clay, and organic matter, *Soil Sci. Soc. Amer. Proc.* 31, 749–52.
PONNAMPERUMA, F. N. (1972) The chemistry of submerged soils, *Adv. Agron.* 24, 29–96.
PONNAMPERUMA, F. N., MARTINEZ, E. and LOY, T. (1966) The influence

of redox potential and partial pressure of carbon dioxide on pH values and the suspension effect of flooded soils, *Soil Sci.* 101, 421–31.

PRATT, P. F. (1961) Effect of pH on the cation exchange capacity of surface soils, *Soil Sci. Soc. Amer. Proc.* 25, 96–98.

UNAMBA–OPARAH, I. (1972) Exchangeable reserve potassium and other cation reserves in the Agodi and Adio soils, *Soil Sci.* 113, 394–409.

6.4 INTERACTION OF ORGANIC MATTER WITH METALS, OXIDES AND CLAYS

BAKER, W. E. (1973) Role of humic acids from Tasmanian podzolic soils in mineral degradation and metal mobilisation, *Geochim. Cosmochim. Acta* 37, 269–81.

HINGSTON, F. J. et al. (1967) Specific adsorption of anions, *Nature* 215, 1459–61.

HODGSON, J. F., LINDSAY, W. L. and TRIERWEILER, J. F. (1966) Micronutrient cation complexing in soil solution: II. Complexing of zinc and copper in displaced solution from calcareous soils, *Soil Sci. Soc. Amer. Proc.* 30, 723–26.

KODAMA, H. and SCHNITZER, M. (1972) Dissolution of chlorite minerals by fulvic acids, *Can. J. Soil Sci.* 53, 240–43.

MORTLAND, M. M. (1970) Clay–organic complexes and interactions, *Adv. Agron.* 22, 75–117.

ORLOV, D. S., PIVOVAROVA, I. A. and GORBUNOV, N. I. (1973) Interaction of humic substances with minerals and the nature of their bond—a review, *Soviet Soil Sci.* 568–81.

SCHNITZER, M. and KODAMA, H. (1976) The dissolution of micas by fulvic acid, *Geoderma* 15, 381–91.

6.5 IRON, ALUMINIUM, AND SILICA

ABRUNA-RODRIGUES, F. et al. (1970) Crop responses to soil acidity factors in Ultisols and Oxisols: I. Tobacco, *Soil Sci. Soc. Amer. Proc.* 34, 629–35.

AKHTYRTSEV, B. P. (1968) Removal and accumulation of oxides in soils of broadleaf forests in the Central Russian Forest steppe, *Soviet Soil Sci.* 1423–34.

ARISTOVSKAYA, T. V. (1975) Role of micro-organisms in the mobilization and fixation of iron in soils, *Soviet Soil Sci.* 215–19.

ARISTOVSKAYA, T. V. and KUTUZOVA, R. S. (1968) Microbiological factors in the extraction of silicon from slightly-soluble natural compounds, *Soviet Soil Sci.* 1653–59.

BACHE, B. W. and SHARP, G S. (1976) Characterization of mobile aluminum in acid soils, *Geoderma* 15, 91–101.

BECKWITH, R. S. and REEVE, R. (1964) Studies of soluble silica in soils, *Aust. J. Soil Res.* 2, 33–45.

COLLINS, J. F. and BUOL, S. W. (1970) Effect of fluctuations in the Eh-pH environment on iron and/or manganese equilibria, *Soil Sci.* 110, 111–18.

CURTIS, C. D. (1970) Differences between lateritic and podzolic weathering, *Geochim. Cosmochim. Acta* 34, 1351–53.

EKPETE, D. M. (1972) Assessment of the lime requirements of eastern Nigeria soils, *Soil Sci.* 113, 363–72.

FOX, R. L. et al. (1967) Soil and plant silicon and silicate response by sugar cane, *Soil Sci. Soc. Amer. Proc.* 31, 775–79.

JONES, L. H. P. and HANDRECK, K. A. (1967) Silica in soils, plants, and animals, *Adv. Agron.* 19, 107–49.

MACKNEY, D. (1961) A podzol development sequence in oakwoods and heath in central England, *J. Soil Sci.* 12, 23–40.

MARION, G. M. et al. (1976) Aluminium and silica solubility in soils, *Soil Sci.* 121, pp 76–85.

MILLER, R. J. (1968) Electron micrographs of acid-edge attack of kaolinite, *Soil Sci.* 105, 166–71.

O'CONNOR, G. A., LINDSAY, W. L. and OLSEN, S. R. (1971) Diffusion of iron and iron chelates in soil, *Soil Sci. Soc. Amer. Proc.* 35, 407–10.

RAUPACH, M. (1963) Solubility of simple aluminum compounds expected in soils, *Aust. J. Soil Res.* 1, 28–54.

ROSS, G. J. and TURNER, R. C. (1971) Effect of different anions on the crystallization of aluminum hydroxide in partially neutralized aqueous aluminium salt systems, *Soil Sci. Soc. Amer. Proc.* 35, 389–92.

SIEVER, R. (1957) The silica budget in the sedimentary cycle, *Amer. Mineral* 42, 821–41.

SMITHSON, F. (1958) Grass opal in British soils, *J. Soil Sci.* 9, 148–54.

TAKKAR, P. N. (1969) Distribution of iron and manganese in meadow forest, non-calcic brown and grey-brown podzolic soils of the Himachal Pradesh (India), *Geoderma* 3, 215–22.

TANAKA, A. and HAYAKAWA, Y. (1975) Comparison of tolerance to soil acidity among crop plants. Part 2. Tolerance to high levels of aluminum and manganese–studies on comparative plant nutrition, *J. Sci. Soil Manure (Japan)* 46, 19–68.

6.6 POTASSIUM, CALCIUM AND MAGNESIUM

ALSTON, A. M. (1972) Availability of magnesium in soils, *J. Agric. Sci.* 79, 197–204.

AVAKYAN, N. O. (1969) Soil adsorption complex and potassium nutrition of plants, *Soviet Soil Sci.* 433–41.

DOLCASTER, D. L. *et al.* (1968) Cation exchange selectivity of some clay-sized minerals and soil materials, *Soil Sci. Soc. Amer. Proc.* 32, 795–98.

KARAMANOS, R. E. and TURNER, R. C. (1977) Potassium-supplying power of some northern-Greece soils in relation to clay-mineral composition, *Geoderma* 17, 209–18.

KARIM, A. (1954) A mineralogical study of the colloid fractions of some great soil groups with particular reference to illites, *J. Soil Sci.* 5, 140–44.

NEMETH, K., MENCEL, K. and CRIMME, H. (1970) The concentration of K, Ca and Mg in the saturation extract in relation to exchangeable K, Ca and Mg, *Soil Sci.* 109, 179–85.

PRINCE, A. L., ZIMMERMAN, M. and BEAR, F. E. (1947) The magnesium-supplying powers of 20 New Jersey soils, *Soil Sci.* 63, 69–94.

RICH, C. I. and BLACK, W. R. (1964) Potassium exchange as affected by cation size, pH, and mineral structure, *Soil Sci.* 97, 384–90.

ST. ARNAUD, R. J. and HERBILLON, A. J. (1973) Occurrence and genesis of secondary magnesium-bearing calcites in soils, *Geoderma* 9, 279–98.

SALMON, R. C. (1963) Magnesium relationships in soils and plants, *J. Sci. Fd Agric.* 14, 605–10.

SALOMONS, W. and MOOK, W. G. (1977) Isotope geochemistry of carbonate dissolution and reprecipitation in soils, *Soil Sci.* 122, 15–24.

WELTE, E. and NIEDERBUDDE, E. A. (1965) Fixation and availability of potassium in loess-derived and alluvial soils, *J. Soil Sci.* 16, 116–20.

UNAMBA-OPARAH, I. (1972) Exchangeable reserve potassium and other cation reserves in the Agodi and Adio soils, *Soil Sci.* 113, 394–409.

6.7 NITROGEN, PHOSPHORUS AND SULPHUR

ADEPETU, J. A. and COREY, R. B. (1976) Organic phosphorus as a predictor of plant-available phosphorus in soils of Southern Nigeria, *Soil Sci.* 122, 159–64.

BASAK, M. N. and BATTACHARYA, R. (1962) Phosphate transformations in rice soil, *Soil Sci.* 94, 258–62.

BLASCO, M. L. and CORNFIELD, A. H. (1967) Effect of soil moisture content during incubation on the nitrogen-mineralizing characteristics of the soils of Colombia (South America), *Geoderma* 1, 19–25.

BLOOMFIELD, C. (1972) The oxidation of iron sulphide soils in relation to the formation of acid sulphate soils and of other deposits in field drains, *J. Soil Sci.* 23, 1–16.

BOLLE-JONES, E. W. (1964) Incidence of sulphur deficiency in Africa. A review, *Emp. J. exp. Agric.* 32, 241–48.

BORNEMISZA, E. and LLANOS, R. (1967) Sulfate movement, adsorption and desorption in three Costa Rican soils, *Soil Sci. Soc. Amer. Proc.* 31, 356–60.

BOSWELL, F. C. and ANDERSON, O. E. (1970) Nitrogen movement comparisons in cropped versus fallowed soils, *Agron. J.* 62, 499–503.

CHAO, T. T., HARWARD, M. E. and FANG, S. C. (1962) Soil constituents and properties in the adsorption of sulphate ions, *Soil Sci.* 94, 276–83.

COLEMAN, R. (1966) The importance of sulphur as a plant nutrient in world crop production, *Soil Sci.* 101, 230–39.

CONNELL, W. E. and PATRICK, W. H. (1969) Reduction of sulfate to sulfide in waterlogged soil, *Soil Sci. Soc. Amer. Proc.* 33, 711–15.

COOPER, G. S. and SMITH, R. L. (1963) Sequence of products formed during denitrification in some diverse Western soils, *Soil Sci. Soc. Amer. Proc.* 27, 659–62.

ENWEZOR, W. O. (1976) Sulphur deficiencies in soils of southeastern Nigeria, *Geoderma* 15, 401–11.

GILMORE, A. R. (1972) Liming retards height growth of young shortleaf pine, *Soil Sci.* 113, 448–52.

GOTOH, S. and YAMAHITA (1966) Oxidation–reduction potential of a paddy soil *in situ* with special reference to the production of ferrous iron, manganous manganese, and sulfide, *Soil Sci. Plant Nutrn (Tokyo)* 12, 230–38.

HANLEY, P. K. MCDONNELL, P. and MURPHY, M. D. (1965) Phosphorus status of Irish soils: 1, *Irish J. agric. Res.* 4, 81–92.

HANLEY, P. K. and MURPHY, M. D. (1970) Phosphate forms in particle size separates of Irish soils in relation to drainage and parent materials, *Soil Sci. Soc. Amer. Proc.* 34, 587–90.

HARWARD, M. E. and REISENAUER, H. M. (1966) Reactions and movement of inorganic soil sulphur, *Soil Sci.* 101, 326–35.

HERRON, G. M. et al. (1968) Residual nitrate nitrogen in fertilized deep loess-derived soils, *Agron. J.* 60, 477–82.

HESSE, P. R. (1962) Phosphorus fixation in mangrove swamp muds, *Nature* 193, 295–96.

ISLAM, A. and MANDAL, R. (1977) Amounts and mineralisation of organic phosphorus compounds and derivatives in some surface soils of Bangladesh, *Geoderma* 17, 57–68.

IVARSON, K. C. (1973) Microbiological formation of basic ferric sulphates, *Can. J. Soil Sci.* 53, 315–23.

JOHANSSON, O. (1959) On sulphur problems in Swedish agriculture, *Kgl. Landbr. Ann.* 25, 57–169.

JUO, A. S. R. and EHTS, B. G. (1968) Chemical and physical properties of iron and aluminium phosphates and their relation to phosphorus availability, *Soil Sci. Soc. Amer. Proc.* 32, 216–21.

KHALID, R. A., PATRICK, W. H. and DE LAUNE, R. D. (1977) Phosphorus sorption characteristics of flooded soils, *Soil Sci. Soc. Amer. Proc.* 41, 305–10.

KINJO, T. and PRATT, P. F. (1971) Nitrate adsorption. I. In some acid soils of Mexico and South America. *Soil Sci. Soc. Amer. Proc.* 35, 722–25.

LARSEN, S., GUNARY, D. and SUTTON, C. D. (1965) The rate of immobilisation of applied phosphate in relation to soil properties, *J. Soil Sci.* 16, 141–46.

LIONNET, J. I. C. (1952) Rendzina soils of coastal flats of the Seychelles, *J. Soil Sci.* 3, 172–81.

MacLEAN, A. J. (1977) Movement of nitrate nitrogen with different cropping systems in two soils, *Can. J. Soil Sci.* 57, 27–33.

MCCLUNG, A. C. and DE FERITAS, L. M. M. (1959) Sulphur deficiency in soils from Brazilian Campos, *Ecology* 40, 315–17.

MCKERCHER, R. B. and ANDERSON, G. (1968) Content of inositol penta- and hexaphosphates in some Canadian soils, *J. Soil Sci.* 19, 47–55.

MCLACHLAN, L. D. (1955) Phosphorus, sulphur, and molybdenum deficiencies in soils from eastern Australia, *Aust. J. agric. Res.* 8, 673–84.

MAHTAB, S. K. *et al.* (1971) Phosphorus diffusion in soils: I. The effect of applied P, clay content, and water content, *Soil Sci. Soc. Amer. Proc.* 35, 393–97.

MATTINGLY, G. E. G. (1975) Labile phosphate in soils, *Soil Sci.* 119, 369–75.

NYE, P. H. and BERTHEUX, T. (1957) The distribution of phosphorus in forest and savannah soils of the Gold Coast and its agricultural significance, *J. Agric. Sci.* 49, 141–59.

OLSEN, S. R. and WATANABE, F. S. (1966) Effective volume of soil around plant roots determined from phosphorus diffusion, *Soil Sci. Soc. Amer. Proc.* 30, 598–602.

OMOTOSO, T. I. (1971) Organic phosphorus contents of some cocoa-growing soils of Southern Nigeria, *Soil Sci.* 112, 195–99.

OZANNE, P. G., KIRTON, D. J. and SHAW, T. C. (1961) The loss of phosphorus from sandy soils, *Aust. J. Agric. Res.* 12, 409–23.

PATRICK, W. H. and WYATT, R. (1964) Soil nitrogen loss as a result of alternate submergence and drying, *Soil Sci. Soc. Amer. Proc.* 28, 647–53.

PEVERILL, K. I. and DOUGLAS, L. A. (1976) The use of undisturbed cores of surface soil for investigating leaching losses of sulphur and phosphorus, *Geoderma* 16, 193–99.

ROBINSON, J. B. (1963) Nitrification in a New Zealand grassland soil, *Plant Soil* 19, 173–83.

ROBINSON, J. B. (1968) A simple available nitrogen index. II. Field crop evaluation, *J. Soil Sci.* 19, 280–90.

ROBINSON, J. B. D. (1967) Soil particle size fractions and nitrogen mineralization, *J. Soil Sci.* 18, 109–17.

RODRIGUES, G. (1954) Fixed ammonia in tropical soils, *J. Soil Sci.* 5, 264–74.

SAUNDERS, W. M. H. (1965) Phosphate retention by New Zealand soils and its relationship to free sesquioxides, organic matter and other soil properties, *N. Z. J. Agric. Res.* 8, 30–57.

SAVANT, N. K. and ELLIS, R. (1964) Changes in redox potential and phosphorus availability in submerged soil, *Soil Sci.* 98, 388–94.

SIMPSON, K. (1961) Factors influencing uptake of phosphorus by crops in southeast Scotland, *Soil Sci.* 92, 1–14.

SINGH, B. R. and KANEHIRO, Y. (1969) Adsorption of nitrate in amorphous and kaolinitic Hawaiian soils, *Soil Sci. Soc. Amer. Proc.* 33, 681–83.

SMITH, S. J., YOUNG, L. B. and MILLER, G. E. (1977) Evaluation of soil nitrogen mineralization potential under modified field conditions, *Soil Sci. Soc. Amer. Proc.* 41, 74–76.

STANFORD, G. and EPSTEIN, E. (1974) Nitrogen mineralisation water relations in soils, *Soil Sci. Amer. Proc.* 38, 103–7.

STARKEY, R. L. (1966) Oxidation and reduction of sulphur compounds in soils, *Soil Sci.* 101, 297–306.

SYERS, J. K. et al. (1971) Phosphate sorption parameters of representative soils from Rio Grande do Sul, Brazil, *Soil Sci.* 112, 267–75.

SYERS, J. K., SHAH, R. and WALKER, T. W. (1969) Fractionation of phosphorus in two alluvial soils and particle-size separates, *Soil Sci.* 108, 283–89.

VAIDYANATHAN, L. V. and TALIBUDEEN, O. (1968) Rate-controlling processes in the release of soil phosphate, *J. Soil Sci.* 19, 342–53.

WALKER, T. W. and SYERS, J. K. (1976) The fate of phosphorus during pedogenesis, *Geoderma* 15, 1–19.

WESTERN, S. (1972) The classification of arid zone soils I. An approach to the classification of arid zone soils using depositional features, *J. Soil Sci.* 23, 266–78.

WETSELAAR, R. (1961) Nitrate distribution in tropical soils, *Plant Soil* 15, 110–33.

WILLIAMS, E. G. and SAUNDERS, W. M. H. (1956) Distribution of phosphorus in profiles and particle-size fractions of some Scottish soils, *J. Soil Sci.* 7, 90–108.

WULLSTEIN, L. H. (1969) Reduction of nitrite deficits by alkaline metal carbonates, *Soil Sci.* 108, 222–26.

YAMANE, I. (1969) Reduction of nitrate and sulphate in submerged soils with special reference to redox potential and water-soluble sugar

content of soils, *Soil Sci. Plant Nutrn* 15, 139–48.
YOSHIDA, T. and ANCAJAS, R. R. (1973) Nitrogen fixing activity in upland and flooded rice fields, *Soil Sci. Soc. Amer. Proc.* 37, 42–46.
YOUNG, J. L. and MCNEAL, B. L. (1964) Ammonia and ammonium reactions with some layer-silicate minerals, *Soil Sci. Soc. Amer. Proc.* 28, 334–39.

6.8 MICRONUTRIENTS

ANDERSON, A. J. (1970) Trace elements for sheep pastures and fodder crops in Australia, *J. Aust. Inst. Agr. Sci.* 36, 15–29.
CLARKE, A. L. and GRAHAM, E. R. (1968) Zinc diffusion and distribution coefficients in soil as affected by soil texture, zinc concentration and pH, *Soil Sci.* 105, 409–18.
GOTOH, S. and PATRICK, W. H. (1972) Transformation of manganese in a waterlogged soil as affected by redox potential and pH, *Soil Sci. Soc. Amer. Proc.* 36, 738–42.
GUPTA, U. C. (1968) Relationship of total and hot-water soluble boron and fixation of added boron, to properties of podzol soils, *Soil Sci. Soc. Amer. Proc.* 32, 45–48.
HODGSON, J. F., LINDSAY, W. L. and TRIERWEILER, J. F. (1966) Micronutrient cation complexing in soil solution: II. Complexing of zinc and copper in displaced solution from calcareous soils, *Soil Sci. Soc. Amer. Proc.* 30, 723–26.
KARAMANOS, R. E., BETTANY, J. R. and STEWART, J. W. B. (1976) The uptake of native and applied lead by alfalfa and bromegrass from soil, *Can. J. Soil Sci.* 56, 485–94.
MCPHAIL, M., PAGE, A. L. and BINGHAM, F. T. (1972) Adsorption interactions of monosilicic and boric acid on hydrous oxides of iron and aluminium, *Soil Sci. Soc. Amer. Proc.* 36, 510–14.
MARSHALL, A. (1972) 'Desert' becomes 'Downs'; the impact of a scientific discovery, *Aust. Geogr.* 12, 23–34.
MORTVEDT, J. J., GIORDANO, P. M. and LINDSAY, W. L. (eds) (1972) *Micronutrients in agriculture* (Soil Sci. Soc. Amer., Madison, Wisconsin).
NAVROT, J. and RAVIKOVITCH, S. (1969) Zinc availability in calcareous soils: III. The level and properties of calcium in soils and its influence on zinc availability, *Soil Sci.* 108, 30–37.
OBUKHOV, A. I. (1968) Behaviours of microelements during weathering and soil formation in the tropics and subtropics of Burma, *Soviet Soil Sci.* 1894–1901.
PATRICK, W. H. and TURNER, F. T. (1968) Effect of redox potential on manganese transformation in waterlogged soil, *Nature* 220, 476–78.

RADOV, A. S. and KOR CHAGINA, Y. D. (1967) Boron and manganese content in the chestnut soils of the Volgograd Oblast, *Soviet Soil Sci.* 647–55.
RANDHAWA, N. S. and KANWAR, J. S. (1964) Zinc, copper and cobalt-status of Punjab soils, *Soil Sci.* 98, 403–7.
RANDHAWA, N. S., KANWAR, J. S. and NIJHAWAN, S. D. (1961) Distribution of different forms of manganese in the Punjab soils, *Soil Sci.* 92, 106–12.
REISENAUER, H. M., TABIKH, A. A. and STOUT, P. R. (1962) Molybdenum reactions with soils and the hydrous oxides of iron, aluminum, and titanium, *Soil Sci. Soc. Amer. Proc.* 26, 23–27.
SVIRIDOV, A. S. *et al.* (1969) Content of available minor elements in the Chernozems of Tamblov Oblast, *Soviet Soil Sci.* 557–61.
TANAKA, A. and NAVASERO, S. A. (1966) Manganese content of the rice plant under water culture condition, *Soil Sci. Plant Nutrn (Tokyo)* 12, 67–72.
TAYLOR, R. M. and GILES, J. B. (1970) The association of vanadium and molybdenum with iron oxides in soils, *J. Soil Sci.* 21, 203–15.
TAYLOR, R. M. and MCKENZIE, R. M. (1966) The association of trace elements with manganese minerals in Australian soils, *Aust. J. Soil Res.* 4, 29–39.
TYLER, G. (1974) Heavy metal pollution and soil enzyme activity, *Plant Soil* 41, 303–11.
WHITEHEAD, D. C. (1973) Studies of iodine in British soils, *J. Soil Sci.* 24, 260–70.
YAALON, D. H., BRENNER, I. and KOYUMDJISKY, H. (1974) Weathering and mobility sequence of minor elements on a basaltic pedomorphic surface, Galilee, Israel, *Geoderma* 12, 233–44.

6.9 SODIUM CHLORIDE AND ASSOCIATED SALTS

AGRAWAL, R. P. and RAMAMOORTHY, B. (1970) Morphological and chemical characteristics of alkali and normal soils from black and red soils of India, *Geoderma* 4, 403–15.
AYERS, A. D. *et al.* (1960) Saline and sodic soils of Spain, *Soil Sci.* 90, 133–38.
CLARIDGE, G. G. C. and CAMPBELL, I. B. (1977) The salts in Antarctic soils, their distribution and relationship to soil processes, *Soil Sci.* 123, 377–84.
CONACHER, A. J. and MURRAY, I. D. (1973) Implications and causes of salinity problems in the Western Australian Wheatbelt: The York-Mawson area, *Aust. Geog. Studies* 11, 40–61.
GREENLEE, G. M., PAWLUK, S. and BOWSER, W. E. (1968) Occurrence of

soil salinity in the dry lands of southwestern Alberta, *Can. J. Soil Sci.* 48, 65–75.

MILJKOVIC, N., AYERS, A. D. and EBERHARD, D. L. (1959) Salt-affected soils of Yugoslavia, *Soil Sci.* 88, 51–55.

ORLOV, D. S. (1967) Activities of ions and salts in soils and their significance for the theory of soil formation and soil fertility, *Soviet Soil Sci.* 1789–99.

SEHGAL, J. L., HALL, G. F. and BHARGANA, G. P. (1975) An appraisal of the problems in classifying saline-sodic soils of the Indo-Gangetic plain in N. W. India, *Geoderma* 14, 7591.

SLAVNYY, Y. A., TURSINA, T. V. and KAURICHEVA, Z. N. (1970) Genesis of saline soils in the Caspian region, *Soviet Soil Sci.* 537–42.

SZABOLCS, I. (1974) *Salt affected soils in Europe* (Nijhoff, The Hague), 63 pp.

WIEGAND, C. L., LYLES, L. and CARTER, D. L. (1966) Interspersed salt-affected and unaffected dryland soils of the Lower Rio Grande Valley: II. Occurrence of salinity in relation to infiltration rate and profile characteristics, *Soil Sci. Soc. Amer. Proc.* 30, 106–10.

Chapter 7 Soil mechanical properties

7.1 BULK DENSITY

ADAMS, W. D. (1973) The effect of organic matter on the bulk and true densities of some uncultivated podzolic soils, *J. Soil Sci.* 24, 10–17.

AVERY, B. W. *et al.* (1969) The soils of Broadbalk, *Rothamsted Exp. Sta. Rep. 1968* 2, 63–115.

BUNTING, B. T. and HATHOUT, S. A. (1971) Physical characteristics and chemical properties of some High-Arctic organic materials from southwest Devon Island, Northwest Territories, Canada, *Soil Sci.* 112, 107–15.

CAMP, C. R. and GILL, W. R. (1969) The effect of drying on soil strength parameters, *Soil Sci. Soc. Amer. Proc.* 33, 641–44.

KUZNETSOV, M. S. (1967) Influence of cohesion among soil grains on the erosion resistance of the light-chestnut soils of the Yergeni Upland, *Soviet Soil Sci.* 1748–53.

KYUMA, K., SUH, Y-S. and KAWAGUCHI, K. (1977) A method of capability evaluation for upland soils. I. Assessment of available water retention capacity, *Soil Sci. Plant Nutrn* 23, 135–49.

NIELSEN, D. R., BIGGAR, J. W. and COREY, J. C. (1972) Application of flow theory to field situations, *Soil Sci.* 113, 254–63.

LOWRY, F. E., TAYLOR, H. M. and HUCK, M. G. (1970) Growth rate and yield of cotton as influenced by depth and bulk density of soil pans, *Soil Sci. Soc. Amer. Proc.* 34, 306–9.

POKOTILO, A. S. (1967) Some physical properties of the separates of Ordinary and Southern Chernozems, *Soviet Soil Sci.* 1862–67.

SCHUURMAN, J. J. (1965) Influence of soil density on root development and growth of oats, *Plant Soil* 22, 352–74.

TABATABAI, M. A. and HANWAY, J. J. (1968) Some chemical and physical properties of different-sized natural aggregates from Iowa soils, *Soil Sci. Soc. Amer. Proc.* 32, 588–91.

WILLIAMS, R. F. B. (1971) Relationships between the composition of soils and physical measurements made on them, *Rothamsted Exp. Sta. Rep. for 1970* 2, 5–35.

ZYUZ', N. S. (1968) Bulk density and hardness of the hillocky sands of the Middle Don, *Soviet Soil Sci.* 1769–76.

WARNCKE, D. D. and BARBER, S. A. (1972) Diffusion of zinc in soil: II. The influence of soil bulk density and its interaction with soil moisture, *Soil Sci. Soc. Amer. Proc.* 36, 42–46.

7.2 SOIL STRENGTH

AUDRIC, T. and BOUQUIER, L. (1976) Collapsing behaviour of some loess soils from Normandy, *Quart. J. Eng. Geol.* 9, 265–77.

BAKHTIN, P. U. et al. (1969) Changes in resistance of some soils to plowing and deformations in relation to their moisture content, *Soviet Soil Sci.* 592–602.

BARDEN, L. (1972) The relation of soil structure to the engineering geology of clay soil, *Quart. J. Eng. Geol.* 5, 85–102.

BRADFORD, J. M. and PIEST, R. F. (1977) Gully wall stability in loess-derived alluvium, *Soil Sci. Soc. Amer. Proc.* 41, 115–22.

CHANDLER, R. J. (1972) Lias Clay: weathering processes and their effect on shear strength, *Geotechnique* 22, 403–31.

FOUNTAINE, E. R. (1954) Investigations into the mechanism of soil adhesion, *J. Soil Sci.* 5, 251–63.

KOLOSKOVA, A. V. and BURKALOV, A. A. (1969) Adhesion of some soils in the Tatar ASSR, *Soviet Soil Sci.* 448–55.

LUMB, P. (1965) The residual soils of Hong Kong, *Geotechnique* 15, 180–94.

MCINTYRE, D. S. (1976) Subplasticity in Australian soils. I. Description, occurrence, and some properties, *Aust. J. Soil Res.* 14, 227–36.

RITCHIE, J. T. et al. (1972) Water movement in undisturbed swelling clay soil, *Soil Sci. Soc. Amer. Proc.* 36, 874–79.

SHIYATYY, Y. I., LAVROVSKIY, A. B. and KHMOLENKO, M. I. (1972) Effect of texture on the cohesion and wind resistance of soil clods, *Soviet Soil Sci.* 105–12.

SLOVIKOVSKIY, V. I. (1968) Resistance of some sandy deposits and soils of the Middle Dneiper region, *Soviet Soil Sci.* 832–35.

TAYLOR, H. M. and RATLIFF, L. F. (1969) Root elongation rates of cotton and peanuts as a function of soil strength and soil water content, *Soil Sci.* 108, 113–19.

TAYLOR, R. K. (1972) The functions of the engineering geologist in urban development, *Quart. J. Eng. Geol.* 4, 221–40.

WARKENTIN, B. P. and MAEDA, T. (1974) Physical properties of allophane soils from the West Indies and Japan, *Soil Sci. Soc. Amer. Proc.* 38, 372–77.

7.3 SWELLING AND DISPERSION

AGRAWAL, R. P. (1970) Critical sodium limits and indices of degree of dispersion in soils, *Trop. Agric.* 47, 67–72.

BARKER, A. C., EMERSON, W. W. and OADES, J. M. (1973) The comparative effects of exchangeable calcium, magnesium and sodium on some physical properties of red-brown earth subsoils, *Aust. J. Soil Res.* 11, 143–50.

DAN, J., YAALON, D. H. and KOYUMDJISKY, H. (1968) Catenary soil relations in Israel, 1. The Netanya catena on coastal dunes of the Sharon, *Geoderma* 2, 95–120.

GROENEWEGEN, H. (1959) Relation between chloride accumulation and soil permeability in the Mirrool irrigation area, New South Wales, *Soil Sci.* 87, 283–88.

HAMID, K. S. and MUSTAFA, M. A. (1975) Dispersion as an index of relative hydraulic conductivity in salt-affected soils of the Sudan, *Geoderma* 14, 107–14.

HARKER, D. B., WEBSTER, G. R. and CAIRNS, R. R. (1977) Factors contributing to crop response on a deep-plowed Solonetz soil, *Can. J. Soil Sci.* 57, 279–87.

MUKHTAR, O. M. A., SWOBODA, A. R. and GODFREY, C. L. (1974) The effect of sodium and calcium chlorides on structure stability of two vertisols: Gezira clay from Sudan, Africa, and Houston Black Clay from Texas, USA, *Soil Sci.* 118, 109–19.

SCHAIK, J. C. van (1967) Influence of adsorbed sodium and gypsum content on permeability of glacial till soils, *J. Soil Sci.* 18, 42–46.

TALSMA, T. and LELIJ, A. van der (1976) Infiltration and water movement in an *in situ* swelling soil during prolonged ponding, *Aust. J. Soil Res.* 14, 337–49.

YAALON, D. H. and KALMAR, D. (1972) Vertical movement in an undisturbed soil: continuous measurement of swelling and shrinking with a sensitive apparatus, *Geoderma* 8, 231–40.

7.4 SHRINKAGE AND FISSURING

BUEHRER, T. F. and DEMING, J. M. (1961) Factors affecting aggregation and permeability of hard spot soils, *Soil Sci.* 92, 248–62.

MCGOWN, A., SALDIVAR-SALI, A. and RADWAN, A. M. (1974) Fissure pattern and slope failures in till at Hurlford, Ayrshire, *Quart. J. Eng. Geol.* 7, 1–26.

VAL'KOV, V. F. (1968) Description of the Very Compact soils of Cuba, *Soviet Soil Sci.* 23–29.

WHITE, E. M. (1970) Giant desiccation cracks in central South Dakota, *Soil Sci.* 110, 71–73.

ZEIN EL ABEDINE, A. and ROBINSON, G. H. (1971) A study on cracking in some Vertisols of the Sudan, *Geoderma* 5, 229–41.

ZEIN EL ABEDINE, A., ROBINSON, G. H. and COMMISSARIS, A. (1971) Approximate age of the vertisols of Gezira, Central clay plain, Sudan, *Soil Sci.* 111, 200–7.

7.5 CONSISTENCY OR ATTERBERG LIMITS

AHMED, S., SWINDALE, L. D. and EL-SWAIFY, S. A. (1969) Effects of adsorbed cations on physical properties of tropical red earths and tropical black earths. I. Plastic limit, percentage stable aggregates and hydraulic conductivity, *J. Soil Sci.* 20, 255–68.

CHAUDHARY, T. N. and GHILDYAL, P. B. (1969) Aggregate stability of puddled soil during rice growth, *J. Indian Soc. Soil Sci.* 17, 261–65.

COLEMAN, J. D., FARRAR, D. M., and MARSH, A. D. (1964) The moisture characteristics, composition and structural analysis of a red clay soil from Nyeri, Kenya, *Geotechnique*, 14, 262–76.

DUMBLETON, M. J. and WEST, G. (1966) Some factors affecting the relation between the clay minerals in soils and their plasticity, *Clay Min. Bull.* 6, 179–93.

FARRAR, D. M. and COLEMAN, J. D. (1967) The correlation of surface area with other properties of nineteen British clay soils, *J. Soil Sci.* 18, 118–24.

GILL, W. R. and REAVES, C. A. (1957) Relationship of Atterberg Limits and cation-exchange capacity to some physical properties of soil, *Soil Sci. Soc. Amer. Proc.* 21, 491–94.

HARRIS, C. (1977) Engineering properties, groundwater conditions, and the nature of soil movement on a solifluction slope in North Norway, *Quart. J. Eng. Geol.* 10, 27–43.

KUBOTA, T. (1971) Effects of organic manure on Atterberg Limits with relation to the stability of soil structure, *J. Sci. Soil Manure (Japan)* 42, 7–11.

MCGOWN, A. and ILEY, P. (1973) A comparison of data from agricultural soil surveys with engineering investigations for roadworks in Ayrshire, *J. Soil Sci.* 24, 145–56.

ODELL, R. T., THORNBURN, T. H. and MCKENZIE, L. J. (1960) Relationships of Atterberg Limits to some other properties of Illinois soils, *Soil Sci. Soc. Amer. Proc.* 24, 297–300.

SANCHEZ, P. A. (1973) Puddling tropical rice soils: I. Growth and nutritional aspects, *Soil Sci.* 115, 149–58.

SOANE, B. D., CAMPBELL, D. J. and HERKES, S. M. (1972) The characterization of some Scottish arable topsoils by agricultural and engineering methods, *J. Soil Sci.* 23, 93–104.

TOWNER, G. D. (1974) A note on the plasticity limits of agricultural soils, *J. Soil Sci.* 25, 307–9.

7.6 COMPACTION

BATEMAN, H. P. (1963) Effect of field machine compaction on soil physical properties and crop response, *Trans. Amer. Soc. Agric. Engs* 6, 19–25.

DAVIES, D. B., FINNEY, J. B. and RICHARDSON, S. J. (1973) Relative effects of tractor weight and wheelslip in causing soil compaction, *J. Soil Sci* 24, 399–409.

HANRAHAN, E. T. (1954) An investigation of some physical properties of peat, *Geotechnique* 4, 108–23.

ISHII, K. and TOKINAGA, Y. (1972) Studies on the soil compaction caused by tractor traffic (Part 5). Distribution of the soil pressure produced in the soil due to tractor traffic, *J. Sci. Soil Manure (Japan)* 43, 1–35.

ROSENBERG, N. J. (1964) Response of plants to the physical effects of soil compaction, *Adv. Agron.* 16, 181–96.

SHELDON, J. C. (1974) The behaviour of seeds in soil III. The influence of seed morphology and the behaviour of seedlings on the establishment of plants from surface-lying seeds, *J. Ecol.* 62, 47–66.

SIDES, G. R. and BARDEN, L. (1970) The times required for the attainment of air-water equilibrium in clay soils, *J. Soil Sci.* 21, 50–62.

SKEMPTON, A. W. (1970) The consolidation of clays by gravitational compaction, *Quart. J. Geol. Soc.* 125, 373–411.

SOANE, B. D. (1973) Techniques for measuring changes in the packing state and cone resistance of soil after the passage of wheels and tracks, *J. Soil Sci.* 24, 311–23.

STEINBRENNER, E. C. and GESSEL, S. P. (1955) The effect of tractor logging on physical properties of some forest soils in southwestern Washington, *Soil Sci. Soc. Amer. Proc.* 15, 372–76.

TAYLOR, H. M., ROBERSON, G. M. and PARKER, J. J. (1967) Cotton seedling taproot elongation as affected by soil strength changes induced by slurrying and water extraction, *Soil Sci. Soc. Amer. Proc.* 31, 700–4.

7.7 BRITTLENESS

BRADFORD, J. M. and BLANCHER, R. W. (1977) Profile modification of a Fragiudalf to increase crop production, *Soil Sci. Soc. Amer. Proc.* 41, 127–31.

CRAMPTON, C. B. (1963) The development and morphology of iron pan podzols and Mid and South Wales, *J. Soil Sci.* 14, 282–302.

GROSSMAN, R. B. and CARLISLE, F. J. (1969) Fragipan soils of the eastern United States, *Adv. Agron.* 21, 237–79.

HARLAN, P. W., FRANZMEIER, C. B. and ROTH, C. B. (1977) Soil formation on loess in southwestern Indiana: II. Distribution of clay and free oxides and fragipan formation, *Soil Sci. Soc. Amer. Proc.* 41, 99–103.

MOORE, T. R. (1976) Sesquioxide-cemented soil horizons in northern Quebec: their distribution, properties and genesis, *Can. J. Soil Sci.* 56, 333–44.

7.8 CEMENTATION OF IRON CONCRETIONS

AHMAD, N. and R. L. JONES (1969) A Plinthaquult of the Aripo savannas, North Trinidad: II. Mineralogy and genesis, *Soil Sci. Soc. Amer. Proc.* 33, 765–68.

CLARK, J. S., BRYDON, J. E. and FARSTAD, L. (1963) Chemical and clay mineralogical properties of the concretionary brown soils of British Columbia, Canada, *Soil Sci.* 95, 344–52.

DOBROVOL'SKIY, G. V. and TERESHINA, T. V. (1970) Ferromanganiferous new formations in southern Taiga soils, *Soviet Soil Sci.* 665–74.

OJANUGA, A. G. and LEE, G. B. (1973) Characteristics, distribution and genesis of nodules and concretions in soils of the Southwestern Upland of Nigeria, *Soil Sci.* 116, 282–91.

RUSANOV, G. V., TSYPANOVA, A. N. and BUSHUYEV, Y. N. (1975) Content and some properties of concretions in the podzolic soils of the central taiga subzone of the Komi ASSR, *Soviet Soil Sci.* 272–80.

7.9 INDURATION

ALEXANDER, L. G. and CADY, J. G. (1962) Genesis and hardening of laterite in soils, *U.S. Dept. Agric. Tech. Bull. 1282*.

BROWN, C. N. (1956) The origin of caliche on the northeastern Llano Estacado, Texas, *J. Geol.* 64, 1–16.

DURY, G. H. (1966) Duricrusted residuals on the Barrier and Cobar

Pediplains of New South Wales, *J. Geol. Soc. Aust.* 13, 299–307.

FLACH, K. W. *et al.* (1969) Pedocementation: induration by silica, carbonates, and sesquioxides in the Quaternary, *Soil Sci.* 107, 442–53.

GILE, L. H. (1961) A classification of *ca* horizons in soils of a desert region, *Soil Sci. Soc. Amer. Proc.* 25, 52–61.

GILE, L. H., PETERSON, F. F. and GROSSMAN, R. B. (1965) The K horizon: a master soil horizon of carbonate accumulation, *Soil Sci.* 99, 74–82.

GOUDIE, A. (1973) *Duricrusts in tropical and subtropical landscapes* (Oxford University Press, London) 174 pp.

HAMMING, E. (1968) On laterites and latosols, *Prof. Geogr* 20, 238–41.

HUTTON, J. T. *et al.* (1972) Composition and genesis of silcrete skins from the Beda valley, southern Arcoona Plateau, South Australia, *J. Geol. Soc. Aust.* 19, 31–39.

JESSUP, R. W. (1960) Laterite soils of the Australian arid zone, *J. Soil Sci.* 11, 106–113.

LIONNET, J. F. G. (1952) Rendzina soils of coastal flats of the Seychelles, *J. Soil Sci.* 3, 172–81.

MCFARLANE, M. J. (1976) *Laterite and landscape* (Academic Press, London) 151 pp.

MAUD, R. R. (1965) Laterite and lateritic soils in coastal Natal, South Africa, *J. Soil Sci.* 16, 60–72.

MULCAHY, M. J. (1960) Laterites and lateritic soils in south-western Australia, *J. Soil Sci.* 11, 206–26.

PATON, T. R. and WILLIAMS, M. A. J. (1972) The concept of laterite, *Ann. Ass. Amer. Geogrs*, 62, 42–60.

SHREVE, R. and MALLORY, T. D. (1933) The relationship of caliche to desert plants, *Soil Sci.* 35, 99–113.

STEPHENS, C. G. (1971) Laterite and silcrete in Australia: a study of the genetic relationships of laterite and silcrete and their companion materials, and their collective significance in the formation of the weathered mantle, soils, relief and drainage of the Australian continent, *Geoderma* 5, 5–52.

STUART, D. M., FOSBERG, M. A. and LEWIS, G. C. (1961) Caliche in southwestern Idaho, *Soil Sci. Soc. Amer. Proc.* 25, 132–35.

WETHERBY, K. G. and OADES, J. M. (1975) Classification of carbonate layers in highland soils of the Northern Murray Mallee, S. A., and their use in stratigraphic and land-use studies, *Aust. J. Soil Res.* 13, 119–32.

Chapter 8 Soil colour

BENNEMA, J., JONGERIUS, A. and LEMOS, R. C. (1970) Micromorphology of some oxic and argillic horizons in south Brazil in relation to weathering sequences, *Geoderma* 4, 333–55.

KRISHNAMURTI, G. S. R. and SATYANARAYANA, K. V. S. (1971) Influence of chemical characteristics in the development of soil colour, *Geoderma* 5, 243–8.

O'NEAL, A. M. (1923) The effect of moisture on soil color, *Soil Sci.* 16, 275–9.

PENDLETON, R. L. and NICKERSON, D. (1951) Soil colors and special Munsell soil color charts, *Soil Sci.* 71, 35–43.

RUSSELL, J. S. and MOORE, A. W. (1967) Use of a numerical method in determining affinities between some deep sandy soils, *Geoderma* 1, 47–68.

SHIELDS, J. A., PAUL, E. A., ARNAUD, R. J. ST. and HEAD, W. K. (1968) Spectrophotometric measurement of soil color and its relation to moisture and organic matter, *Can. J. Soil Sci.* 48, 271–80.

SINGH, S. (1954) A study of the black cotton soils with special reference to their coloration, *J. Soil Sci.* 5, 289–99.

SINGH, S. (1956) The formation of dark-coloured clay-organic complex in black soils, *J. Soil Sci.* 7, 43–58.

SOILEAU, J. M. and MCCRACKEN, R. J. (1967) Free iron and coloration in certain well drained coastal plain soils in relation to their other properties and classification, *Soil Sci. Soc. Amer. Proc.* 31, 248–55.

Appendix: soil names

The ashy grey colour of a characteristic soil in northern Europe is so widely observed that the traditional Russian word 'podzol' has passed readily into general usage. However, as with the 'black earth', or Chernozems, the significance of colour is lost in the phonetic rendering. In the case of podzol, there is also ambiguity, as the key element -zol (Russian *zola*, ashes) has been widely taken up as a convenient suffix to mean 'soil'. Furthermore, whilst the formative element 'spod' (Greek *spodos*, wood ashes) has passed into the U.S.D.A. Soil Taxonomy terminology, the 'spodic' horizon designates the brightly coloured subsoil in which iron and aluminium oxides accumulate together with illuviated organic matter, not the ash-coloured podzolic eluvial horizon. This single instance epitomizes the problems in devising, translating, adopting and adapting names for soil types. It must therefore be stressed that the relatively new U.S.D.A. Soil Taxonomy is designed specifically for specialists already familiar with preceding systems. In view of the sheer number, linguistics, and uncertainties in shades of meaning in soil names, the terms which have been in use longest are generally preferred in the present account in which the priority is the properties rather than the partitioning of the soil. However, many instances of changes in soil properties within the soil profile are cited under 'depth' in the Index, and some information on soil classification and the use of soil names is inevitably incorporated at points in the text where the most characteristic property of a soil horizon or soil type is under discussion. For instance, in discussing soil organic matter, Chapter 3 opens with a description of the characteristics and subdivisions of soil A_0 (organic) horizons (p. 70). The thickness of the humified Chernozem profile is illustrated (p. 119) and a complete podzol profile contrasted with the mull profile of a Brown

Earth (p. 96). Finally, the style in which the U.S.D.A. terminology is structured is illustrated with reference to organic soils or 'Histosols' (Greek *histos*, tissue) on pages 97–8.

A brief description of the more commonly encountered names follows, with sufficient illustration of the U.S.D.A. terminology to prepare readers at least to recognize the source of other similar terms which might be encountered in further reading. The classical climatic sequence is incorporated into Figure 1.19c on p. 38.

Andosols (F.A.O. World Soil Map key, derived from the Japanese *an*, dark, and *do*, soil). Dark soils which develop on fresh volcanic ash and lava, with abundant amorphous mineral material. The U.S.D.A. equivalent is Andept (p. 180). Here 'ept' is the formative element taken from one of the 10 soil orders, Inceptisol (Latin *inceptum*, beginning). Inceptisols include one or more diagnostic horizons which formed quickly but do not represent significant illuviation, eluviation, or extreme weathering.

Brown Earth (West European type), developed under deciduous woodland in humid temperate climates on parent materials relatively rich in bases. Since illuviation is not pronounced, there is no great compaction in the B horizon (pp. 95–6), and a subsoil may not be distinguishable between the organic surface horizons and the C horizon.

Chernozems are soils with a maximum development of humified organic matter (p. 119). Soil reaction is neutral to alkaline throughout the profile and carbonate accumulation in deeper horizons may reach a maximum of 15 per cent. In the Thick Chernozem, the $A_1 + A_2$ horizon is 1 m thick, decreasing to 70 cm in the Ordinary Chernozem, in which the A_1 horizon is reduced to less than half the total A horizon thickness. The proportion is much less than half in the Southern Chernozem, in which a greyish shade is characteristic. The Northern or Leached Chernozem is also greyish, approaching the colour of a podzol. Most Chernozems are equivalent to some subdivision of the U.S.D.A. Mollisol order, from which their -oll suffix is derived. Thus, the suborder name, the second level in the classification hierarchy, is Boroll, the element 'bor' (Greek *boreas*, northern) meaning cool. At the third level of subdivision, probable 'Great group' equivalents to the Ordinary Chernozem include Vermiborolls, with the prefix 'verm' (Latin base of *vermes*, worm) referring to bioturbation, especially by earthworms. An equivalent to Leached Chernozem is Argiboroll, with the formative element 'arg' being derived from 'argillic horizon' (Latin *argilla*, white clay).

Chestnut soils lie on the cooler margin of semi-arid zone soils, characterized by a dark brown colour, similar to that of a ripe chestnut. In the A_1 horizon, the upper 5–7 cm is laminated, relatively friable and lighter in colour than the lower horizons. Vertical cracking induces prismatic structures in the lower horizons where gypsum and calcium carbonate accumulate. These soils are equivalent to U.S.D.A. suborders Xeroll, Ustoll, and Boroll, with the prefixes 'xer' (Greek *xeros*, dry) indicating an annual dry season, and 'ust' (Latin base of *ustus*, burnt) indicating a dry climate with a hot summer.

Gley soils (British Isles) have a distinct humic or peaty topsoil and a diagnostic gleyed horizon, g, within 40 cm of the surface. A podzol horizon, E, may underlie the organic mat. The diagnostic feature of Eg or Bg horizons is the rusty or ochreous mottles. Greyish ped faces may also be recognizable. In U.S.D.A. terminology peaty gleys are Histic Humaquepts. The formative element 'aqu' (Latin *aqua*, water) indicates characteristics associated with wetted soil, 'hum' (Latin *humus*, earth) indicating the presence of humus.

Grey Forest soils have a thin A_1 horizon over a light-coloured bleached A_2 horizon, with a high percentage of bases and clay illuviation in the B horizon. They are transitional soils with a range of characteristics intermediate between those of Leached Chernozems and podzols. One U.S.D.A. equivalent would be an Argiaquoll.

Histosol (U.S.D.A., formerly Bog soils, Organic soils, peat, muck) are identified where a soil is more than 30 per cent organic matter to a depth of 40 cm (pp. 97–8).

Lateritic soils are tropical soils from which alkalies and alkaline earths, together with the silica from silicates, have been removed. Aluminium silicate clays have accumulated as have, to a greater or lesser extent, hydrated sesquioxides. There is some overlap with the U.S.D.A. Oxisols, in which the 'oxic' horizon defines a highly weathered subsoil as having less than 10 meq/100g of exchangeable bases.

Mollisol (U.S.D.A.) has a thick dark surface horizon in which base saturation with bivalent cations exceeds 50 per cent and in which structure is moderately to strongly developed. It includes Chestnut and Chernozem soils of earlier classifications but is unrelated to the 'mollisol', sometimes a term describing the active layer in periglacial environments.

Podzols have a surface mat of largely undecomposed organic matter (p. 96) with a distinctive, ashy grey colour in the A_2 horizon, since it is poorer in bases and sesquioxides than the parent material. A fragipan is often present in the B horizon, with a higher content of sesquioxides, manganese oxide and organic matter attributable to illuviation. Gley podzolic soils are transitional with organic soils, with leaching of the A horizon and accumulation in the B horizon being less pronounced. Podzols are virtually synonymous with the Spodosols of the U.S.D.A. terminology.

Red Earth (p. 106) is a sub-tropical and tropical soil with lateritization incomplete. It therefore retains more clayey aluminium silicates than laterite. In the U.S.D.A. system it could be described as an Oxide Palehumult, the element 'pale' (Greek *paleos*, old) indicating long development.

Red Yellow Podzols are sub-tropical soils which are usually moist but dry out during summer. They are low in bases and there is illuviation and accumulation of clay in the subsoil. This soil group is largely covered by the subdivisions of the U.S.D.A. Ultisol order.

Sierozems are semi-desert grey earths, with a surface structure determined largely by earthworms and by burrowing insects and reptiles. Since the humus content is only 1 per cent, the upper 10 cm are grey to greyish brown, but the 10–20 cm horizon is browner and more friable due to the presence and activity of plant roots. The carbonate horizon includes concretions, and there are transitional forms ranging to the calcretes of desert soils. Sierozems fall within the U.S.D.A. order of Aridisols (Latin *aridus*, dry) for which the formative element which terminates the name of lower categories of the classification is 'id'. Thus, at the second level of subdivision, typical Sierozems fall within the suborder of Orthids (Greek *orthos*, true), in which the element 'orth' signifies 'the common ones'. The nature of the subsoil is incorporated into the Great Group names at the third level of subdivision, with the 'cambic' in Camborthids indicating a subsoil horizon formed by the weathering of minerals within the horizon, rather than being the result of illuviation. In Calciorthids, the carbonate subsoil is prominent.

Solods are leached saline soils, having a pale A_2 horizon and a degraded, fine-textured B horizon. One U.S.D.A. example is the Albaqualf, in which 'alb' (Latin *albus*, white) expresses the white appearance of the soil.

Solonchaks are light-coloured, structureless, saline soils in which the concentration of soluble salts is high. They occur in sub-humid and semi-arid areas, developed beneath a sparse growth of halophytic plants. In the U.S.D.A. terminology they are Salorthids (Latin base of *sal*, salt).

Solonetz are sub-humid or semi-arid grassland soils. Salts formerly present have been leached out, and the characteristic dark-coloured, hard, highly alkaline B horizon has a distinctive cloddy prismatic or columnar structure. This soil type includes the U.S.D.A. Natralbolls and Natrargids, in which the element 'natr' is derived from *natrium*, sodium.

Tundra soils have a mere 3 cm surface layer of little-decomposed organic matter overlying a thin horizon of yellowish-grey loam. Beneath, the characteristic horizon is a uniform, very sticky loam which is grey-blue in colour and which flows readily. This horizon is about 10 cm thick, but is more prominent in wetter areas. One U.S.D.A. tundra soil is the Cryaquept (Greek *kryos*, cold).

Vertisol (U.S.D.A., F.A.O., and French ORSTOM system) is a term (Latin *verto*, turn) which covers a range of earlier soil names (p. 119) where swelling clays crack upon drying (p. 210). The associated shrinkage induces shearing and turbation in the soil. This diagnostic characteristic (p. 212) is observed in lower latitudes with distinct wet and dry seasons.

Towards the finest degrees of sub-division within soil classifications, Soil Series are named after a geographical feature near the area where the soil was first mapped. The location thus gives a significant guide to the weathering regime under which the soil has developed. In addition, different soil types within a Series are usually distinguished by the description of the texture of the surface horizon. This combination of information on regional setting and local characteristics means that several specific soil types can be quoted without needing to digress into describing full details of individual cases.

Subject index

A horizon, 9–10, 47, 70, 75, 96, 103, 158, 210–12, 227; A_0, 70, 169; A_p (plough), 75, 109, 112, 132, 157, 159, 182, 195, 203, 210; A_1, 9, 96, 108, 122–3, 128, 205; A_{12}, 70, 96; A_2, 9, 96, 108, 159
acid soils, 65, 84, 96, 160–7, 177, 185, 190–1, 196–7, 213, 221
acidity, 82, 160–6, 169, 184, 193, 202
acids, 82–3, 165, 173, 194
adhesion, 13, 107, 207–10; intermolecular, 126
adsorption, 33, 57–8, 84–6, 126–9, 155, 161–7, 173–6, 179, 181–3, 188, 192–4, 197–8, 200–2, 211
aeration, 31, 93, 109, 152, 166
aerobic conditions, 150–2, 181, 189–91, 196
afforestation, 13, 26–7, 222
aggregates, 82, 102, 145–6, 149, 166, 202, 208, 213–15; cementing agent, 105–8, 222; coarse, 30, 33, 68, 112, 143, 150, 226; water-stable, 91, 102, 105–6, 136, 219
aggregation, 5, 54, 81, 101, 105–7, 118, 213, 218; micro-, 105, 109, 112, 211
agricultural: labourers, 21; engineer, 5; processing plant, 32; soils (*see also* arable soils, cultivation, ploughing, tillage), 130–1, 160, 214, 218
agriculture, 1, 7, 15–28, 33–5, 41, 45, 59, 98–102, 105, 110–12, 116–17, 155, 179, 215, 218, 228; no-tillage, 21, 23
akiochi, 151
alcohols, 88–90, 154, 158
algae, 3, 71, 93, 181

aliphatic (open-chain) structures, 91
alkaline soils, 65, 84, 166, 173, 184, 190–1
alkaline solution, 83–4, 90–2
alkalinity, 53, 82, 160, 164–5
alkalization, 200–2
allophane, 67, 157, 165, 186, 207
alluvial: deposits, 7, 11, 13, 52–4, 59, 110, 207; soils, 17, 49–52, 120–3, 132–4, 195, 220
alluvium, 16, 48, 54, 60, 112, 200, 207
alpine environments, 54, 67, 71
aluminium, 32, 47–8, 55–7, 61–7, 156–8, 162–81, 188, 191–3, 197, 219–21, 225; hydroxide, 33, 58, 66–8, 164–5, 172–3, 192–5, 215; oxide, 46, 107, 161, 167, 173, 180, 187, 192–4, 215, 221
alumino-silicates, 49, 170, 173, 179, 194, 223
aminization, 183
amino: acids, 82–4, 89, 90–1, 95, 149, 183; sugars, 82, 106, 158
ammonia, 85, 90, 166, 179, 184, 187, 191
ammonification, 183, 191
ammonium, 96, 115, 141, 149, 166–7, 174–5, 178–9, 182–8, 191, 211, 213
anaerobic: conditions, 30, 93, 115, 145–52, 170, 189, 194, 200, 227; organisms, 151, 154, 169; respiration, 168, 181
Ångström unit, 55
animals, 3, 11–14, 19, 22, 69, 74–8, 81, 86, 113–15, 119, 206, 218
anion, 55–6, 59, 83–4, 155–7, 166–7, 172–3, 178–9, 186, 198–200; exchange, 161–2, 167, 186–7, 193–6
Antarctic soils, 136, 200
ants, 73, 76–7, 114, 128

277

278 Subject index

apatite, 47, 50, 53, 186
arable soils, 72, 74, 78, 93, 118, 133, 137, 148, 203
Arctic environments, 54, 71, 136–7
argillans, 103, 168
arid-zone soils, 33, 59, 157, 174, 198–9, 222–3
arid zones, 45, 55, 66, 75, 93, 159–60, 169, 197, 199, 227
aromatic structures, 86, 91, 94
arsenic, 31, 59
arthropods, 76–7, 81
ash: content, 27, 72, 98, 206; volcanic, 13, 52, 61, 67, 98, 179–80
atmosphere, 144–6, 148, 150–1, 178
atomic radius, 55–7, 175, 219
atrazine, 21
Atterberg limits, 4, 214–15; liquid, 29; shrinkage, 207
augite, 46–53
available water, 54, 59–60, 126–7, 130–5; capacity (AWC), 42, 130–6, 206

B horizon, 9–10, 30, 38, 96, 101, 103, 108, 110, 124, 160, 169–70, 200, 202, 212, 218–19, 225; B_1, 9, 47, 205; B_2, 9, 39, 47, 96, 108, 123, 205, 212
bacteria, 30, 76, 81, 85, 89, 111–13, 168, 179, 181, 191–4; autotrophic, 183–5; nitrifying, 115, 183–6; parasitic, 179; pathogenic, 31
barium, 175, 192
bases, 63, 65–6, 82, 97, 107, 111, 160–2, 165, 191
bedrock, 5, 15, 17, 44, 184, 220
benzene rings, 86–7
bicarbonate, 163, 199–201
biocides, 3, 21, 154
biological oxygen demand (BOD), 31, 184
biomass, 24–5, 69–72
biotite, 49–50, 64–5
boron, 69, 192–7
boulders, 6, 44
brittleness, 113, 218–19
brucite, 57–8, 62, 65–7
bulk density, 97–8, 115, 132, 203–10, 213–18
burrowing, 109, 113–14, 119, 206, 218

C horizon, 9, 47, 96, 210, 212
Chroma, Munsell, 224–5, 228
calcareous soils, 9, 14, 158, 160, 166, 190, 194–7, 219
calcium, 11, 33, 40, 48, 50, 63, 65, 72, 78, 96–7, 107, 117, 135, 147, 160–2, 167–8, 172–7, 179, 186–8, 191, 198, 200, 210–15; carbonate, 53–4, 78, 96, 197–8, 202, 223–5; chloride, 199–201; sulphate, 53, 198–201
calcrete, 7, 222–3
capillarity, 124–9, 198, 215, 223
capillary: fringe, 127; pores, 107, 111–12, 126, 131, 136; rise, 122–3, 126, 131, 144, 169; -ruptured moisture content, 126
carbohydrates, 80–2, 85–6, 179
carbon, 78–82, 86–7, 90–3, 115, 135, 151, 156, 159, 178–83, 186–7, 204, 215; dioxide, 5, 79, 85, 90–2, 137, 145–50, 163, 166, 174–6, 183, 186, 194
carbonates, 35, 53, 78, 174, 198–202, 219, 222–3
carbonic acid, 147, 164, 176
carbonyl group, 79–80, 87
carboxylic acid group, 79, 82–4, 87, 90, 158, 163–6, 187
cation exchange, 59, 65–7, 154, 162, 186, 196; capacity (CEC), 154–64, 186, 211, 215, 226
cations, 47–8, 55, 58–9, 64–5, 83–4, 106, 135, 154–8, 160–6, 172, 179, 198, 201
cattle, 11, 15, 81, 99, 214; bulls, 200
cellulose, 76, 80–2, 86, 95, 97
ceramic industry, 60
channels, 113, 123, 128, 172, 221
chelation, 82, 166, 168
chloride, 197–200, 226
chlorine, 50, 199
chlorite, 47, 50, 62, 65–7, 215
chlorophyll, 176, 178
cities, 17, 28–31
civil engineer, 30, 50, 62, 99
classification, see Appendix
clay, 1–5, 9, 13, 15, 26, 29, 33, 43–4, 53–61, 75–6, 86–7, 93–5, 105–10, 118–26, 131–5, 139, 154–61, 164–7, 172–7, 180, 186–7, 195–6, 203–6, 210–14, 219–20, 225–6, 228; coarse, 92, 160; fine, 110, 157, 159–60, 208; heavy, 25, 59, 117; particles, 82, 156, 167, 211; sandy, 9, 43, 118, 215; silty, 9, 43; soils, 53, 98, 112, 120, 136, 160
clay crystal, 112, 179, 211; framework, 56–7, 59, 86, 175–7
clay crystal structure: aluminium-oxygen octahedra, 56–8, 62; amorphous, 67–8, 179; basal spacing, 63–4, 66; expansion, 58, 65–7, 154, 157, 176–7, 179, 186, 211; 'holes', 56–7, 64, 174; interlayer material, 58, 62, 157, 159, 165, 177; magnesium-oxygen octahedra, 57–8, 62, 65; silicon-oxygen tetrahedra, 56–8, 62, 65, 175
clay mineralogy, 55–6, 61, 164, 168–9, 171, 181, 187, 196, 212, 222

Subject index 279

clay minerals: 1:1-layer silicates, 58, 177, 182, 187; 2:1-layer silicates, 58, 64–7, 174, 176–8, 187; 2:2-layer silicates, 67
clay surfaces, 135, 167, 169, 188, 198, 211, 217
climate, 11, 20, 35, 39, 45, 94–5, 115, 136, 192, 197, 200, 221–2
climatic change, 9, 100, 200
clods, 105, 118, 208, 213
coatings, 219, 222, 227; clay-skin, 103, 109–10, 168, 213, 222; flood, 110; humus, 213; iron (ferrans) 108; oxide, 161; sesquioxide, 165
cobalt, 22, 84, 151, 192–6
cohesion, 13, 59, 108, 169, 208–10, 213
collembola, 70, 73, 77, 96
colloids, 84, 107, 140–1, 154, 158, 167–70, 173, 194, 198, 206, 215, 226
compaction, 3, 5, 23, 30–2, 53, 117–18, 120, 184, 206, 213–19
complexing, 166, 172, 194, 196
concrete, 162
concretions, calcareous, 14, 202, 223; ferromanganiferous, 219–21
condensation (biochemical), 83–4, 87, 90, 92, 94, 141
condensation (vapour), 144
consistency, 101, 214–15, 223
consolidation, 60, 99, 145, 217
constructions, 15, 18, 28, 30, 99, 216, 223
cool environments, 71, 98, 137–8, 141, 148, 159, 219–20, 227
copper, 18, 22, 84, 98, 151, 166, 192–3, 196–7, 228
corrosion, 30
cracking, 3, 30–1, 104–5, 107–9, 115, 121, 169, 210–13
crops, field: general, 1, 3, 19–23, 27, 50, 54, 64, 69, 72, 74, 78, 93, 95, 116–17, 129, 131–2, 146, 154, 178, 181, 184, 191, 193, 195, 199, 222; varieties: alfalfa, 138, 213; barley, 20; corn, 20–1, 74, 218; cotton, 19, 59, 205–6, 218; legumes (*see also* pasture), 19, 179; maize, 137; millet, 218; oats, 20; peas, 20, 74; potatoes, 20, 111, 23; rye, 20, 23; sugar beet, 218; turnip, 20; wheat, 17, 20, 74–5, 129, 218
crystallization, 180, 220; calcite, 53, 174; iron, 221, 227; sesquioxide, 181
cultivation (*see also* ploughing, tillage), 10, 15, 18, 20–1, 54, 75, 95, 98, 101, 104–5, 108, 112, 115, 141, 181, 210; shifting, 93
cumulative frequency curves, 43, 46–7
cutans (*see also* coatings), 103, 108–10

decomposition, biochemical, 5, 33, 72, 74, 78, 83, 89, 93, 95, 141, 168, 206
deficiencies, nutrient, 19, 21–2, 25, 27, 33, 98, 161, 168–9, 178, 193, 195–6
deflocculation, 107
dehydration, 80, 83, 94, 107, 220
denitrification (nitrogen reduction), 148–50, 152, 154, 184, 186, 188
deoxyribonucleic acid, 78
depths, soil, 1, 3, 5, 9, 15–17, 24–6, 29–30, 32–3, 37–40, 49, 61, 97, 110, 112–16, 119, 122–3, 128, 130, 132, 136–42, 145, 152, 160, 167, 184, 211, 213–18, 220–2
desert, 36, 42, 45, 54, 75, 92; flora, 11; pavements, 45; sands, 18; soils, 3, 45, 49, 51–3, 147, 181, 213
desiccation (*see also* drying), 104, 115, 213, 220
diatoms, 71, 173
diffusion: gaseous, 144–50, 217; solutes, 36, 110, 168, 185, 188, 205–6; thermal, 137, 139, 140
dispersion, 5, 91, 107, 111, 120–1, 155, 167, 202, 210–13, 226; ratio, 110
downpour, 124, 130, 208
drainage, 13, 15, 24, 28, 31, 99, 115, 143, 159, 206, 218, 227; free, 75, 93, 95–6, 124, 127, 147, 149, 153, 167; gravitational, 111, 126–7, 130–1, 143; impeded, 53, 124, 170, 181, 227
drought, 15, 22–7, 55, 118, 130–1, 222
drumlin, 9, 39, 45
drying, 3, 5, 12, 35, 92, 103, 107–12, 118, 127–8, 137, 142–4, 168–9, 174, 177, 188, 208, 213, 219
dust, 53–5, 98, 148

earthworms, 25, 70–1, 76–8, 96–7, 109, 113–14, 121, 127–8, 143; casts, 109
ecology, 10
edaphic factor, 11, 13
effluent, 30–1
Eh values, 152–4, 165
electrolyte, 155, 198, 209
electrons, 55–6, 79–80, 146, 151–4, 165, 167
energy, 69, 71, 81, 85, 88, 92, 105, 129, 146, 169, 179, 182–3, 186–7, 198, 217, 226
engineering (geotechnical), 4, 28, 30, 33, 35, 42, 50, 60, 68, 99, 110, 117, 154, 162, 206–11, 214–18, 222
enzymes, 59, 81–4, 89, 95, 162, 169, 179, 182
epidote, 47, 50–2
erodibility, 43, 105, 110, 205, 208
erosion, 1, 5, 7, 9, 15, 17–18, 26–7, 32,

37, 41, 45, 59, 69, 104, 109, 118, 160, 184, 208, 214
evaporation, 110, 126, 137, 140, 144–5, 169, 174, 197–9, 215, 222
evapotranspiration, 1, 7, 27, 123, 129, 184, 197, 222
exchange sites, 135, 166–7, 169–72, 176–7, 179, 187, 211
excrement, 31, 70, 73–8, 97, 109

fabric, 101, 219; K-, 223
farmers, 17, 23, 214
feldspar, 48–9, 52–3, 61–2, 158
fermentation, 146, 168–9; horizon (F), 70, 95–6
ferric and ferrous compounds, see iron
fertility, 17, 19, 20–2, 27–8, 46, 52–4, 62, 94, 98, 105, 126, 151, 154, 162, 199, 227
fertilization, 25–7, 32–3, 50, 161, 177, 181, 184, 216
fertilizers, 19–23, 27, 32, 45, 50, 56, 64, 74, 99, 149, 178–81, 184–7, 191–3
fibres, 97–8, 204, 206
field capacity, 127–8, 130–5, 177, 184, 201, 208
fissures, 31, 53, 110, 115, 213–14
fixation, 32, 178, 194, 196; nitrogen, 179, 181, 186, 191; potassium, 64, 174–7; phosphorus, 67, 181, 188, 191
flocculation, 82, 107, 139, 168, 176
flood, 1, 3, 48–9, 53, 123, 148, 165–6, 188, 193–6, 211
flooded soils (*see also* paddy soils), 146–7, 152, 179, 188–9, 193
floodplain, 7, 8, 52; soils, 7, 35, 48, 110, 157
fodder, 13, 23
food, 15, 25, 74, 78, 84, 93; chain, 13; production, 19
forage, 11, 74, 78
forest, 11, 14, 26, 72–4, 96, 113, 116, 118, 137, 140, 191–3, 222; clearance, 118, 137; fire, 27; management, 25, 27; soils, 25–8, 36, 75, 106, 118, 142–3, 147, 163; -steppe zone, 22, 71, 170
fragipans (hardpan), 110, 115, 218–19
freeze-thaw, 3, 44, 54, 137, 140
freezing, 2–5, 33, 91–2, 107, 136–43, 206
fulvic acids, 90–2, 166, 170, 226
fungi, 70, 76–7, 85, 89, 105, 174, 183, 193

garnet, 47, 51–2
gas, 31, 99, 111, 126, 143–4, 149, 166, 170, 182–4
geomorphology, 4–5
gibbsite, 57–8, 62–7, 108, 164
gilgai, 120

glacial drift, 7, 9, 16, 18, 32, 44–5, 49, 51, 54, 113, 121, 147, 158, 224
glaebules, 101, 104, 108
gleying, 168–72, 219–20, 227
glucose, 79–81, 85, 95, 146
goethite, 63, 168, 221, 225–7
grass, 13, 15, 19–20, 59, 72, 88, 104, 116–17, 146, 153, 173, 186
grassland (*see also* savanna), 13–14, 74, 77–8, 93–5, 106, 114, 147–9, 203
gravel, 5–6, 8, 12, 16, 30, 33, 42, 45, 53, 128, 139–40, 208
grazing, 12–13, 18, 41, 45, 59, 71, 74, 142
ground surface, 29, 34, 45, 53, 124, 136, 140, 147, 200; depressions, 13–14, 197, 200, 210; micro-relief, 35, 115, 210
groundwater, 1, 7, 31, 123, 173, 186, 199–200, 221–3
gypsum, see calcium sulphate

habitats, 10, 12, 15–17, 36
haematite, 225, 227
halloysite, 62–4, 196, 215, 222
hardening, 219–21, 223; frost, 139
hardpan (*see also* fragipan), 30, 115, 124, 218
harvesting, 25, 210, 217
hayfields, 99
heaths, 71, 96
heavy minerals, 47–52
hemi-cellulose, 81, 94–5, 97
herbicides, 21, 87, 154
hillslopes, 1, 7, 200, 208
horizons (*see also* A, B, C, fermentation (F), humification (H), induration (K), and litter (L) horizons), 46–7, 51, 70, 108, 110, 138, 170, 218, 221; charcoal, 9; carbonate, 9; indices, 108; lower, 1, 91, 107, 113, 117–18, 148, 168; surface, 1, 3, 152–3, 226; upper, 91, 97–8, 118, 124, 141, 168, 214, 219
hornblende, 47, 50–3
horses, 53
'hotspots', 41
hue, Munsell, 224–5
humic acids, 84, 88–94, 98, 107, 141, 158, 166–7, 188, 194, 196, 222, 226
humidity, 94, 111, 144
humification, 89, 94–5, 158, 160, 226; horizon (H), 70, 112
humins, 90–2, 206, 226
humus, 17, 25, 54, 69, 78, 81, 89–92, 95, 104, 107, 110, 140, 154–5, 158, 160, 167, 170, 176, 187, 196, 213–14, 227; raw, 27, 96, 169
hydration, 57, 63, 155, 161, 211, 225
hydraulic conductivity (K), 28, 53–4, 60,

99, 119, 122–3, 126–8, 131, 142, 206, 213, 223
hydrocarbons, 86–7
hydrogen, 57, 62–5, 78, 80–7, 90–1, 94, 98, 149, 151, 153, 160–72, 178–9, 183, 193, 202; bonding, 62, 167; ion concentration, 162–5, 190; sulphide, 21, 23, 149, 170, 191
hydrology, 1–3
hydrolysis, 80–5, 89, 164, 169, 183
hydromicas, 64–5, 158, 186
hydroxyl, anion, 49, 57–62, 66, 162–5, 194–8, 202; groups, 80, 89–90, 94, 157–8, 161, 182, 187
hysteresis, 128

ice, 2, 34, 53, 108, 139, 141–2
illite, 62–8, 120, 123, 135, 157, 174–5, 177
illuvial horizon, 38, 96, 104, 202, 228
illuviation, 70, 108–10, 124, 212, 223
ilmenite, 47, 227
induration, 30, 113, 220–3
industry, 1, 3, 31
infertility, 25, 32, 44, 50, 75, 140
infiltration, 1–3, 27, 53, 102, 105, 110, 116–21, 124, 129, 143, 197, 208, 217
insects, 19, 22, 25, 31, 76–8, 81, 121, 174
interflow, 4, 200
iodine, 21, 194
ion (*see also* anion, cation, diffusion, electrolyte, soil solution), 55–6, 66, 144, 155, 206–8, 211, 215; exchange, 154–62, 164, 166; layers, 155, 161
ionic radius, 55–6, 58, 178, 211, 219
ionization, 55, 82–3, 158, 162, 164
iron, 7, 21, 32, 47–9, 54, 63–5, 84, 99, 107, 110, 151–3, 157, 116–71, 178, 191, 193, 197, 219–20, 225–7; accumulation, 221; availability, 168–9; compounds, 113, 154, 161, 167, 181, 228; ferric, 152–3, 168–9, 188, 226; ferrous, 63, 137, 153, 165, 168–70, 188, 191, 225, 227; gel, 107, 168, 221; hydroxide, 33, 66–7, 113, 165–70, 188, 192–5, 215, 219; mobility, 167–8, 220–1; oxide (extractable), 40, 46, 60, 63, 67, 107, 153, 156–7, 168, 170, 173, 187, 192–7, 211, 215, 220–1, 225–7; pipes, 113; pyrites, 193
irrigation, 1, 3, 19, 22–3, 30, 45, 53–4, 99, 111, 117–20, 123–4, 129–31, 143, 151, 158, 169, 200–1, 214
isomorphous substitution, 63–6, 154, 157, 178

K-: fabric, 223; horizon, 10, 222–3
kaolin, 68, 135

kaolinite, 61–6, 157, 160–1, 174, 177, 182, 186–8, 196, 207, 212, 215, 222, 226
krotovinas, 113

lakes, 3, 7, 18, 32, 38
land: drainage, 117; levelling, 32; surface, 5, 7, 28, 33, 41, 97, 221–2; use, 25, 28, 39, 95, 104, 118, 148, 222–3
landfill, 31
landscaping, 32
landslip, 4, 5, 28, 209
larvae, 70, 77–8
laterite, 6, 7, 17, 220–2; silicified, 6; pisolithic, 221
lateritic soils, 30, 63, 65, 68, 102, 107–8, 117, 128, 157, 192–3, 206, 215
latitudes: high, 33, 35, 44, 53, 67, 97, 138, 148, 159, 161–2, 204, 219; mid-, 24, 33, 44, 95–6, 111, 113, 141–2, 159–61, 177, 181; tropical, 30, 54, 59, 75, 78, 94–7, 99, 101, 118, 137–8, 159–60, 200, 208, 227
layers, 1–3, 70, 96, 120, 123, 222; active, 34, 142; frozen, 3, 138
leaching, 13, 27, 33, 35, 39, 50, 63–6, 70, 91–4, 110, 141, 147, 152, 164–6, 170, 183–7, 193–4, 198–9, 202, 206, 220, 227
leaf: fall, 71–2; fragments, 70, 76–7
leaves, 77–8, 86, 115, 131, 135, 170, 173
lessivage, 169
lichens, 71
ligands, 166–7
lignin, 84–90, 94–5
liming, 33, 160–1
linkages (biochemical), 80–1, 85, 107, 178, 188
linuron, 87
litter, 27, 70–2, 77–8, 84, 93–5, 118, 135, 137, 142, 145, 169, 177; layer (L) 27, 70, 81, 86, 96
livestock, 15, 74
load, 15, 217
loam, 2, 9, 13–14, 121, 126, 132–4, 175, 217; clay, 43, 47, 75, 92–3, 117, 127, 131, 133, 144, 149, 203, 205; coarse, 43, 205; fine clay, 112, 120; fine sandy, 133–4, 217; medium, 43, 76, 205; sandy, 9, 43, 76, 92–3, 123, 132–7, 169, 212–13, 216; sandy clay, 43, 133, 212; sandy silt, 43; silty, 9, 14, 43–4, 104, 132–3, 190–1, 195, 213; silty clay, 14, 43, 123, 133, 205
loess, 9, 15–16, 40, 51–2, 54, 112, 184
logging, 28

macrofauna, 12, 73, 76–8, 109, 113–14
macropores, 112–13, 121, 123, 127, 129, 143

282 Subject index

magnesium, 11, 21, 46–50, 57–8, 63–7, 84, 107, 147, 160–2, 167–8, 172–9, 191, 198, 200; aluminosilicate, 64; bi-carbonates, 200; carbonate, 200; chloride, 199–201; sulphate, 199–201
magnetite, 46–7
malnutrition, 17, 178
man, 3, 10–11, 18, 41, 45, 53, 68, 100, 222; Masai, 15; maya, 15, 17
management practices, 19, 112, 161, 218
manganese, 32, 84, 110, 151–4, 162, 169, 174, 192–7, 219, 225; oxide, 211, 219
manure, 135, 149; poultry, 31–2; farmyard (FYM), 30, 74, 132, 134; green, 33, 94
mapping, 36–9, 227
marine deposits, 18, 50, 60, 113, 193, 210
market gardening (truck farming), 50, 98; crops: asparagus, 50; celery, 98; lettuce, 98; onions, 98, tomatoes, 23, 146
mass flow, 145, 177
meadows, 30, 99, 147
mechanization (see also vehicles), 19, 23, 25
mellowness, 101
metals, 82, 84, 91, 107, 155, 162, 166, 168
mica, 48–9, 53, 60–1, 64–5, 68, 158, 174–5
micro-biological processes, 82, 89, 106, 113, 145, 153, 162, 168, 173, 182–3, 191
micro-organisms, 12, 27, 31, 59, 78, 81–4, 92–3, 105–6, 113, 138, 141, 144, 147, 151, 166–9, 173, 178–9, 181–4, 186–8, 191
micromorphology, 103, 108
micron, 42
mites, 70, 77, 96, 109
moder, 96–7
moisture, soil, 1–4, 20, 22, 27, 60, 116, 129–43, 148, 153, 168, 177, 188–9, 191, 199, 201, 205–10, 213–15, 218, 222, 227; minimum capacity, 130; maximum content, 130, 188; hygroscopic, 214; potential, 129; retention, 123, 134; tension, 119, 123, 128, 134
mole, 76, 113; hill, 41, 116, 119
molybdenum, 18, 69, 162, 194–7
monosaccharides, 79–82
monosilicic acid, 172–4, 193, 222
montmorillonite, 62, 65–7, 86, 111, 120–3, 135, 157, 174, 177, 186–7, 212, 215, 226
mor, 95–7
mosses, 71, 137–8, 142, 147–8
mottling, 37, 170, 172, 220
mulch, 123, 128, 145
mull, 95–7
Munsell colour code, 224–5
mycorrhizae, 146

needles, pine, 70, 72, 93, 95

negative charge, 55–6, 64–5, 82–4, 126, 154–7, 161, 164–5, 187, 208, 211
nematodes, 77–8
nitrate ions, 115, 179–80, 182–3
nitrates, 3, 96, 98, 123, 141, 149, 152–4, 161, 184–6, 189–90, 197–200
nitric acid, 148, 152, 164
nitrification, 3, 141, 183, 186–8, 191
nitrite, 166, 184, 190–1
nitrogen, 3, 19, 22, 26–7, 32, 69, 72, 74, 78, 80, 82, 84, 91–8, 115, 149–53, 178–87; available, 93, 149, 182–3, 187; cycles, 183; dioxide, 148, 182; loss, 183–4, 186, 188, 191
nitrogenous substances, 80–1, 85, 90, 93, 182, 185
nitrous oxide, 148–52, 184
nodules: calcareous, 223; ironstone and aluminous, 194, 219–21
nutrients, 11, 21, 24–7, 51, 54–5, 59, 60, 64–9, 72, 74, 76, 78, 81, 96–8, 105, 144, 151, 155, 161–2, 177–8, 197, 215; cycling, 94, 169, 177; micro-, 22, 32, 49, 59, 69, 162, 166, 174, 192–6; uptake, 110, 115, 137, 141

olivine, 46, 50, 52
organic acids, 61, 87, 137, 147, 154, 166, 169–70, 194, 227
organic complexes, 95, 168, 188
organic compounds, 3, 31, 71, 79–82, 86, 89, 161, 168, 183, 187, 220; bonds, 80–8, 166–7
organic (O) horizon, 70, 75, 95, 97, 124, 194, 219
organic matter, 3, 5, 22, 32–3, 44, 59, 71–5, 81, 86–7, 102, 105–10, 131–6, 141–2, 148–9, 153, 157–60, 162, 165–73, 177–9, 182–3, 187–8, 191–6, 202–4, 209, 211, 214, 219, 221, 226–7; comminution, 54, 76–8, 80, 96; decomposition, 3, 5, 59, 69, 71–2, 76–7, 82–4, 92–9, 105, 141, 149, 152, 163–4, 181–4, 189, 215; residues, 69, 74–5, 89, 93–4, 153; synthesis, 78, 80, 82, 84, 89–91, 95
organic molecules, 66, 80, 82, 90, 94, 166
organic soils (Histosols), 7, 97–8, 187, 196
organo-: clay compounds, 159, 167, 226; mineral compounds, 7, 102, 107, 166, 170, 194; metal complexes, 102, 107, 166–7, 170, 193, 196
ortstein, 219–20
osmosis, 129, 201
oxidation, 153, 165; iron, 113, 168, 170, 221, 226; organic matter, 31, 75, 79–80, 85, 87–90, 94, 97, 99, 141, 147, 149, 151, 160, 183, 185; sulphides, 32, 199;

microbiological, 31, 80, 189, 191–4; plant root, 146
oxidation-reduction (redox) processes, 7, 85, 151, 194, 219
oxygen, 78, 80, 82, 85–7, 90, 94, 98, 111–13, 144–52, 167–8, 170, 177, 181, 184, 186, 189, 191, 196; anion, 55–8, 62, 64; diffusion, 153, 188; diffusion rate (ODR), 146, 153, 186

paddy soils, 147, 151, 165, 168, 179, 181, 188, 191, 211
paleosols, 5–7, 9, 63, 228
paraquat, 21
parent materials, 12, 44, 48–51, 61, 67, 124, 160, 164, 178, 194, 200
particle: characteristics, 8–9, 45; shape, 209; sizes (see also texture), 5, 29, 33, 42–3, 53, 55, 60, 65, 75, 86, 93, 111, 121–3, 126, 140–1, 157, 186, 196; sorting, 8
pasture, 19, 25, 74, 78, 95, 99, 105, 113, 132, 142, 179, 200; clover, 22, 179, 199; tropical, 23, 25
peat, 32, 71, 97–100, 136–7, 142, 168, 170, 172, 196, 206, 227; bogs, 137–8, 146, 170; high moor, 94; horizons, 97, 99–100, 172; -lands, 97–9; low moor, 99
peaty: gleys, 25; soils, 74, 85, 98–9, 136, 147, 166, 168, 204
peds (see also aggregates, coatings, soil crumbs), 101, 104, 107–11, 208, 220, 222
percolation, 31, 121, 123–4, 129–30, 136, 145, 150, 162, 169, 176, 184, 215
permafrost, 34–5, 53, 84, 113, 137–8, 141–2
permanent wilting point, 130–1, 201
permeability, 1, 3–5, 30, 32, 53, 93, 99, 111, 116, 118, 120–4, 136, 201, 206, 215, 227; air, 145
pesticides, 41, 60, 149, 166
pF, 122, 125, 127–9
pH, 33, 35, 37, 90, 96, 162–6, 169–73, 190, 195–7, 200, 202
phenolic groups, 87, 158, 163, 166, 187
phosphates, 32, 50, 59, 147, 161, 166, 173, 178, 180–2, 185, 187–91, 194, 197
phosphorus, 3, 11, 21–2, 26, 32–3, 50, 72, 74, 78, 82, 85, 97, 110, 178, 180, 185–90, 193, 215; availability, 151, 180, 185, 190–1; fixation, 181, 186; immobilization, 92, 187
photosynthesis, 71, 129, 147–8
phytoliths, 171, 173
pipelines, 30
plant: breeding, 20–1; cover, 13, 15, 45, 129, 147; germination, 137, 217–18; growth, 24, 32, 41–2, 46, 59, 110, 115, 117, 129–41, 146–7, 176–8, 182, 184–5, 188, 199, 201, 217–18; materials, 76, 89, 183; moisture requirements, 129–32; mortality, 217; nutrition, 19, 93–4, 166, 176, 179; physiology, 19, 21, 23, 130–1, 146; residues, 23, 91, 95, 97, 106, 117, 214; spacing, 19; uptake of nutrient, 179, 182–3, 185, 188, 191, 194, 196; uptake of water, 120
plantations, 24, 27, 50, 59, 72, 99, 113, 120, 130, 177, 223
plants, 11, 13, 19, 21–2, 25, 32–3, 45, 54, 64, 74, 78, 80, 82, 85–6, 98, 112–13, 129–31, 139, 146, 166, 169, 172–4, 177, 179, 194, 201, 218
plasma, soil, 103–4, 108
plasticity, 30, 33, 60, 215
ploughing, 30, 33, 38, 60, 108, 117, 207, 214–15, 217; deep, 44
podzolic soils, 26, 59, 63, 80, 91–2, 95–6, 99, 121, 124, 137, 153, 160–1, 163, 167, 169, 173
podzolization, 166, 169
polar areas, 67, 71, 74
polder, 99, 213
pollen, 9, 11, 77, 100
pollution, 25, 31, 123
polygons, 142
polymers, 81, 87, 89–92, 102, 165
polyphenols, 84, 89–90, 93, 170
polysaccharide, 79, 81–2, 84, 91, 106; soil, 82, 106, 158
pores (see also krotovinas, macropores), 1–3, 53, 75, 102–3, 111–14, 117–21, 123, 126–9, 131, 135–6, 139, 142–5, 155, 167–8, 188, 206, 213, 215–16, 218; clogging (see also coatings), 143, 221
porosity, 1, 5, 31, 99, 105, 111–14, 118, 120–1, 128, 130–1, 135, 140, 142–3, 188, 203, 208, 213, 215, 221
positive charge, 55–6, 84, 155, 157, 161, 181
potassium, 20–1, 40, 46–9, 63–7, 72, 74, 98, 110, 160, 174–8, 198–200, 211, 213; availability, 176; immobilization, 174; sulphate, 199
poultry, 31–2
prairie (see also grassland), 114
precipitation, chemical, 162, 174, 180–1, 193–4, 199, 219–22, 227
precipitation (see also rainfall), 1, 105, 129, 184
pressure membrane apparatus, 125, 130
profile: soil, 4, 7, 9, 45, 70, 75, 91, 95, 97, 103, 123–4, 129, 135, 138–40, 143, 145,

147, 152, 158, 160, 162, 166–72, 198, 201, 220, 228; differentiation (*see also* depths), 4, 14, 108, 117, 119, 151
profile subdivisions (*see also* A, B, C, and K horizon): lettered subscripts, 10, 96, 108; numerical subscripts, 9, 70
proteins, 66, 82–6, 95, 98, 115, 149, 178–9, 182–3
protons, 58, 82–3, 161–2, 169
protoplasm, 82, 84, 129
puddling, 32, 112, 215

quartz, 46, 48, 50–4, 61, 102, 172–3, 219, 222, 226
Quaternary, 5, 12, 46, 49, 99
quick clays, 200, 209
quinones, 84, 87, 89–91

radiation, 23, 71, 136, 139, 148, 198
radical, organic, 82–3, 87, 89–90, 94, 165, 183
radiocarbon datings, 9, 10, 100
railway construction, 99
raindrop impact, 1, 18, 118, 120
rainfall (*see also* downpour), 3, 13, 22–3, 55, 98, 102, 116, 118, 120–1, 124, 130, 141, 147–8, 170, 186–7, 197, 199, 212, 223
rainforest, 36, 71–2, 95, 147
ratio: C:H, 94; C:N, 92–4, 98; HA/FA, 91–2; O:H, 94; quartz-feldspar, 51–2
re-cycling, 71–4, 81
re-vegetation, 28
reaction (*see also* hydrogen ion concentration, pH), 162
reclamation, 99, 117
redox, phenomena (*see also* oxidation-reduction), 153–4, 170, 219; potential, 24, 151–3, 165, 170–1, 188, 190, 194–5
reduction, 151, 165, 168–70, 181, 184–90, 194–6, 221, 227
resistance, soil, 28, 204–5, 207–8
respiration, micro-biological, 181; plant root, 5, 145–8
respiratory quotient (R.Q.), 150
rhizobium, 179, 191
rhizosphere (*see also* roots), 152, 168, 185
rice, 19, 21–2, 59, 99, 137, 146, 152, 179, 181, 191, 193–4, 197; soils (*see also* paddy soils), 23, 151, 206, 211, 215
ring structures, carbon, 80, 82, 86–7, 89–90
roads, 28, 30, 41, 53, 219; construction, 28, 34, 99; corrugations, 30; cuts, 32
rocks: basic, 52, 158, 221, 225; basalt, 46, 147, 158, 192, 220; chalk, 5; granite, 113; intermediate, 65; Jurassic strata, 4; lava, 13; limestone, 15, 44–5, 50; metamorphic, 61; quartzite, 222; rhyolite, 208; sandstone, 46; sedimentary, 60–1; shales, 18; silicified conglomerate, 12; Tertiary formations, 48; tuff, 13; volcanic, 158; volcanic glass, 65
root, crops (*see also* crops), 25, 210; development, 32, 110–12, 115, 137, 185, 217–18, 222–3; infection, 214; nodules, 146; penetration, 44, 60, 11–12, 130, 206–8; pattern, 13, 35; zone (rhizosphere), 184
roots, 24, 27, 35, 59, 69–70, 72, 77, 81, 84–6, 103–5, 109, 113, 115, 121, 126, 130, 136–9, 144–7, 151–2, 168, 172, 177, 179, 184, 188, 194, 214, 218–22
rotations, 19, 23, 26
runoff, 1, 27, 105, 116, 118, 120–1, 134, 184, 208

s-matrix, 103–4
saline: -alkali soils, 14, 200–2; soils, 3, 155, 199–202, 226; water, 22, 35, 117
salinity, 3, 13, 201–2
salt-affected soils, 11, 111, 117, 197, 199–202, 228
sand, 2, 5–8, 12–13, 17, 26, 29, 37, 43–54, 75, 104, 108, 112, 120–1, 124, 131–4, 140, 147, 157, 175, 186, 206, 208, 211, 222, 228; blows, 18; coarse, 8, 42, 51, 122, 133, 158, 180; fine, 8, 42, 49, 158, 163, 175, 180, 186; loamy, 48, 133–4, 139–40, 186, 205; silty, 175; very fine, 43, 133
sandy soils, 15, 17, 22, 32, 35, 46, 50, 55, 70–1, 93, 98, 101, 126–8, 132, 137, 140, 144, 153, 160, 186–7, 202, 206, 215, 226, 228
savanna (*see also* tropical grassland), 15, 41, 75–6, 186, 192–3, 220–2
season, 139, 142; dry, 13, 30, 141, 208, 221; growing, 146, 183, 185; spring, 3, 35, 50, 140; summer, 7, 23–4, 35, 142, 144, 148, 168, 197, 212; wet, 15, 141, 208, 212, 227; winter, 23–4, 143
seasonal change, 4, 23, 138, 140, 168, 197, 212, 227
seaweed, 17
sediments, 17–18, 49–51, 60; suspended, 52–3
seeds, 19, 81, 137; bed, 105; growth, 137, 217
selenium, 22, 31
semi-arid areas, 33, 93, 120, 134–5, 144, 197, 199; soils, 3, 9, 110, 222
sensitivity, 31
septic tank, 31
sesquioxides (*see also* iron and aluminium oxides), 63, 103, 106–8, 159–61, 165,

169, 171, 180–2, 187, 192–5, 220, 227
settlements, 16, 29; dry-point, 15–16, 45
sewage: disposal, 30, 41; sludge, 31, 74
shear strength, 28, 60, 99, 207–8, 211
sheep, 11, 45
shelter, 12, 15, 45; -belts, 27–8, 123
shrinkage, 30, 63, 99, 103, 107, 109–10, 121, 210–14, 218, 222
shrubs, 11, 193; Nuttall saltbush, 11; sagebush, 11; shadscale, 11; winterfat, 11; roses, 228
side chains, organic, 82–3, 88–91
silage, 74
silcrete, 5–7, 45, 222
silica, 48, 63, 65, 67, 167, 169, 171–4, 219–20, 222–3; crystallization, 172–3; mobilization, 173
silicate: sheets, 56–8, 62–7, 157; units, 57
silicon, 21–2, 55–7, 63–5, 157–8, 172–4, 192–3
silt, 9, 29, 37, 42–3, 51–5, 75, 104, 110, 121, 124, 131–3, 135, 140, 156–8, 180, 186, 195, 205, 219; coarse, 52, 75, 203; fine, 51–2, 75, 126, 132, 134; pan, 218; sandy, 175; loamy, 43
silty soils, 15, 51–5, 135
silviculture, *see* forestry
slaking, 60, 101, 106, 111
slickensiding, 107, 210, 212
slopes (*see also* hillslopes), 13–15, 28, 200, 207; deposits, 10, 44
slurries, 31–2
smectites, 65
snow: cover, 137, 142; melt, 130
sodic soils, 3, 201–2, 211
sodium, 48, 63, 65, 111, 117, 160, 173, 179, 198–202, 210–11, 215, 226; bicarbonate, 200, 202; carbonate, 199, 202, 226; chloride, 174, 197, 199–201; nitrate, 174; percentage, exchangeable (ESP), 198, 201–2, 211, 213; silicate, 202; sulphate, 174, 199–201
soil, dry, 144–5, 154, 208, 210, 212
soil air, 5, 30, 115, 124, 131, 134, 143–50, 155, 210, 217, 227; circulation, 60, 75, 97, 113, 143–6, 170, 181, 218; 'respiration', 147, 164
soil classification, *see* Appendix
soil colour, 224–8
soil crumbs (*see also* peds), 60, 101–2, 107–10, 118, 141–2, 145, 152, 154, 161, 210
soil fauna, 25, 41, 44–5, 93, 96–7, 109, 113–14
soil formation (*see also* weathering), 5, 7, 27, 81, 110, 123, 151, 227–8
soil heterogeneity, 36–9, 115

soil mapping, 5, 36, 39, 43; boundaries, 36–7, 39, 41; surveys, 17, 36, 42, 214
soil matrix, 101
soil mechanics, 5, 28, 216
soil solution, 54, 64, 82, 117, 123, 137, 145, 155, 160–2, 164, 166, 168, 170–7, 185, 188–9, 193–4, 196, 198, 200–2, 209, 211, 219, 222–3
soil surface, 1, 44–5, 78, 99, 105, 118, 121, 123, 130, 138, 148, 201, 226; bare, 145, 147, 184; blows, 18; conditioner, 102; creep, 14; crusts, 1, 53, 93, 118, 142, 181, 199–201, 221–2, 226; drifting, 12; exposure, 26, 32, 118, 136–8, 160, 220; sealing, 105, 118
soil water (*see also* moisture), 1, 5, 22, 24, 27, 29–30, 115, 117–43, 147–8, 152, 167, 170, 177, 188–91, 196–7, 200–1, 203, 205, 207–10, 217; pressure, 110, 125–8, 142
solodization, 202
solonization, 201–2
sphagnum, 97, 138, 148
spoil, mining, 32
stability: aggregate, 102, 105–11, 120–1, 166, 211; ground surface, 28, 30, 32, 53, 99, 207–8, 213, 216
starch, 59, 81–2, 85
stem, 26, 36, 81; flow, 36
steppe (*see also* grassland), 22, 28, 92, 94, 114, 116, 119, 121, 132
stickiness, 209–10
stony soils, 13, 17, 42, 44–5, 141, 143
stresses, 28, 203, 208–9, 212
structural units (*see also* peds), 109, 123; angular blocky, 124; columnar, 122, 202; 212; double-wedge, 109; granular (*see also* soil crumbs), 142, 221; honeycomb, 142; needle-ice, 142; platy, 109; polygonal, 110; poor, 3, 120, 121; prismatic, 109, 202; solonetzic, 111, 202
structure, 32, 39, 54, 69, 101–11, 132, 135–6, 141, 167, 210–11, 213, 215, 226
stubble, 72
sub-humid areas, 22, 28, 123, 129, 211, 224
subsoil, 1, 4, 28, 30, 32, 53, 115, 124, 141, 150, 152, 178, 186, 201, 204, 214, 216, 218, 220, 227
suction (non-alkaline), soil (*see also* pF), 125–8, 142, 144, 208
sugars, 80, 82, 89, 94, 158, 181
sulphates, 32, 152, 154, 161, 178, 182, 185–9, 191, 193, 198–200, 226
sulphides, 32, 141, 151, 154, 185, 189, 191
sulphur, 21–2, 64, 78, 80, 82, 149, 151, 153, 178, 180, 182, 185–9, 191, 193;

Subject index

dioxide, 149; immobilization, 182; mineralization, 185–6, 188
sulphuric acid, 32, 164, 185, 193
surface area, colloidal, 31, 55, 134, 156–7, 161, 188, 226
surface tension, 129
suspensions, 107, 155, 168
swampy areas, 14–15, 35, 85, 212, 228
swelling, 4–5, 30, 63, 66, 106–7, 109–10, 118, 121, 124, 139, 206–7, 210–13, 218, 222

taiga, 28, 71, 91, 94, 137
temperature: air, 23–5, 34, 67, 72, 94, 114, 136–43, 153; soil, 1–3, 5, 12, 34, 115, 136–44, 148, 203
tension, soil water, 131, 135–6, 209
termites, 35, 41, 76, 114
texture, 1, 3, 13, 31, 39, 43, 54, 92, 105, 111, 115, 131–3, 135–6, 142, 169, 175–7, 186, 196, 203, 205–6, 209, 212, 226; coarse, 7, 22, 121, 131–2, 139–40, 142, 145, 187; fine, 45, 123, 138, 144, 157, 186
thawing, 3, 5, 7, 34, 53, 137–8, 140, 142–3
thermal: conductivity, 139–40; gradients, 138–40, 142–3
thermonuclear fallout, 154
thin-sections, 101, 109, 221
thiobacilli, 185–6, 189, 191
tillage (*see also* ploughing), 105, 118, 208, 216, 228
time, 2–5, 26–7, 34, 94, 102, 105, 121, 124, 131, 139, 163
tissue, 86, 145, 178–9, 181–2
titanium, 192, 225
topsoil, 4, 32, 74, 140, 148, 160, 184, 202
toxic compounds, 31, 41, 99, 151, 155, 162, 166, 191, 193, 199
toxicity, 3, 21, 23, 25, 31–2, 59, 67, 151, 154, 162, 169, 171, 174, 195
trace elements, 22, 91, 151, 166, 193–7
traffic: automobile, 30, 54; pedestrian, 32; tractor, 216–18; wagon, 53
translocation, 47, 91, 202, 212
transpiration, 174, 198
tree: growth, 24; bark, 71; branches, 26, 71; foliage, 26; logs, 76, 99; trunks, 71; twigs, 77
tree crops: cacao, 177; coconut, 223; coffee, 113; fruit, 137; rubber, 59, 99, 112; sugar cane, 120, 174, 199, 218; vines, 112, 131, 218
trees, 11, 26–8, 44, 71, 76, 86, 88, 93, 111–13, 129, 173, 179, 218–19; acacia, 179; black spruce, 44; Douglas fir, 109, 148; eucalyptus, 11, 72; jack pine, 44; Loblolly pine, 26; lodgepole pine, 72; oaks, 17, 70–1, 137; pines, 130, 140, 148, 160, 206; red pine, 72; slash pine, 24; Sitka spruce, 25
triazine, 154
tropical soils, 14, 33, 36, 49, 61, 68, 118, 135–8, 141, 157, 161–3, 175, 177, 179–80, 192–3, 215, 220–2, 227
tundra, 71, 113, 137
tunnels (*see also* macropores), 109, 113–14, 121, 127–8, 143
turbation, 110
turgidity, 129

underground gas leakage, 30
United States Department of Agriculture (U.S.D.A.) Soil Taxonomy (*see also* Appendix), 39–40, 97–8, 108, 119, 180, 213
urbanization, 1, 3, 28, 30–1, 41, 50, 143
urea, 85
uronic acid, 79, 82, 95, 158

valley, 3, 10, 17, 54, 120
value, Munsell, 224–6
van der Waals force, 66, 107
vapour, 140, 143–4, 149, 197; movement, 139, 142, 144–5; loss, 131; pressure, 129, 144, 149
vegetation, 9–10, 13–14, 32, 34, 39, 71–2, 74, 77, 97, 136, 140–2, 169, 173, 201
vehicle weight, 60, 216–18
vermiculite, 46–7, 62, 64–5, 67, 157, 165, 174–5, 186
vertisols, 13–14, 76, 109, 119, 124, 210, 213
viruses, 31
viscosity, 142, 217
voids (*see also* pores), 103–9, 113, 124, 168, 203; Ratio, 206, 209

waste: disposal, 3, 30–2, 60; treatment, 31
water: acid, 62; adsorption, 126–9, 211–12, 217; alkaline, 62; balance, 120; initial content, 116, 124; drinking, 3; films, 145, 147, 155, 206, 208–9, 214; -holding capacity, 13, 50, 69; levels, 124; loss, 123, 215, 222; molecules, 57, 63–5, 80, 83–4, 90, 106, 126–7, 155, 162, 167, 208, 214; movement, 5, 99, 105, 109, 115, 117, 120, 126, 129, 139, 142, 145; quality, 3, 22, 32, 35; retention, 59, 99, 105, 126, 129, 134–5; stagnant, 109; storage capacity, 55, 109, 135; supply, 15; supply to plants, 111, 132, 137, 218
waterlogging, 7, 13, 35, 93, 97, 124, 143, 149, 151–3, 170, 188, 219–20, 222
weather, 22–3, 27, 114–15, 200

weathering, 7, 13, 32, 36, 44–52, 54, 56, 61, 63–7, 81, 109, 112, 159–63, 169, 174, 177, 179, 181, 188, 199, 215, 219–22, 227; index minerals, 51; ratios, 40, 51
weeds, 21
wells, 28
wetted soil, 144, 147, 149, 165, 169, 189, 210
wetting, 5, 106–12, 118, 120–1, 123–4, 128, 130–1, 145, 219; fingers, 121, 123–4; front, 1, 112, 121, 145
wilting, 22, 27; point, 128, 131–2, 134–5
wind, 27, 113, 137, 145, 198; erosion, 15, 18, 37, 53, 59, 109, 202, 208; -throw, 44; transport, 7, 16, 40, 43, 54
wood, 25, 81, 86, 99; production, 11, 44, 71
woodland, 11, 14, 70–2, 76, 93, 95–6, 113–14, 118, 135, 146, 160; clearance (*see also* forest clearance), 10, 100, 137

X-ray diffraction, 60, 67

yields, 17, 20, 22–3, 59, 74, 105, 129, 202, 205

zinc, 21, 84, 166, 193, 195–7

3 2311 00069 080 3

/553.6P692G>C1/

553.6 P692g

Pitty, A. F.

Geography and soil
 properties.

A. C. BUEHLER LIBRARY
ELMHURST COLLEGE
Elmhurst, Illinois 60126